ARCHITECTURE AND PRINCIPLES OF SYSTEMS ENGINEERING

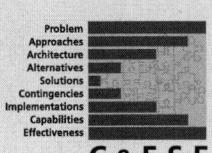

COMPLEX AND ENTERPRISE SYSTEMS ENGINEERING

Series Editors: Paul R. Garvey and Brian E. White

The MITRE Corporation

www.enterprise-systems-engineering.com

Designing Complex Systems: Foundations of Design in the Functional Domain
Erik W. Aslaksen
ISBN: 978-1-4200-8753-6
Publication Date: October 2008

Architecture and Principles of Systems Engineering
Charles Dickerson and Dimitri N. Mavris
ISBN: 978-1-4200-7253-2
Publication Date: May 2009

Model-Oriented Systems Engineering Science: A Unifying Framework for Traditional and Complex Systems
Duane W. Hybertson
ISBN: 978-1-4200-7251-8
Publication Date: May 2009

Enterprise Systems Engineering: Theory and Practice
George Rebovich, Jr. and Brian E. White
ISBN: 978-1-4200-7329-4
Publication Date: October 2009

Leadership in Decentralized Organizations
Beverly G. McCarter and Brian E. White
ISBN: 978-1-4200-7417-8
Publication Date: October 2009

Complex Enterprise Systems Engineering for Operational Excellence
Kenneth C. Hoffman and Kirkor Bozdogan
ISBN: 978-1-4200-8256-2
Publication Date: November 2009

Engineering Mega-Systems: The Challenge of Systems Engineering in the Information Age
Renee Stevens
ISBN: 978-1-4200-7666-0
Publication Date: December 2009

Social and Cognitive Aspects of Engineering Practice
Stuart S. Shapiro
ISBN: 978-1-4200-7333-1
Publication Date: March 2010

RELATED BOOKS

Analytical Methods for Risk Management: A Systems Engineering Perspective
Paul R. Garvey
ISBN: 978-1-58488-637-2

Probability Methods for Cost Uncertainty Analysis: A Systems Engineering Perspective
Paul R. Garvey
ISBN: 978-0-8247-8966-4

AUERBACH PUBLICATIONS

www.auerbach-publications.com
To Order Call: 1-800-272-7737 • Fax: 1-800-374-3401
E-mail: orders@crcpress.com

ARCHITECTURE AND PRINCIPLES OF SYSTEMS ENGINEERING

C.E. Dickerson
D.N. Mavris

CRC Press
Taylor & Francis Group
Boca Raton   London   New York

CRC Press is an imprint of the
Taylor & Francis Group, an **Informa** business
AN AUERBACH BOOK

Complex and Enterprise Systems Engineering Series

Auerbach Publications
Taylor & Francis Group
6000 Broken Sound Parkway NW, Suite 300
Boca Raton, FL 33487-2742

© 2010 by Taylor and Francis Group, LLC
Auerbach Publications is an imprint of Taylor & Francis Group, an Informa business

No claim to original U.S. Government works

Printed in the United States of America on acid-free paper
10 9 8 7 6 5 4 3 2

International Standard Book Number: 978-1-4200-7253-2 (Hardback)

Library of Congress Cataloging-in-Publication Data

Dickerson, Charles, 1949-
   Architecture and principles of systems engineering / Charles Dickerson, Dimitri N. Mavris.
      p. cm. -- (Complex and enterprise systems engineering)
   Includes bibliographical references and index.
   ISBN 978-1-4200-7253-2 (hardcover : alk. paper)
   1. Systems engineering. 2. Computer architecture. I. Mavris, Dimitri. II. Title.

   TA168.D643 2009
   620.001'171--dc22   2009007601

Visit the Taylor & Francis Web site at
http://www.taylorandfrancis.com

and the Auerbach Web site at
http://www.auerbach-publications.com

This book is dedicated to our families who have sacrificed their time with us so that we could finish this important work.

Contents

SECTION 2 MODELING LANGUAGES, FRAMEWORKS, AND GRAPHICAL TOOLS

SECTION 3 USING ARCHITECTURE MODELS IN SYSTEMS ANALYSIS AND DESIGN

SECTION 4   AEROSPACE AND DEFENSE SYSTEMS ENGINEERING

SECTION 5   CASE STUDIES

Acknowledgments

The material in this course is from a collaborative effort by Professor C.E. Dickerson of Loughborough University and Professor D.N. Mavris of the Georgia Institute of Technology. We each wish to express special thanks to our Ph.D. students who have generously devoted their personal time to support the creation of this material: Phil Johnson, Hernando Jimenez, Kelly Griendling, and William Engler.

Each has directly contributed to the writing of this book, as noted by their authorship and co-authorship of individual chapters. And their overall technical contributions and insights on the course have been invaluable.

This course is based on but substantially extends the Advanced Systems Design M.Sc. course for aerospace engineering at the Georgia Institute of Technology and the previous Systems Architecture M.Sc. course taught at Loughborough University on the subject of defense systems architecture and frameworks.

The defense systems architecture course at Loughborough University had been taught by Professor Alex Levis and Dr. Lee Wagenhals, both of George Mason University in Virginia. Course materials from all three universities have been integrated into this book.

The authors have been fortunate to receive key contributions from two leading subject matter experts, one from the U.S. and one from the U.K.

Dr. Russell Peak, a Senior Research staff at the Georgia Institute of Technology, has contributed the chapter on the application of the Systems Modeling Language (SysML) to modeling and simulation. Dr. Peak is the Director of the Modeling and Simulation Lab at Georgia Tech and represents the University to the OMG SysML Task Force.

Mr. Andrew Daw, Chief System Engineer at BAE Systems for Capability Development, is a recognized thought leader for Through Life Capability Management. He has contributed the chapter on the 'wicked' aspects of acquiring defense capabilities. Mr. Daw is the past president of the U.K. Chapter of the International Council on Systems Engineering (INCOSE).

List of Principles Used for Model-Based Architecture and Systems Engineering

Conceptual Integrity and the Role of the Architect

Conceptual integrity is the most important consideration in system design. The architect should be responsible for the conceptual integrity of all aspects of the product perceivable by the user.

The Principle of Definition

One needs both a formal definition of a design, for precision, and a prose definition for comprehensibility.

Model Transformation

Model transformations relate to system design and should preserve the relationships *between the parameters being modeled.*

Reflection of Structure in System Design

The solution should reflect the inherent structure of the problem.

Modular Structured Design

Systems should be comprised of modules, each of which is highly cohesive but collectively are loosely coupled.

Structured Analysis

The specification of the problem should be separated from that of the solution.

The Authors

Professor C.E. Dickerson is the Royal Academy of Engineering Chair of Systems Engineering at Loughborough University in the United Kingdom. His research program at the university is focused on model-driven architecture and systems engineering. He has authored numerous papers as well as an internationally recognized book on military systems architecture, and has co-authored key government reports. As a member of IEEE and OMG, and as the Chair of the INCOSE Architecture Working Group, he works with the systems engineering community on systems architecture practice and standards.

Before joining Loughborough University, he was a Technical Fellow at BAE Systems, providing corporate leadership for architecture-based and system of systems engineering. Previously, as a member of MIT Lincoln Laboratory, he conducted tests and research on electromagnetic scattering. As an MIT IPA, he served as Aegis Systems engineer for the U.S. Navy Theater-Wide Program and then as the Director of Architecture for the Chief Engineer of the U.S. Navy. His aerospace experience includes air vehicle survivability and avionics design at the Lockheed Skunk Works in the Burbank (California) facility and the Northrop Advanced Systems Division, and operations analysis at the Center for Naval Analyses. He received a Ph.D. degree from Purdue University in 1980.

Dimitri Mavris is the Boeing Professor of Advanced Aerospace Systems Analysis at the Guggenheim School of Aerospace Engineering, Georgia Institute of Technology. Throughout his academic career he has authored and co-authored over 60 refereed publications, more than 300 conference papers, and more than 100 technical reports. Since his initial appointment as academic faculty in 1996, Dr. Mavris has graduated more than 110 master's degree students and 50 doctoral students.

In addition to his academic duties, Dr. Mavris is the founder and director of Georgia Tech's Aerospace Systems Design Laboratory (ASDL), a unique academic organization home to 200 researchers and graduate-level students actively pursuing research in the areas of multidisciplinary analysis, design, and optimization; MDO/A; and nondeterministic design theory. Since its inception in 1992, ASDL has been named a Center of Excellence in Robust Systems Design and Optimization under the General Electric University Strategic Alliance (GE USA), and a NASA/DoD University Research Engineering Technology Institute (URETI) on Aeropropulsion and Power Technology (UAPT). In addition, ASDL is a member of the Federal Aviation Administration's Center of Excellence under the Partnership for Air Transportation Noise and Emissions Reduction (PARTNER).

Dr. Mavris is an associate fellow of the American Institute of Aeronautics and Astronautics (AIAA) and fellow of the National Institute of Aerospace. He also serves as Deputy Director for AIAA's Aircraft Technology Integration and Operations Group and as a Chair for the AIAA Energy Optimized Aircraft and Equipment Systems Program Committee.

List of Figures

3. Concepts, Standards, and Terminology

4. Structure, Analysis, Design, and Models

SECTION 2   MODELING LANGUAGES, FRAMEWORKS, AND GRAPHICAL TOOLS

5. Architecture Modeling Languages

6. Applications of SysML to Modeling and Simulation

SECTION 3 USING ARCHITECTURE MODELS IN SYSTEMS ANALYSIS AND DESIGN

14. Capabilities Assessment

15. Toward Systems of Systems and Network-Enabled Capabilities

SECTION 4   AEROSPACE AND DEFENSE SYSTEMS ENGINEERING

25.   Remote Monitoring

FOUNDATIONS OF ARCHITECTURE AND SYSTEMS ENGINEERING

1

Chapter 1

Introduction

Why does the community need another book on architecture and systems engineering? How will this book help you personally gain a better understanding of these subjects and develop new skills that can help you in the practice or research of systems architecture and engineering?

Today, the systems engineering community is actively rethinking its concepts and practices as it undergoes dramatic growth. However, tensions exist across the various systems disciplines. In the domain of software engineering and information technology, Moore's law leads to an order-of-magnitude improvement in technical capability every four to six years, a timeframe that aerospace and defense systems enterprises can require to develop a single new system, such as an air vehicle. The development of major new systems can take even longer.

The emergence of Model Driven Architecture (MDA™) and recent initiatives for model-based systems engineering (MBSE) will play an important role in determining how the practice of architecture and systems engineering evolves over the next several years. How will this affect you as an architect or engineering professional? Major changes have occurred in the practice of software engineering over the past two decades. The times will demand that major changes occur in systems engineering in the years to come.

This book will give students and readers the foundation and elementary methods to step into the domain of model-based architecture and systems engineering practices. A special attractiveness of the approach in this book is that it is as widely applicable as the interests and needs of the practitioner, and it can be inserted at any point into any organization's practice of systems architecting and engineering. The concepts, standards, and terminology provided in this book embody the emerging model-based approaches, but are rooted in the long-standing practices of

3

engineering, science, and mathematics. Each of the authors brings substantial experience from academic, government, and commercial research and development.

Fundamental questions are addressed. What is systems architecture? How does it relate to systems engineering? What is the role of a systems architect? How should systems architecture be practiced?

The expectation of the authors is to change how you think about architecture and systems. Our goal is to give you new skills that you can take back to your workplace or research program. However, your ability to reason in new ways from this book is more important than the details of any information that you might gain from it.

Readers of this book should ask themselves a basic question that is embodied in a comparison of an ancient Greek philosopher and one of the greatest 20th-century inventors: Diogenes and Thomas Edison. They might appear to have been on two separate paths. However, are these paths in conflict or do they have a common root that makes them complementary?

Diogenes was a Greek philosopher, who was a second-generation disciple of Socrates, one of the greatest seekers of truth in the history of Western civilization. Diogenes has been characterized as an old man with a lantern who used its light to look for truth. In fact, he chose to live a life of poverty on the streets of ancient Greece, sleeping in a large tub. He was probably considered an annoyance because, as a philosophical cynic, he always asked the hard questions that most of us do not want to answer.

The question "What is truth?" is thousands of years old. However, the pursuit of the answer to this question crosses many boundaries. In scientific terms, the answer to this question by Western philosophers, ancient and modern, necessarily leads to the consideration of mathematical logic and the role of models in logic and science, which are discussed in detail in Chapter 2 ("Logical and Scientific Approach") and Chapter 5 ("Architecture Modeling Languages").

Thomas Edison, on the other hand, was a great inventor in modern times. Remembered for inventing the electric light bulb, he became a wealthy and recognized technology leader of his time. His pursuit of "truth" two millennia after Diogenes, in the modern times of technology, could be considered the pursuit of valuable intellectual capital.

He did not live on the streets as did Diogenes, but did he have anything in common with Diogenes? Do you have anything in common with either of these great minds?

At the age of 22, Edison received his first patent: an electronic vote recorder to be used by legislative bodies. It was a simple concept but very advanced for its time. Votes could be recorded by the flip of a switch. However, legislators preferred to cast their votes by voice; Edison could not *sell* the invention. The inventor's response was:

> Anything that won't sell, I don't want to invent. Its sale is a proof of *utility*, and utility is success.

In the years to come, he applied this principle to his engineering practice of invention. He became very wealthy and some would say that he gave up the passion of technology for the pursuit of money. However, others might consider that he valued the utility of his inventions and focused on the utility of the technologies emerging in his day rather than the technologies. It might be considered that he was an early "venture capitalist." His viewpoint coupled with his passion for technology could not help but make him a very wealthy businessman.

Edison himself said that invention was 99 percent perspiration and 1 percent inspiration. It has been said of him that often he was found late at night sleeping on one of the laboratory benches, in between running experiments. Is that so different from sleeping on the streets, as did Diogenes?

Who best fits your personal model? If Diogenes and Edison seem too remote, then consider two contemporary great business leaders, Steve Jobs and Bill Gates. Some might consider that Steve Jobs is technology driven, a revolutionary (e.g., iPod/iTunes) who charges premium prices, and a thought leader who is focused on aesthetics. They might also consider that Bill Gates, on the other hand, is market driven, evolutionary, charges commodity prices, and is a business leader who is more focused on the utility of his products.

It is not a coincidence that the two contemporary leaders were both from the computer industry. This industry has seen dramatic technological growth and generated opportunities for wealth over the past decades, but it has also been the scene of dramatic failures. The evolution of the computer business architecture from the years 1980–2000 saw dramatic, if not devastating, changes for the major computer companies of the time. At the beginning of the 1980s, the industry was organized vertically. Each major company in the computer industry had its own sales and distribution network, application software, operating system, computing platform, and even its own microchips. A scant 20 years later, the industry had been turned on its head. All but one of the major companies that were prosperous in the 1980s had gone the way of the dinosaurs. What happened? Driven by open architecture, the industry became characterized by horizontal integration. The industry had also shifted to a commodity market.

Given the fast cycle of Moore's law, the computer and software industry is necessarily compelled to adapt to change much faster than many other industries. Will the systems engineering practices of the next decade see the same dramatic changes as did the computer and software engineering industry? Shouldn't the systems engineering community be carefully considering what has happened and is happening in the computer and software engineering communities? Does the systems engineering community hold the same narrow views that the dinosaurs of the computer industry held through the 1980s and 1990s?

Multiple views and viewpoints are part of everyday life, but not all of us recognize this perspective. Legitimate but differing points of view abound. If you observe the daily world around you, especially during travel through airports, you will see many signs of the commercial realization of the importance of viewpoints in the form of advertisements. Using pictures of everyday people and objects, challenging questions are asked. For example, what is a full moon? It may be a symbol of lunacy to one person while another may view it as a symbol of romance. Which is the "correct" point of view?

Those readers who are engineers or have a background in science may be thinking that these types of point–counterpoint contrasts belong to nontechnical domains, where regions of grayness reside. In science and engineering, we think that we know what is and what is not. However, serious differences can occur in science and engineering too!

In science the quest for determinism failed in the 20th century with the advent of quantum theory. But Albert Einstein, for example, intellectually struggled against the concepts of quantum theory, saying that he did not believe God played with dice (in the design of the universe). Why did Einstein have a viewpoint that was so different from that of his contemporaries?

In Chapter 3 ("Concepts, Standards, and Terminology"), a few of the myriad concepts and terminologies of architecture and systems engineering will be reviewed. There are over 100 definitions of architecture alone. With so many viewpoints abounding in the precepts of architecture and systems engineering, there is a challenge as to how broadly the concepts and terminology of architecture and systems engineering should be treated in this book. The approach to the material in this book is to keep it simple. There are many points of view. The viewpoints of the authors will focus on just one role of an architect and a commonly practiced role of a systems engineer. Your situation will undoubtedly be different, but the approach presented in this book will give you a way of reasoning and a common thread end to end that you can apply to your own situation. Although this book will provide a standard introduction to the methods and practice of systems architecture, it is important to recognize that new ways of reasoning are more important than the details of practice.

These views are based on the authors' successful professional experiences. The architecture viewpoint is based on personal experience as the Director of Architecture (2000–2003) for the (1st–3rd) Chief Engineer of the U.S. Navy. In this role, the architect was positioned between the acquisition authority (the Assistant Secretary of the Navy) and the government leaders of system development (the system commands). This is not unlike the position that many systems architects find themselves in aerospace and commercial practice.

The systems engineering viewpoint is based on the personal experience of the director of a large academic aerospace systems engineering laboratory at the Georgia Institute of Technology. Often facing ambiguous and ill-defined problems from which clear systems understanding and crisp conclusions must be produced, this role embodies a balancing act between spearheading the dynamic interaction with industry and government customers, and leading the analytical efforts that yield relevant systems knowledge.

However, this book is not just a reiteration of past achievements. It is our view of the path to the future, based on what we know has worked and the power of model-based approaches to architecture and systems engineering.

What is the role of a systems architect in systems engineering? The view taken in this book is succinctly expressed in *The Mythical Man Month* (Brooks 1995):

> Conceptual integrity is the most important consideration in system design. The architect should be responsible for the conceptual integrity of all aspects of the product perceivable by the user.

This view is supported by personal professional experience. It is the basis of this book from the viewpoint of the systems architect.

Key to conceptual integrity are modeling and modeling languages, architecture frameworks, and systems engineering standards. The practice of systems architecture requires communication skills, tools, experience, and knowledge of case studies.

The material in this book has been taught as a one-semester course in Systems Architecture at the M.Sc. and Ph.D. level. However, there is sufficient material to support two semesters of study, one being more focused on systems architecture but with an exposure to software engineering and the other being more focused on aerospace and defense systems engineering. If the material is spread over two semesters, it is recommended that the instructor spend additional time on laboratory practicum so that the students can gain greater proficiency in one or more of the software tools available today for the practice of systems architecture.

Reference

Brooks, F.P. 1995. *The Mythical Man Month*. Boston: Addison Wesley Longman.

Chapter 2

Logical and Scientific Approach

What are the formal languages and methods of systems engineering? The answer to this question is at the heart of the approach described in this chapter. Mathematics, science, and the engineering disciplines each use formal languages, methods, and models. Logic and science provide a sound foundation for a model-based approach to architecture and systems engineering.

Key Concepts

Formal languages and methods
Integrity and consistency of terms
Scientific models
Logical models

Motivation and Background

Modern mathematics, science, and engineering enjoy the benefit of thousands of years of human thought, experience, and practice. And over the past century, with the advent of large-scale systems that today are commonplace, systems engineering has emerged as a distinct engineering discipline, but from any historical perspective, it must be considered still young and maturing. When compared with mathematics, science, and the historical academic disciplines of engineering, we see that

- Mathematics uses
    - Formal logic (the predicate calculus) for description
    - Logical methods and mathematical induction for reasoning

- Science uses
  - Mathematics (a formal language) for description
  - Models and experimental methods for reasoning
- Engineering uses
  - Science and mathematics for description and reasoning
  - Methods, tools, and prototypes for design and decision

The languages, tools, and formalized methods of systems engineering are beginning to emerge. Because the systems viewpoint seeks to reconcile the differences between stakeholders, collaborative systems engineering and design methods are becoming more prevalent. Figure 2.1 depicts the Collaborative Visualization Environment (CoVE) at the Georgia Institute of Technology, which has enjoyed significant success as an academic environment for research in collaborative systems engineering and design. Most large aerospace companies today have collaborative systems engineering environments with substantial visualization capabilities.

But what are the formal languages of systems engineering? Without the precision of a formal language, large-scale tools cannot be commercially developed. Systems engineering has undergone substantial growth over the past decades, but broadly accepted formal languages for systems engineering have not. By comparison, the rise of electronic computation was accompanied by the development of several formal languages, which in recent years have converged in the broadly accepted Unified Modeling Language (UML) and certain computer languages, the standards for which are managed by the Object Management Group (OMG), an international open group. And over the past decade, the OMG Model Driven Architecture (MDA™) has emerged as a new model-based approach to software design and development.

Figure 2.1 The Collaborative Visualization Environment (CoVE).

Therefore, as with the comparison to mathematics, science, and the academic disciplines of engineering made earlier, we see that software engineering indeed does have formal languages, methods, and tools. Specifically:

- Software engineering uses
  - The Unified Modeling Language (UML)
  - Computer-aided software engineering (CASE) tools
- The OMG Model Driven Architecture (MDA) uses
  - UML to create models independent of technology
  - Model transformations for design and development

Efforts to apply UML to systems engineering have been ongoing over the past decade. Figure 2.2 illustrates the meta model for the UML profile for two defense systems architecture frameworks: the Department of Defense Architecture Framework (DoDAF) and the Ministry of Defence Architecture Framework (MODAF). The UML Profile for DoDAF and MODAF (UPDM) is a recently approved OMG standard that is a significant step towards establishing architecture standards for defense systems engineering. The UPDM also takes advantage of a recently approved systems engineering extension to UML, which is called the Systems Modeling Language (SysML). Hatley (2000) has also developed a UML diagram, which was a model of his process for requirements engineering and

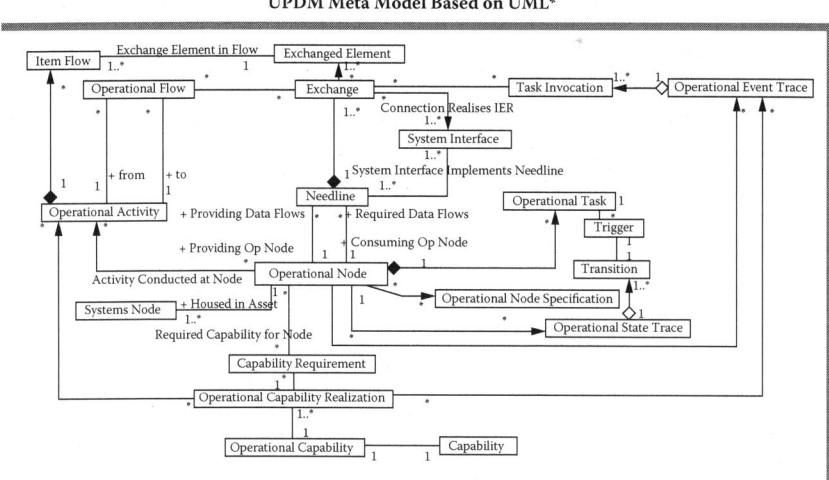

*From UPDM submission to OMG, March 2007

Figure 2.2   Modeling and formalizing systems engineering.

system architecture. This model was developed while UML was in its early stages but has been put into practice commercially.

We see then that although systems engineering currently lacks a broadly accepted formal language, it could use UML and the Systems Modeling Language (SysML) and recent advances in architecture frameworks and emerging tools to take steps forward that would better formalize the systems engineering discipline.

Where can all of this be expected to lead the practitioners of systems engineering and the businesses that rely on its practice? This is difficult to say, but given that the world of computer systems has developed at a faster pace than that of large-scale systems, it might be worthwhile to leverage the advances in software engineering over the past decade, especially the use of system models and model transforms.

Will the systems engineering community undergo these types of changes? This is also difficult to say, but when the languages, commercially available tools, and methods of systems engineering are firmly established and broadly accepted, the practice of systems engineering cannot be expected to be the same as it is today. If trends in the software engineering industry, such as UML and MDA, are any indication, then we should expect that systems engineering will become a model-based discipline over the next decade, rather than the document-based discipline that it is today.

Scientific Basis of Engineering

If systems engineering is to be properly understood, then it must be understood in the general context of engineering. What is engineering? The following definition will be used for the purposes of this book:

> *Engineering* is the most primitive level of concept realization where the relationships between function, behavior, and structure for the purpose of solving a problem can be described using the laws of science.

Although there are no doubt many other definitions of this common term, this definition is well suited to the architecture and systems engineering approach taken by this book.

A simple example from structural engineering can be used to illustrate the relationship between science and engineering:

- Problem: How to best use bricks to build a bridge with a planar surface.
- Purpose: The bridge enables transport across a gap in the terrain.
- The underlying physics:
  - Concept of a moment in mechanics.
  - Principle of dispersion of energy in a structure.

- ■ Implications for the engineering of a bridge:
  - The bridge must transfer sufficiently large moments of energy from the plane of transport to the base of the bridge on the ground.
  - If the bricks are arranged in an arch, then the moments of energy will be evenly dispersed through the structure and transferred to the base.
  - Among the benefits of the design is that the number of bricks required is greatly reduced and the resulting structure can be visually attractive.

It has been considered (Finch 1951) that in the evolution of any engineering discipline in the context of Western civilization, the discipline always starts as a craft but when demands for large volumes of production cannot be met by craftsmen, the practice of the craft must evolve. Successful (and profitable!) large-scale production of goods and services relies on science, for its depth of understanding, precision, and repeatability. The honing of a craft using science to achieve commercial development results in the practice of engineering.

This short intuitive description of engineering should not lull the reader into a false sense that good engineering comes easily. Notwithstanding the commercial and organizational problems that can challenge the engineer, reliably understanding and predicting the relationships between function, behavior, and structure is a serious technical problem. The ubiquitous engineering term *emergent behavior* can work both for as well as against the engineer. Murphy's law is always at work.

For those who may consider modern science and engineering to be absolutes, the catastrophic collapse of the Tacoma Narrows Bridge in 1940 stands as one of the great counterexamples (University of Washington 2006). Harmonic oscillations, which are well understood in science and engineering, were the cause of the collapse. More recently, the Millennium Bridge in London suffered a similar design flaw, which fortunately was corrected before a catastrophic event occurred (but after the bridge went into service). See, for example, Arup 2008.

Experimental and Logical Basis of Science

What is science? If engineering in general and systems engineering in particular are to be founded on science, then there needs to be an agreement on what is meant by *science*. The following perspective is offered as a simple aid to understanding the term:

- ■ The Latin word *scientia* means knowledge.
- ■ The English word *science* refers to any systematic body of knowledge, but more commonly refers to one that is based on the scientific method.
- ■ The scientific method consists of
  - Characterization of observables (by definition and measurement).
  - Formulation of hypotheses (i.e., interpretations of models), which are tested by comparing:

  – Predictions (based on the hypotheses).
  – Experimentation (which must be repeatable).

It will be important to understand the difference between principles and laws. In physics, principles are deeper insights from which laws can be derived. Consider the following:

■ Examples of principles:
  – The conservation of energy
  – The equivalence of mass and energy
■ Examples of laws:
  – Newton's laws of dynamics
  – The law of gravity

Science generally seeks to develop models of its principles and laws. Consider Newton's three laws for describing the dynamics of bodies of matter:

■ Law of inertia: An object will stay at rest or continue at a constant velocity unless acted upon by an external force.
■ Law of acceleration: The net force on an object is equal to the mass of the object multiplied by its acceleration.
■ Law of reciprocal actions: Every action has an equal and opposite reaction.

These laws are couched in natural language, but they were modeled using algebra and calculus. The power of Newton's calculus allowed many physical systems to be modeled and their behavior predicted. The simple law of acceleration, when properly modeled with Newton's calculus, leads to Hamiltonian mechanics, where systems of large numbers of particles and fields are described by partial differential equations. Hamiltonian mechanics is surprisingly universal. It can even be used to formulate problems in both classical and quantum mechanics.

If this is sounding too easy, you are right! What makes it hard? The problem is easier to model than the solution! For example, the equations for a system of three bodies of mass remain unsolved even today. The orbits of the planets in our solar system would be unpredictable if it were not for the dominant mass of the sun!

So, the recently maturing discipline of systems engineering has something to learn from the historical disciplines of science and engineering. The power of formal language modeling is just one thing to be learned from science and engineering. However, a new level of precision for the description of systems engineering problems will not be a *panacea*. Hard problems and unsolvable problems will still be with us. And reliable prediction of emergent behavior is not always attainable. Wicked problems abound and are discussed in detail for defense system acquisition and development in Chapter 21 ("Wicked Problems").

Logical Modeling of Sentences

Newton sought principles in physics. We should seek *principles* for architecture and systems engineering. The key systems engineering principle for the logical modeling of sentences will be called the *principle of definition*:

> One needs both a formal definition of a design, for precision, and a prose definition for comprehensibility.

This principle was set forth by the pioneers of digital computer architecture (Brooks 1995, p. 234, 6.3) and remains valid today. When this principle is used to reason about the concepts and terminology regarding architecture and systems engineering, the term *design* will be replaced by the term *concept* and the term *prose definition* will be taken to be a natural language definition, which will become the subject of the reasoning. Thus, logical modeling will be used for precision and natural language will be used for comprehensibility. The purpose of the following discussion is to provide a procedure to ensure that the conceptual integrity of the formal definition of a specific term, for example, is preserved by concordance between precision and comprehensibility.

Relationship of Logical Modeling to Science and the Predicate Calculus

If logical models should be used for precision, what then constitutes a good model?

- Stephen Hawking (Hawking 1988) attributes a good model in physics to be
  - Simple
  - Mathematically correct
  - Experimentally verifiable
- By analogy, a good logical model of a natural language sentence should be
  - Simple
  - Logically well formed and consistent
  - Verifiable through logical interpretation

The concept of logical modeling presented here is motivated by the meaning of models in the predicate calculus. See, for example, *Models and Ultraproducts* (Bell and Slomson 1969). The terms *model* and *sentence* have precise meanings in the predicate calculus, which will be discussed in detail in Chapter 5 ("Architecture Modeling Languages").

- The formal languages of mathematical logic are
  - Propositional calculus,
  - Predicate calculus.

- The term *calculus* derives from the Latin for calculation:
  - In logic, *calculus* refers to the calculation of truth.
  - By contrast, in the *calculus* of Newton and Leibnitz, the term refers to the calculation of limits.
- *Propositions* are declarative statements and are represented by propositional variables; that is, the propositions are represented as abstract variables.
- *Predicates* are statements of relationships between variables and are represented by predicate letters; that is, the relationships are represented as abstract predicates that depend on abstract variables, to which the abstract relationships refer.

A *sentence* is a well-formed formula of predicates that is fully quantified; that is, the formula adheres to the constructs of logic and there are no free variables in the formula. Free variables (i.e., variables that have not been given proper scope through quantifiers) admit unbounded interpretation, which is problematic. In natural language, scope can be provided by certain types of adjectives. In the predicate calculus, a *model* is a relational structure (i.e., a set with a family of relations on the set) in which sentences are reasoned about. A relational structure is said to be a *model of a sentence* if the sentence is valid when interpreted into the model.

The *objective* of the logical modeling of sentences is to extract the relations comprising the sentence using a formal language to derive a minimal model that is complete and captures the intended meaning of the sentence. Keywords will be selected that remain undefined to avoid the circularities possible in natural language. This approach follows the axiomatic approach of mathematics. See Dickerson (2008) for the foundational work that links the scientific concept of a model with the concept of model in mathematical logic.

Details of the Procedure for the Recommended Modeling Approach

When deriving a logical model of a sentence, it is important to avoid introducing information that does not belong to the intended meaning of the sentence. The model should capture the exact meaning of the sentence within the limitations of natural language.

The first step in modeling a sentence is to list the keywords that give it meaning:

- The keywords will be undefined and their meaning will be determined by relations between the words, which will be represented graphically, using notations from the Unified Modeling Language (UML).

- The primary type of diagram needed from UML is the class diagram. Approaches similar to this have been used, for example, in the IEEE Standard 1471 (IEEE 2000) and the commercial practice of architecture (Hatley 2000).
- The Systems Modeling Language (SysML) is also used.

The approach described here uses UML and SysML notations with certain conventions for the purpose of tracking keywords, changes, and rationale. Specifically:

- Natural language notation
  - The **defined term** uses the following:
    - Bold font
    - Underlined
    - Not italics
  - The ***keywords*** in the definition use the following:
    - Bold font
    - Not underlined
    - Italics
  - *Other words* in the definition use:
    - Not bold font
    - Not underlined
    - Italics
  - Additional words (i.e., not in the definition) use the following:
    - Not bold
    - Not underlined
    - Not italics
- Graphical notation:
  - Nouns are placed in boxes
  - Verbs and relations are placed on lines
  - Solid boxes and lines are used for keywords
  - Dash-dot graphics are used otherwise

Adjectives are converted to verbs or possessed nouns whenever meaning is sharpened, unless their role gives logical scope. Depending on the readability of the text, bold font will not be used for discussion of the sentence and keywords in textual discussions of diagrams.

The procedure is finished when all of the keywords (both cited from the sentence and necessary for completion of the model) have been identified and all of the relations between the words (both explicit and intended by the natural language) have been captured in the diagram.

The criteria for the goodness of a logical model include the intrinsic consistency and completeness of the model, and the correctness of the model relative to the natural language under accepted interpretations. It is also desirable to have independence between the relationships and independence between the keywords.

Detailed Illustration of the Approach to the Term "System"

This section presents a detailed discussion of the term *system* to illustrate how logical modeling of sentences can be used to compare and reconcile competing definitions. Two definitions of system will be examined. This first is the definition from the INCOSE *Systems Engineering Handbook* (Haskins et al. 2007):

> A system is a combination of interacting elements organized to achieve one or more stated purposes.

In the context of engineering, the term *system* will always mean an engineered system.

Based on the INCOSE definition of system, the term could comprise seven keywords: combination, interacting, elements, organized, achieve, stated, and purpose. Using the notations described in the previous section, the sentence would then be depicted as follows:

> A **system** is a **combination** of **interacting elements organized** to **achieve** one or more **stated purposes**.

The second step in developing the logical model would be to capture the relationships between **system** and the keywords. Figure 2.3 shows the relationship between **system** and *stated purposes* in graphical form, using the notation of a UML abstract class diagram.

The next step in developing the logical model would be to capture any other relationships between **system** and other keywords. Figure 2.4 shows reasonable relationships between **system**, *combination*, and *elements* in graphical form, again using the notation of a UML abstract class diagram.

Now that the direct relationships between **system** and the related keywords have been captured, the next step in developing the logical model is to capture any relationships between the keywords. Figure 2.5 shows a possible relationship between *combinations* and *stated purposes.* The keyword *organized* has been modeled as verb (on a line), which results in the *combinations* becoming organized. Because the natural language definition being modeled is ambiguous as to whether the elements are *organized* or the *combinations* are *organized*, the dash-dot notation has been used to indicate that other interpretations may be possible.

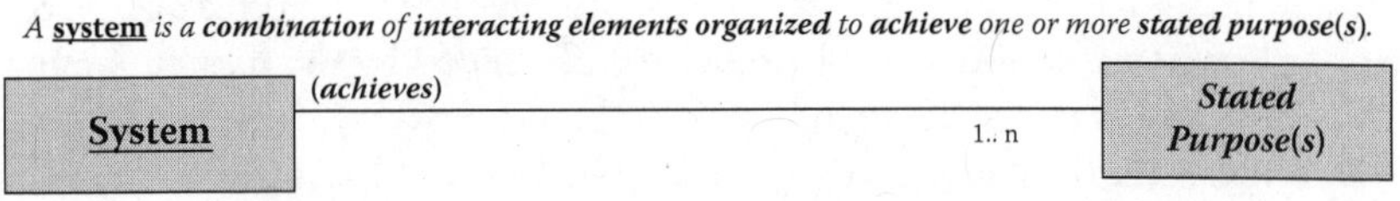

A **system** *is a* **combination** *of* **interacting elements** *organized to* **achieve** *one or more* **stated purpose(s)**.

Figure 2.3 The relationship between <u>system</u> and *stated purposes*.

A <u>*system*</u> *is a combination of interacting elements organized to achieve one or more stated purpose(s).*

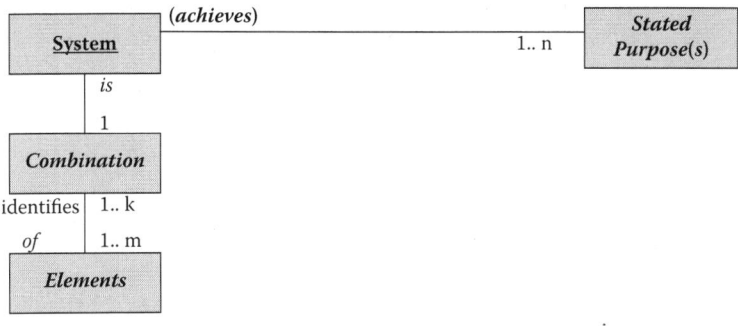

Figure 2.4 **Reasonable relationships between <u>system</u>, *combination*, *elements*; and the relationship between <u>system</u> and *stated purposes*.**

A <u>*system*</u> *is a combination of interacting elements organized to achieve one or more stated purpose(s).*

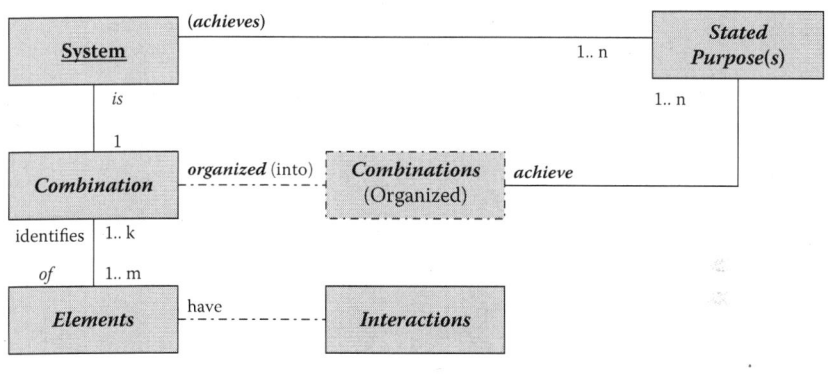

Figure 2.5 **The relationship between *combinations* and *stated purposes*.**

As indicated in Figure 2.6, the diagram of the logical model makes clear that there are relationships missing in the model; for example, how do the ***interactions*** relate to ***achieving*** *one or more* ***stated purposes***? This will lead to the next step in developing the logical model, but before we take that step we will introduce a competing definition of **system**, which will actually give us insight as to how best to proceed.

Definition of **system** (Hitchins 2003):

A **system** is an open set of complementary, interacting parts with properties, capabilities, and behaviors emerging both from the parts and from their interactions.

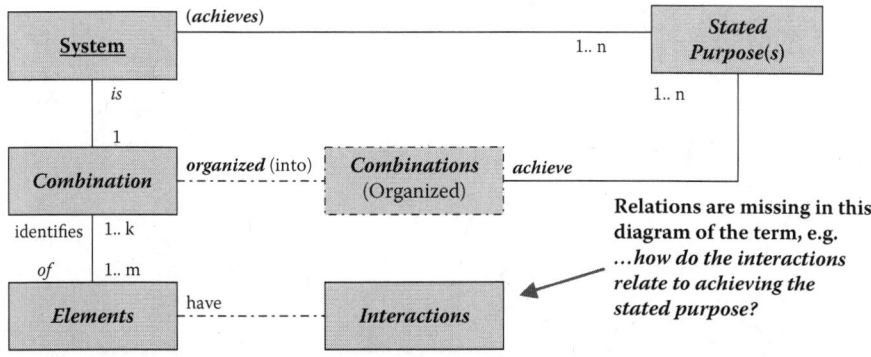

Figure 2.6   Logical assessment of the diagram for <u>system</u>.

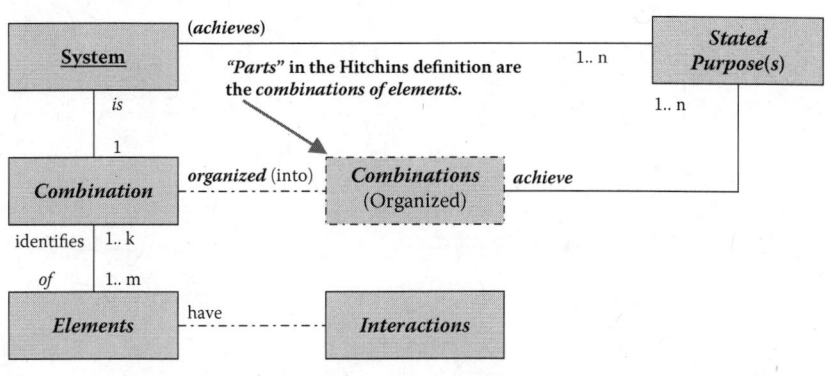

Figure 2.7   Interpretation of *parts* in the Hitchins definition as the *combinations* of *elements* of the INCOSE definition.

Hitchins emphasizes the commonly accepted concept of emergent behavior for systems. It is also clear from Hitchins that both the parts and the interactions between the parts cause emergent properties, capabilities, and behaviors.

Only these two concepts will be used to extend the model of the INCOSE definition of **system** and fill the gaps.

Figure 2.7 indicates how the diagram in Figure 2.6 can be interpreted to include the 'interacting parts' in the Hitchins definition. The 'parts' will correspond to the *combinations of elements*. In Figure 2.8 an insertion point is identified for the 'properties, capabilities, and behaviors' used in the Hitchins definition. The diagram in Figure 2.9 then makes the insertion and recalls the previously identified gap related to the interactions.

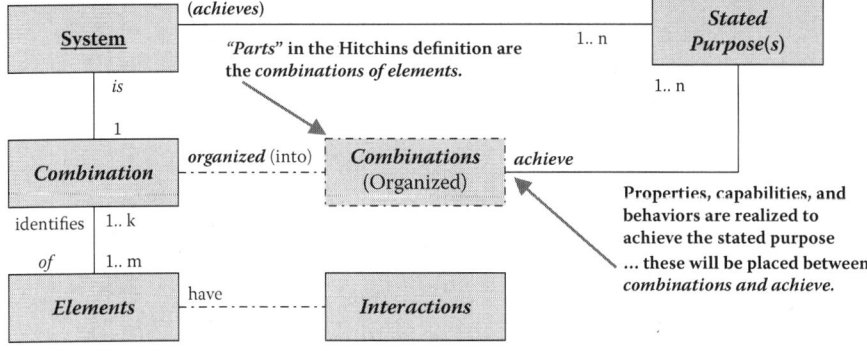

Figure 2.8 The issue of realization of *properties, capabilities,* and *behaviors* from the combinations in the diagram for the INCOSE definition of <u>system</u>.

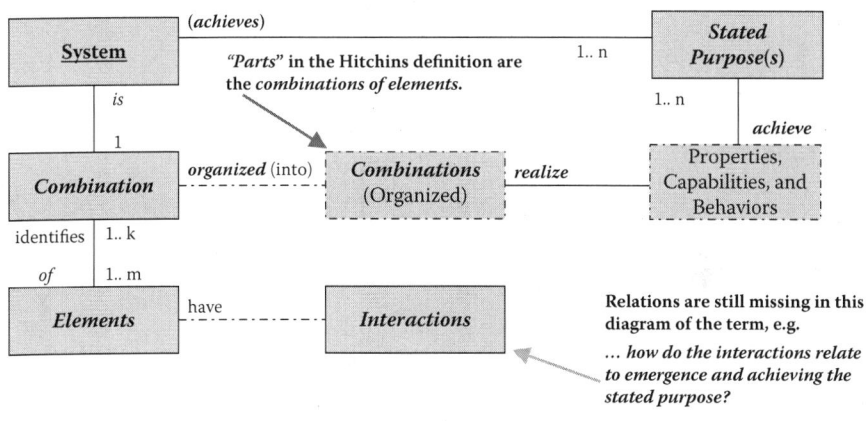

Figure 2.9 The issue of *emergence* from *interactions.*

Figure 2.10 takes a first step in identifying the relationship between the interactions and the properties, capabilities, and behaviors of the system. The SysML symbol for decomposition has also been introduced.

This raises significant questions, as called out in Figure 2.11. The combinations are organized. And the interactions are between the elements of the combinations, so should they not also be organized? And is the organization of the interactions independent of the organization of the combinations? This cannot be true. Therefore, are there many organizing principles or are there one or more overarching organizing principles?

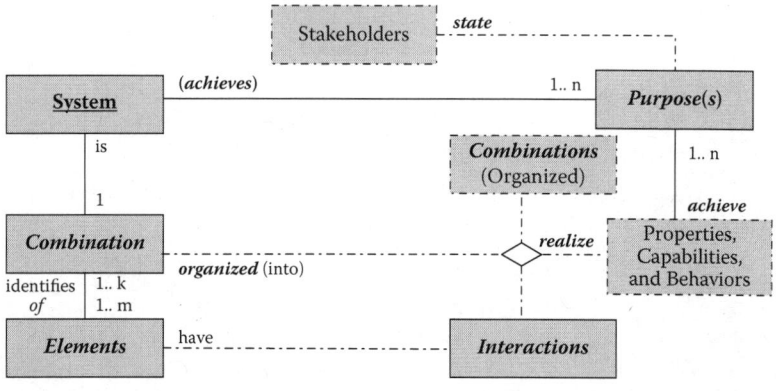

A **system** *is a* **combination** *of* **interacting elements organized** *to* **realize properties, behaviors,** *and* **capabilities** *that* **achieve** *one or more* **stated purpose**(s).

Figure 2.10   Initial extension of the diagram for system.

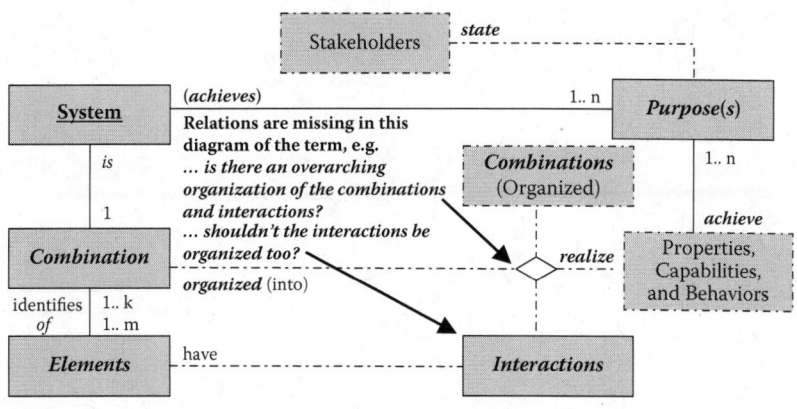

A **system** *is a* **combination** *of* **interacting elements organized** *to* **realize properties, behaviors,** *and* **capabilities** *that* **achieve** *one or more* **stated purpose**(s).

Figure 2.11   Final issues to be resolved in the diagram for system.

Figure 2.12 shows an extension of the diagram in Figure 2.11 that considers the existence of one or more overarching organizing principles that organize both the combinations (i.e., the parts of the system) and the interactions. This captures the Hitchins understanding of the system properties, capabilities, and behaviors caused by both the parts and the interaction between the parts as being with a sense of system wholeness (i.e., there are overarching organizing principles). All dash-dot notations have been dropped in Figure 2.12 because this has been taken as a concluding resolution between the specific details of the two definitions under consideration.

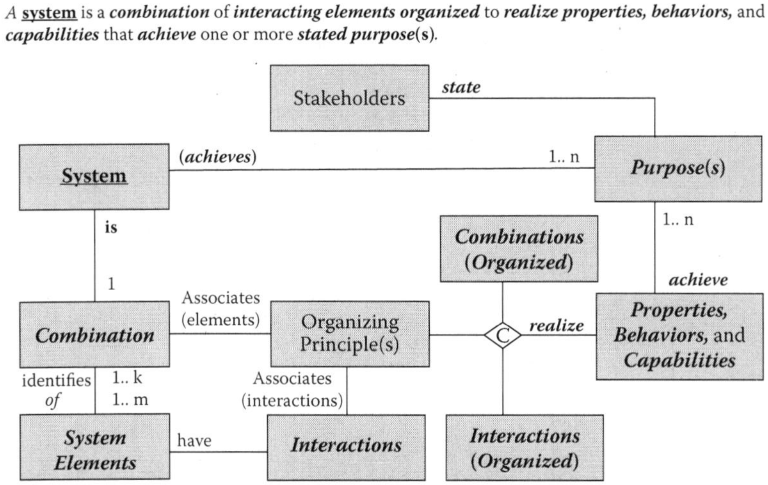
A system is a combination of interacting elements organized to realize properties, behaviors, and capabilities that achieve one or more stated purpose(s).
Stakeholders
state
System
(achieves)
1.. n
Purpose(s)
is
1
1.. n
Combinations (Organized)
achieve
Combination
Associates (elements)
Organizing Principle(s)
C
realize
Properties, Behaviors, and Capabilities
identifies of
1.. k
1.. m
Associates (interactions)
System Elements
have
Interactions
Interactions (Organized)

Figure 2.12 An extension of the diagram for system.

A Suggested Model-Based Definition of Systems Engineering

Now that the definition and model of system has been reconciled against two differing definitions, it would be appropriate to consider other definitions of systems engineering and their logical models. Standards-based definitions will be considered in Chapter 3 ("Concepts, Standards, and Terminology"). In this section, we shall offer a very simple model-based definition and discuss the implications. The approach here is to use the technique of viewpoints and views that are commonly practiced in software engineering.

Recall that a *viewpoint* on a system is a technique for abstraction using a selected set of architectural concepts and structuring rules to focus on particular concerns within that system.

The term *abstraction* is used to mean the process of suppressing selected detail to establish a simplified model. The suppression of detail can be accomplished by deletion, distortion, or generalization (Bandler and Grinder 1975). These three methods may be complete but are not independent.

In the MDA Guide V1.0.1 (Miller and Mukerji 2003), a *viewpoint model* or *view* of a system is a representation of that system from the perspective of a particular viewpoint. The definition in the MDA Guide V1.0.1 is based on IEEE 1471.

A general definition of engineering was presented earlier in this chapter. However, in the following definition of *systems engineering*, any definition of engineering could be used:

> Systems engineering is the practice of engineering from the systems viewpoint.

This possible definition of the term *systems engineering* actually captures two different viewpoints of the engineering community. One of the viewpoints speaks to systems engineering as the engineering of systems. Another viewpoint emphasizes the systems viewpoint of systems engineering. This possible definition of systems engineering has interesting properties. It

- Captures the concept that systems engineers abstract away certain details of design to view a product as a system
- Inherits the concepts and properties that are attributed to system
- Has a syntax that is consistent with the terminology of other engineering disciplines:
  - Chemical engineering could be considered as the practice of engineering from the viewpoint of chemistry.
  - Structural engineering could be considered as the practice of engineering from the structural viewpoint.

The way that the term *engineering* is used in this definition of systems engineering also gives insight into the commonly experienced conflicts in industry over the difference in roles between systems engineers and other engineering disciplines.

Summary

Mathematics, science, and the engineering disciplines each use formal languages, methods, and models. Logic and science provide a sound foundation for a model-based approach to architecture and systems engineering. UML is a widely accepted modeling language for software engineering that can be used to draw logical diagrams of natural language sentences. Thus, logical modeling with UML diagrams can be used for precision, and natural language can be used for comprehensibility in a concordant way.

The methods and procedures in this chapter for creating UML diagrams of natural language sentences were used to model and reason about the term *system*. Ambiguities and gaps in a popular standard definition of system were resolved using UML diagrams of the definition. The definition was also normalized against another well-known definition. In this way, the simplicity and power of logical modeling as an engineering practice (using UML) were clearly demonstrated.

References

Arup. The millennium bridge. http://www.arup.com/MillenniumBridge/ (accessed August 8, 2008).

Bandler, R. and Grinder, J. 1975. *The Structure of Magic I*. Palo Alto: Science and Behavior Books, Inc.

Bell, J.L. and Slomson, A.B. 1969. *Models and Ultraproducts: An Introduction*. Amsterdam: North-Holland Publishing Company.

Brooks, F.P. 1995. *The Mythical Man Month*. Boston: Addison Wesley Longman.

Dickerson, C.E. June 2008. Towards a logical and scientific foundation for system concepts, principles, and terminology. *Proceedings of the Third IEEE International Conference on System of Systems Engineering*.

Finch, J.K. 1951. *Engineering and Western Civilization*. New York: McGraw-Hill.

Haskins, C., Forsbery, K., and Kruger, M. 2007. *Systems Engineering Handbook*. Seattle: INCOSE.

Hatley, D., Hruschka, P., and Pirbhai, I. 2000. *Process for Systems Architecture and Requirements Engineering*. New York: Dorset House Publishing.

Hawking, S.W. 1988. *A Brief History of Time*. New York: Bantam Books.

Hitchins, D.K. 2003. *Advanced Systems Thinking, Engineering and Management*. Boston: Artech House.

Institute of Electrical and Electronics Engineers (IEEE). 2000. IEEE 1471. IEEE Recommended Practice for Architecture Description of Software-Intensive Systems.

Miller, J. and Mukerji, J. 2003. MDA guide version 1.0.1. OMG. http://www.omg.org/docs/omg/03-06-01.pdf (accessed 8 August 2008).

University of Washington. January 24, 2006. History of the Tacoma Narrows Bridge. http://www.lib.washington.edu/specialcoll/exhibits/tnb/.

Chapter 3

Concepts, Standards, and Terminology

Systems engineering terms have different meanings depending on the user. Standards groups seek to normalize terminology across the broad community of users. However, this does provide the precision required by a model-based approach to architecture and systems engineering.

Key Concepts

Engineering standards
Standards groups
Normalization of terms
Logical modeling

Systems Engineering Standards

When dealing with the standards promulgated through the various national and international organizations, it is important to understand that the standardization of concepts and terminology is an agreement between users, practitioners, and other stakeholders of the standard. As such, these standards tend to reflect a consensus of current practice across differing business and technical domains. Furthermore, an international standard must also account for interpretation in multiple languages. It can easily be the case that a single keyword in English, for example, has no corresponding word in one of more of the languages of Europe, much less the many languages and dialects of countries such as India and China.

With so many influences on the choice and use of terminology, it should not be expected that the current concepts and terminology of systems engineering should be standardized to a level that passes the test of logical precision. In the domain of science and mathematics, the use of precision symbolic languages has greatly alleviated this problem, but it has taken centuries for that normalization to occur. Systems engineering in comparison with science and mathematics is a young subject that has a way to go before its concepts and terminology can be normalized to the level of precision and ubiquity that is currently enjoyed by science and mathematics. However, in the modern age of science and information technology, it should be expected that such normalization is a possibility within our grasp.

There are many national and international standards organizations whose work is related to architecture and systems engineering. This chapter will focus on seven of these organizations as the basis for the concepts and terminology used throughout the book:

1. International Organization for Standardization (ISO)
2. Institute of Electrical and Electronics Engineering (IEEE)
3. International Electrotechnical Commission (IEC)
4. Electronics Industries Alliance (EIA)
5. Electronics Components Association (ECA)
6. International Council on Systems Engineering (INCOSE)
7. The Object Management Group (OMG)

The ISO is the largest developer and publisher of a broad range of international standards. The IEEE is a nonprofit organization and is a leading professional association for the advancement of technology, publishing not only standards, but also a wide range of technical journals. The IEC is a leading organization that prepares and publishes international standards for electrical, electronic, and related technologies. The EIA is a trade organization representing and sponsoring standards for the U.S. high-technology community. The ECA is an associate of the EIA and manages EIA standards and technology activities for the electronics industry sector comprising manufacturers and suppliers. INCOSE is a nonprofit membership organization founded to develop and disseminate the interdisciplinary principles and practices to enable realization of successful systems.

The OMG is an international, nonprofit computer industry consortium with open membership that has task forces which develop enterprise integration standards for software-related technologies and industries. Being an open group, the OMG operates under a different business model from the previously mentioned organizations. All of OMG's standards are published without fees on an open Web site as well as a large body of reference material. Unlike professional societies such as the IEEE or INCOSE that raise operating funds through fees for the distribution of journals and standards, the OMG raises operating funds by other mechanisms.

The relationship between the systems engineering standards of the ISO, IEC, IEEE, and EIA can be understood by considering the level of detail provided and the breadth of scope across the system life cycle considered. The ISO/IEC 15288 standard has the greatest breadth of scope but at a lesser level of detail. The IEEE

1220 standard provides a much greater level of detail but at a reduced breadth of scope. The EIA 632 and EIA/IS 632 standards are in between these ISO/IEC 15288 and IEEE 1220 standards in terms of level of detail and breadth of scope.

Standards-Based Definitions of Systems Engineering

It may come as a surprise that most of the aforementioned standards do not offer a formal definition of the term *systems engineering*. This includes the ISO/IEC 15288 standard, both the 1998 and 2005 releases of IEEE 1220, and the 1998 release of the ANSI/EIC-632 standard.

However, in the 1994 release of the IEEE 1220 standard, the following definition was offered:

> Systems engineering is an interdisciplinary collaborative approach to derive, evolve, and verify a life-cycle-balanced system solution that satisfies customer expectations and meets public acceptability.

In the same year, the following definition of systems engineering was given in EIA/IS-632:

> Systems engineering is an interdisciplinary approach encompassing the entire technical effort to evolve and verify an integrated and life-cycle-balanced set of system people, product, and process solutions that satisfy customer needs. Systems engineering encompasses the following:
> - The technical efforts related to the development, manufacturing, verification, deployment, operations, support, disposal of, and user training for, system products and processes
> - The definition and management of the system configuration
> - The translation of the system definition into work breakdown structures
> - The development of information for management decision making.

This definition captures many of the aspects of the term *systems engineering* that are familiar to most practitioners. It is also attractive in that it offers a relatively simple sentence as the definition and separates the details of how systems engineering is accomplished (i.e., the technical and management efforts, etc.) from the definition.

Recall that Chapter 2 ("Logical and Scientific Approach") concluded with a suggested model-based definition of systems engineering: "systems engineering is the practice of engineering from the systems viewpoint." The EIA/IS-632 can be viewed as a straightforward interpretation of the suggested model-based definition. The four bullets of the EIA/IS-632 definition might be regarded as a short statement of how engineering is practiced, that is, an interpretation of the phrase "the practice of engineering" used in the model-based definition. Similarly, the entire statement of the first sentence in EIA/IS-632 might be considered an interpretation of the phrase "from a systems viewpoint" used in the model-based definition.

In contrast with the model-based definition of systems engineering, which established a clear relationship between the terms *engineering* and *system*, it might be considered that the EIA/IS-632 definition has intertwined a definition of engineering with a definition of a system. Most engineering disciplines would claim that the four bullets in the EAI/ES definition are already part of the practice of their discipline. This can lead to confusion, as is sometimes encountered by a new systems engineer who joins a project team only to be told that the engineering team is already performing the requisite "systems" engineering functions.

This is not to say that either definition is right or wrong, or even to suggest that one has any advantages over the other. Rather, each definition has organized the information contained in the concepts and terminology that we call "systems engineering" in different ways for different purposes. And therefore, the different definitions will have different utility in different contexts. However, the organization of the information will definitely influence how the terminology can be applied or extended. This difference will become clear in Chapter 4 ("Structure, Analysis, Design, and Models"), where the concept of model transforms will be first applied to systems engineering in this book. The definition of systems engineering considered at the end of Chapter 2 ("Logical and Scientific Approach") is easily extended to offer a definition of model-based systems engineering as "the practice of systems engineering from the viewpoint of models." This viewpoint will be made clear in a simple and straightforward way in Chapter 4. The reader should ask how the EIA/IS-632 definition could be extended to model-based systems engineering.

The Systems Engineering Vee

EIA/IS-632 also introduced the concept of the Systems Engineering "Vee" model, which has been interpreted and depicted in various ways. Figure 3.1 provides one such depiction. The Vee model is interdisciplinary. At the highest level, it represents repeated applications of definition and decomposition leading to the synthesis of a system design, then repeated applications of integration and verification (against system specifications) into the final product, which is validated against system requirements and customer intent.

On the left-hand side of the Vee, analysis results in increasing detail of specification, starting with decomposition of a system concept into user requirements, which are an agreement between the customer and the developer as to what will be delivered and what will constitute acceptance of the delivery. The analysis process generates system specifications from the concept and requirements, and from those specifications configuration items are specified. The specification process continues in greater level of detail until they are sufficient to generate computer code and be

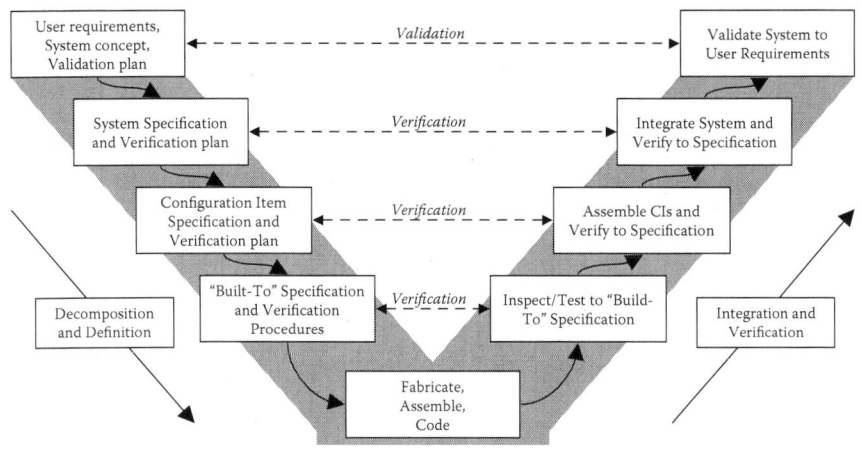

Figure 3.1 The Systems Engineering Vee model. (Adapted from Forsberg and Mooz "vee" model.)

used to either buy parts for the system or build them. Agreement must be reached at every stage of the process and controlled.

The right-hand side of the Vee is where the repeated applications of integration and verification are performed. The left-hand side of the Vee is synchronized with the right-hand side via validation plans at each stage of definition and decomposition in the analysis process. In practice, the Systems Engineering Vee is often implemented as a "waterfall" process, which is serial in nature, requiring significant amounts of time to actually execute. And it can be vulnerable to dramatic programmatic impacts when design flaws fail to be detected until the integration and verification phase of the system engineering process.

Implications of the Vee Model for the Logical Diagram for System

In the discussion of the EIA/IS-632 definition of *systems engineering*, a standards-based definition was compared with our suggested model-based definition of *systems engineering*. It was possible to show how an interpretation of the model-based definition could give the same meaning as the standards-based definition. Although the Vee speaks to the term *system* only indirectly, it does have implications for its definition. What is that implication (Figure 3.2)?

In Chapter 2 ("Logical and Scientific Approach"), two differing definitions of the term *system* were reconciled using logical modeling. The resulting definition became:

A <u>**system**</u> *is a **combination** of **interacting elements organized** to **realize properties, behaviors,** and **capabilities** that **achieve** one or more **stated purpose(**s**)**.*

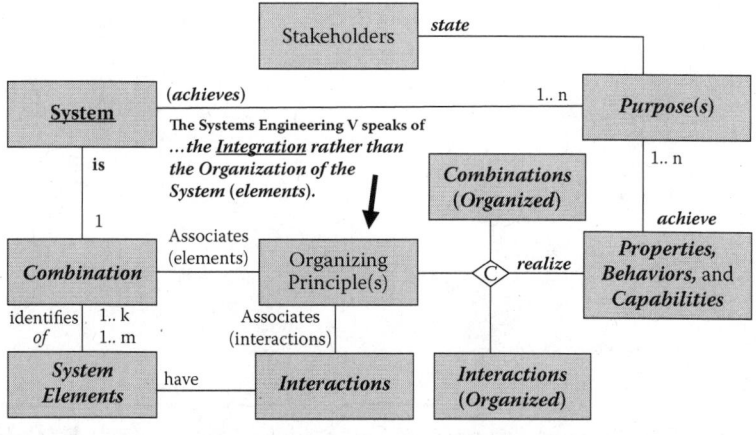

. **Figure 3.2   What is the implication of the Vee for *system*?**

> A system is a combination of interacting elements organized to real-
> ize properties, behaviors, and capabilities that achieve one or more
> stated purpose(s).

The basic principles underlying the Systems Engineering Vee are the repeated application of definition and decomposition followed by the repeated application of integration and verification. Although the term *verification* may be considered separately from the previous discussion of system realization, it is significant that the standards bodies have chosen the term *integration* rather than *organization* to describe the end state of the realized system.

The final modification to the definition of *system* must account for the central role of integration in the Systems Engineering Vee: A system is a combination of interacting elements *integrated* to realize properties, behaviors, and capabilities that achieve one or more stated purpose(s). The previous model for *system* is easily adapted for this change. Figure 3.3 exhibits the simple change.

Definitions and Models of Key Terms

This section will provide a collection of about two dozen key terms needed for the practice of architecture and systems engineering. Commonly accepted definitions will be cited; but when there are choices to be made between terms, we have chosen or modified definitions in a way that provides a collection of terms that support a model-based approach to architecture and systems engineering. This gives the

A <u>system</u> is a *combination* of *interacting elements organized* to *realize properties, behaviors,* and *capabilities* that *achieve* one or more *stated purpose*(s).

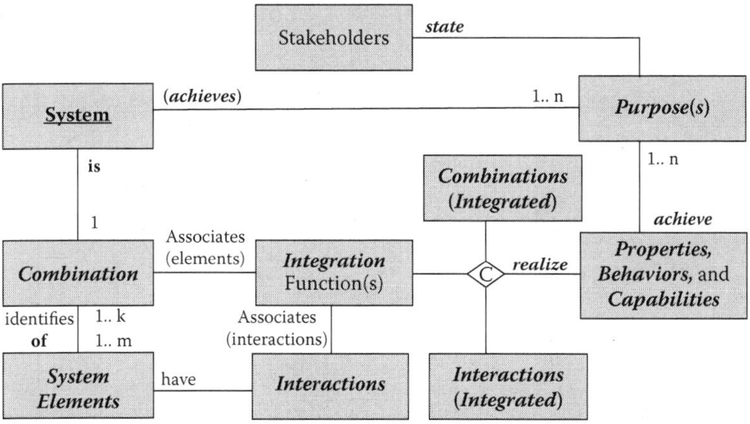

Figure 3.3 Adaptation of the INCOSE and Hitchins definition to the Systems Engineering Vee.

collection conceptual integrity. Consistency between terms has been maintained to the extent possible.

It is appropriate to recall the Principle of Definition:

> One needs both a formal definition of a design, for precision, and a prose definition for comprehensibility.

This principle should be applied to the key terms of systems engineering by defining them in natural language and logically modeling the definitions using UML and SysML. A few key models relating to *system* and *architecture* will be presented in this section. Models of terms not covered are suitable for class exercises and student projects.

A list of key terms follows:

Abstraction	Environment
Model	Context
Relationship/Relations	System
Function	System
Combination	Boundary
Arrangement	Interface
Interaction	Integration
System	Interoperation
System of Systems	Behavior
Family of Systems	Capability

Connection	Requirements
Structure	Engineering
System Architecture	Requirements Engineering

Definitions of each of these terms follow.

Abstraction

Abstraction is the suppression of irrelevant detail. Abstraction is a function of the human mind. The mind constantly seeks to make sense (seek meaning) of perceptions. As indicated in the Chapter 2 ("Logical and Scientific Approach"), abstraction relates to meaning and is accomplished by

Generalization
Deletion
Distortion

These three methods of abstraction may be complete, but they are not independent.

Model

In software and systems engineering practices, a model is typically considered to be an abstraction that represents a system. Modeling languages such as the Unified Modeling Language (UML) are considered to be technologies.

The predicate calculus of mathematical logic is the modeling language that establishes the logical foundation for all concepts of the term *model*. The concept of a model in mathematical logic is precise and well understood:

> A model is a relational structure for which the interpretation of a (logical) sentence in the predicate calculus becomes valid.

The relational structure is referred to as a model of the sentence. A relational structure is a set M and a collection of mathematical relations $\{R_m\}$ on the set. These logical concepts and terms will be discussed in Chapter 5 ("Architecture Modeling Languages").

Relations and Relationships

A relation is simply an assignment (or mathematical function) of one or more arguments (or logical variables) whose range is the set of truth values {true, false}. A relation of only two arguments is called a *binary relation*. An example of a binary relation is the relation "$\_ < \_$" on the set of real numbers. Writing this relation with algebraic variables interprets the relation as "$x < y$", where x and y are unspecified real numbers. A commonly used notation for relations uses a letter, such as R, to represent the relation. The statement $x < y$ is called an instance of the relation R. Instances of R are typically written in one of two ways: $R(x, y)$ or xRy. An instance of a relation, such as xRy, is referred as a relationship between x and y.

Combinations and Arrangements

A combination is a subset of elements selected or identified from a larger set. No reference is made to relationships or order. An arrangement is an ordering or permutation of a combination or combinations of elements.

The distinction between the terms *combination* and *arrangement* can be easily seen by a natural language example using collections of letters. Let A be the set of all letters in the English alphabet. The subset of A denoted by the listing {a, e, p, r} is a combination of letters and, in fact, {e, r, p, a} is the same combination. The following orderings are arrangements of the given combination:

$$[p, e, a, r], [r, e, a, p], [e, r, p, a]$$

The first two have meaning in the English language, when a higher-order relationship between the letters is established by removing the commas from the ordering. The third ordering does not have a natural language meaning but it is still an arrangement of the letters in the original combination.

Arrangements can also repeat letters. The ordering "a peer" is a meaningful arrangement of the two combinations: {e, e, p, r}, {a} derived from the original combination, {a, e, p, r}.

Permutations are a special type of arrangement in which this sort of repetition is not allowed. In strict mathematical terms, a permutation of the set {a, e, p, r} would be a one-to-one mapping of the set onto itself.

Interactions

Interactions are a key term in many definitions of the terms *architecture* and *system*. Interactions involve the exchange (or transport) of energy, material, or information between system elements. See, for example, Hatley (2000), Hitchins (2003), or paragraph 0.2.3 of ISO/IEEE standard15704. Events are the "atomic" level of detail in an interaction (in the logical or Aristotelian sense).

Transport is the movement of energy, material, or information between elements of a system that is associated with the event of an interaction between two elements. Movement is meant in the sense of physics; that is, matter or energy change physical location over a period of time. Although information is an abstraction, in (physical) systems it is transported by energy or material. The term *transport* is logically modeled in Figure 3.4.

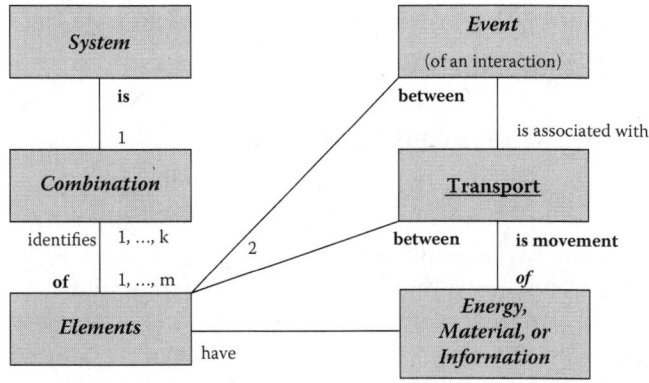

Transport is the *movement of energy, material, or information between elements of a system that is associated with the event of an interaction between two elements.*

Figure 3.4 Diagram of the term *transport*.

System

The term *system* has already been discussed and analyzed extensively. The definition arrived at for the purpose of this book is the adaptation of the definitions of Sowa and the Systems Engineering Vee to the reconciled INCOSE and Hitchins definitions:

> A system is a combination of interacting elements integrated to realize properties, behaviors, and capabilities that achieve one or more stated purpose(s).

The logical diagram for system was given in Figure 3.3.

System of Systems (SoS)

The term *system of systems* has been a topic of active discussion for several years and likely will continue to be in the years to come. From the many definitions to choose from, the following three give a sense of the breadth of interest, authority of sources, and representative definitions:

From the DoD Defense Acquisition Guidebook (Defense Acquisition University 2004):

> A system of systems is a set or arrangement of systems that results when independent and useful systems are integrated into a larger system that delivers unique capabilities.

From the INCOSE Systems Engineering Handbook V.3 (INCOSE 2007):

> System of systems applies to a system-of-interest whose system elements are themselves systems; typically, these entail large-scale interdisciplinary problems with multiple, heterogeneous, distributed systems.

From the U.S. DoD JCIDS Manual (Chairman of the Joint Chiefs of Staff 2007b):

> A system of systems is a set or arrangement of interdependent systems that are related or connected to provide a given capability. The loss of any part of the system will significantly degrade the performance or capabilities of the whole.

Each of these definitions makes the point that an SoS consists of systems. Each definition also tries to attribute system-level properties to the SoS. There is an opinion among many systems engineers and other stakeholders that if the term *system* were correctly defined, then it could be applied to itself to give a meaningful definition for SoS. This is a reasonable test to apply to our model-based definition of system. Applying the model-based definition of system to itself gives the following:

> A system of systems is a combination of interacting systems [i.e., elements of the SoS] integrated to realize properties, behaviors, and capabilities that achieve one or more stated purpose(s).

The logical model for *system* is readily applied to itself to produce a model of SoS that is based on a consistent definition of *system*. This definition does not indicate whether all of the elements are capable of directly achieving the stated purpose of the SoS. Each of the "interacting systems" has properties, behaviors, and capabilities that when integrated achieve one or more (of the) stated purposes. However, some of the "interactions" must play the role of integrating or supporting functions.

There are three key properties of the SoS that give rise to this subtlety. First, each of the systems of interest (SOI) might be independently capable of achieving the stated purpose, in that systems in the SOI are considered to be in some way peers. However, as a system, the SoS has emergent behavior that is realized through the integrating functions of the SoS. The "other elements" of the SoS are infrastructure that either provide connectivity, integration, or support of the SOI.

Family of Systems (FoS)

The term *family of systems* has been discussed for several years, although not as topical as the term *system of systems*, and likely will continue to be in the years to come. However, FoSs are actually more common than SoSs. Military forces, especially joint forces, routinely assemble families of systems for specified mission objectives. The key term here is *assemble*, rather than integrate. The family must have a viable way to interoperate, but this is a more relaxed requirement than to actually integrate the systems. From the many definitions to choose from, the following two give a sense of the breadth of interest, authority of sources, and representative definitions:

From the U.S. DoD JCIDS Manual (Chairman of the Joint Chiefs of Staff 2007b):

> A family of systems is a set of systems that provides similar capabilities through different approaches to achieve similar or complementary effects.

From the U.S. Under Secretary of Defense for Acquisition, Technology and Logistics (Office of the Under Secretary of Defense 2006):

> A family of systems is a set or arrangement of independent (not inter-dependent) systems that can be arranged or interconnected in various ways to provide different capabilities. The mix of systems can be tailored to provide desired capabilities dependent on the situation.

The term *family*, in a strict sense, has a weaker meaning than the term *system*. To use the model-based definition of system to describe a family requires the relaxation of two terms: interacting and integrated (which are distinctions of systems rather than families). The resulting definition is as follows:

> A family of systems is a combination of related systems [i.e., elements] organized to realize properties, behaviors, and capabilities that achieve one or more stated purpose(s).

The diagram of the logical model is easily adapted to accommodate this change in detail.

The key properties of the FoS are that each of the SOI have related properties, behaviors, and capabilities for the purpose of achieving the stated purpose. These relationships could be similar or complementary. As a family that may not be a system, the FoS may not have emergent behavior that is realized through integration. Similar to the SoS properties, the other elements of the FoS other than the SOI are infrastructure that either provide connectivity, interoperability, or support for the SOI but may not be integration elements.

Structure

There are many definitions of the term *structure,* and sometimes structure is confused with arrangement. In the *Oxford English Dictionary*, it is clear that its definition as both a noun and an adjective has a stronger meaning than the term *arrangement*:

Noun: The mutual relation of the constituent parts or elements of a whole as determining its peculiar nature or character; make, frame.

Adjective: Organized or arranged so as to produce a desired result. Also loosely, formal, organized, not haphazard.

Structure is distinguished from arrangement in that it refers not just to the system elements, but also to system properties.

The meaning of the term *structure* has been the subject of discussion among noted philosophers for at least a century. The book *Knowledge Representation* (Sowa 2000) has built on this philosophical history to offer a precise definition of the term for the structure of a system. The following adaptation of Sowa's definition of structure for systems engineering purposes is offered:

> System structure is a stable instance of connecting or binding components of a system that explains how the junctures are organized to realize some property, behavior, or capability of the system.

A simple example from arithmetic serves to illustrate this point. Consider the arrangement 101010 of the combination of digits {0, 1}.

> In base 10, it means 101,010: 100,000 + 1,000 + 10
> In base 2 (binary), it means 42: 32 + 8 +2

Structure is distinguished from "arrangement" in that it refers not just to the relationships of the system elements (digits in this case), but also to system properties (the value of the number).

System Architecture

There are various engineering definitions of this term and the unscoped term *architecture* is reported to have at least 130 definitions around the world. We will focus on two key definitions, one from the electrical engineering community and the other from the software engineering community.

From the IEEE Std 610.12-1990:

> [System] Architecture is the organization of the system components, their relations to each other, and to the environment, and the principles guiding its [i.e., the system's] design and evolution.

From the OMG MDA™ (Miller and Mukerji 2003):

> The architecture of a system is a specification of the parts and connectors of the system and the rules for the interactions of the parts using the connectors.

The key attributes of the historical definitions (i.e., those related to the art and science of designing physical structures) are structure, utility, and beauty. According to Pollio (ca. 27 BC), a good building should satisfy three principles: *Firmitatis, Utilitatis, Venustatis* (durability, utility, and beauty).

Figure 3.5 depicts one possible logical model of the IEEE definition of system architecture. Figure 3.6 depicts one possible logical model of the OMG MDA™ definition of system architecture The architecture of a system is a specification of the parts and connectors of the system and the rules for the interactions of the parts using the connectors. The two logical diagrams can be used to argue that

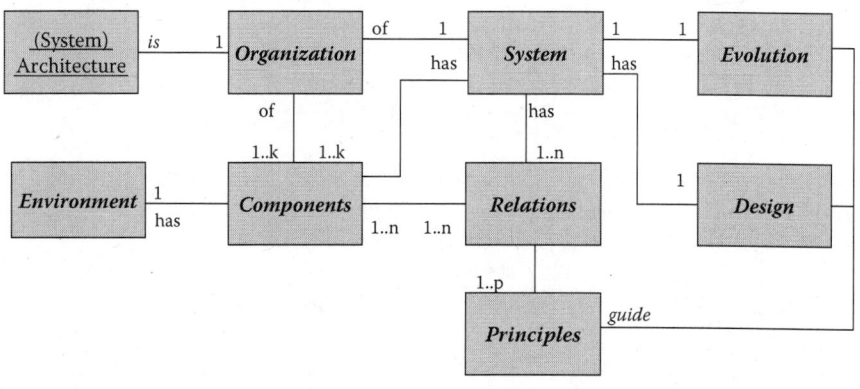

Figure 3.5   A model of system architecture based on IEEE Standard 610.12.

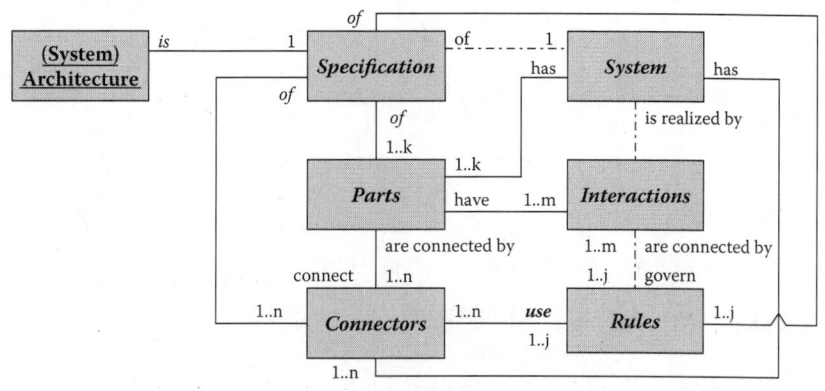

Figure 3.6   A model of the OMG MDA™ definition of system architecture.

the MDA™ definition is a more narrowly focused interpretation (specialization) of the IEEE definition.

System Boundary

There are various definitions of the concept of a system boundary that are rooted in science and engineering. Some are entropy based and others are more deterministic, such as the boundary of an object from which electromagnetic energy is scattered. In systems engineering, system boundaries are sometimes just a rule for what is in the system and what is not. The following proposed engineering definition of *boundary* seeks to normalize these different approaches in a way that lends itself to our model-based concept of a system:

> The system boundary is a demarcation drawn between the system and the environment across which energy, material, or information can be transported between the system and its environment.

Accordingly, a *boundary element of a system* is a system element that interacts directly with the environment and relates to at least one other system element. The *system frontier* is the collection of all of the system boundary elements.

System Interface

Although the following two authoritative definitions derive from military sources, they are well suited to define system interfaces in terms of the previous definitions established.

From the U.S. DoD dictionary (Department of Defense 2008):

> A system interface is a boundary or point common to two or more [compatible] systems across which necessary information [energy or material] flow takes place.

From the Military Handbook 61A (Department of Defense 2001):

> Interfaces are defined in physical and functional characteristics that exist at a common boundary with cofunctioning items that allow systems, equipment, software, and data to be compatible.

System Behavior

System behavior is a well-understood concept in most engineering disciplines. The following succinct definition is representative of the concept:

> System behavior comprises the actions or reactions of a system, in relation to the environment under specified conditions or circumstances.

On the other hand, emergent behavior, which is central to both systems engineering and the model-based definition of system used in this book, is a subject of greater debate. Emergence refers to the way system behavior arises out of a multiplicity of interactions that cannot be attributed simply to the individual actions alone. Examples include a phased array radar, which has emergent behavior (N^2 law) that is at the SoS level (i.e., the radar comprises N smaller "radars" (transmit/receive modules) whose energies are coherently processed on both transmit and receive to yield a nonlinear improvement in system performance.) On the other hand, a mechanical clock also has emergent behavior but not in the same way as a phased array radar. Each part of a mechanical clock, whether a gear, a spring, or a dial, must be carefully synchronized. If not, the clock will not present the correct time to the user. Both synchronization and time are higher-order properties of the system referred to as a clock.

Capability: Military Definitions

The concept of *capability* was first introduced in the late 1990s by the U.K. Ministry of Defence (MOD) as a way to break down the stovepipes of systems tailored to a narrow mission that the Cold War had produced. The U.S. DoD responded favorably to this concept, and over a period of just a few years settled upon the following definition:

From the U.S. DoD Joint Staff (Chairman of the Joint Chiefs of Staff 2005):

> A capability is the ability to achieve a desired effect under specified standards and conditions through combinations of ways and means to perform a set of tasks.

A definition of capability from the U.K. Ministry of Defence (MOD):

> A capability is the ability to execute a specified course of action that is defined by a user and is expressed in non-equipment-based operational terms.

These definitions are actually concordant, and the focus on the user in the MOD is what made successes in the United States possible during the early years of capabilities-based acquisition.

Requirements Engineering

The objective of requirements engineering is the generation, verification and validation, and management of product and service requirements to identify customer needs. The generation of requirements can involve system definition and decomposition, and the development, derivation, and allocation of requirements. Management of requirements involves control and documentation.

The requirements engineering process delineates the major tasks involved, identifies the relative sequence in which they are performed, identifies the inputs and outputs of interfacing entities, and defines the objective evidence required for effective systems engineering.

Meeting the Challenge of Standardizing Systems Engineering Terminology

The challenge of standardizing systems engineering terminology crosses international boundaries of professional societies and standards groups. Systems concepts and principles are currently an active topic of discussion among the larger community. The International Organization of Standardization (ISO) is engaged with issues pertaining to architecture concepts through working groups: JTC1/SC7/WG42 and TC184/SC5/WG1.

INCOSE, the International Council on Systems Engineering, has various initiatives, including special working groups on model-based architecture and systems engineering. And at the INCOSE International Workshop in 2008, a special working group was launched to support the efforts of ISO. For the past eight years, the Object Management Group (OMG) has matured Model Driven Architecture (MDA™), which is linked to systems engineering through INCOSE. And OMG has a special working group to develop next-generation concepts and standards for model transformations, which are at the heart of MDA.

The IEEE has also recently initiated the International Council on Systems of Systems (ICSOS).

Defense interests include the U.S. Office of the Secretary of Defense, which has efforts under way in SoS engineering.

Summary

There are several standards for systems engineering of varying breadth of scope and level of detail. The Systems Engineering Vee is rooted in standards and embodies key concepts of the systems engineering process: definition and decomposition, system synthesis and integration, and verification and validation. It provided insight into the logical model of *system* developed in Chapter 2 ("Logical and Scientific Approach") that has influenced the definition developed and used in this book. The resulting logical model accounts for and normalizes several points of view.

The terms of systems engineering have different meanings depending on the user; but as part of a model-based approach to architecture and systems engineering, it was necessary to establish a baseline of terminology that was consistent and had conceptual integrity. The concept of *model* used for logical modeling in this book is founded on mathematical logic, where the concept and terminology for *model* is precise and is well understood. (The mathematical foundation will be presented in Chapter 5 ["Architecture Modeling Languages"].) The ISO, IEEE, INCOSE, and OMG, as well as DoD and MOD, are working to normalize concepts and terminology.

References

ANSI/GEIA EIA-632. 2003. Processes for Engineering a System.

Chairman of the Joint Chiefs of Staff Manual. 2007. Operation of the joint capabilities integration and development system. http://www.dtic.mil/cjcs_directives/cdata/unlimit/m317001.pdf

Chairman of the Joint Chiefs of Staff. 11 May 2005. Chairman Joint Chiefs of Staff Instruction (CJCSI) 3170.01B.

Defense Acquisition University. 2004. *Defense Acquisition Guidebook.* https://acc.dau.mil/CommunityBrowser.aspx?id=106843

Department of Defense. 2001. MIL-HDBK-61A(SE) Military Handbook 61A (SE) *Configuration Management Guidance.*

Department of Defense. 2008. Joint Publication 1-02 *Dictionary of Military and Associated Terms.* http://www.dtic.mil/doctrine/jel/new_pubs/jp1_02.pdf

Forsberg, K. and Mooz, H. 1991. The relationship of system engineering to the project cycle. Joint Conference sponsored by the National Council of System Engineers and the American Society for Engineering Management, Chatanooga, TN, October 1991.

Hatley, D., Hruschka, P., and Pirbhai, I. 2000. *Process for Systems Architecture and Requirements Engineering.* New York: Dorset House Publishing.

Hitchins, D.K. 2003. *Advanced Systems Thinking, Engineering and Management.* Boston: Artech House.

Institute of Electrical and Electronics Engineers (IEEE). 1990. IEEE Std 610.12-1990 IEEE Standard Glossary of Software Engineering Terminology.

Institute of Electrical and Electronics Engineers (IEEE). 1998. IEEE Std 1220. Standard for the Application and Management of the Systems Engineering Process.

International Council on Systems Engineering (INCOSE). August 2007. *INCOSE Systems Engineering Handbook,* v.3.1. Seattle: INCOSE Central Office.

ISO 15704. 2000. Industrial Automation Systems—Requirements for Enterprise-Reference Architectures and Methodologies.

ISO/IEC 15288. 2008. Systems and Software Engineering—System Life Cycle Processes.

Miller, J. and Mukerji, J. 2003. MDA Guide Version 1.0.1. OMG. http://www.omg.org/docs/omg/03-06-01.pdf (accessed 8 August 2008).

Office of the Under Secretary of Defense (Acquisition, Technology and Logistics), Deputy Under Secretary of Defense (Acquisition and Technology), Director, Systems and Software Engineering. 22 December 2006. *System of Systems Systems Engineering Guide, Version. 9.*

Marcus Vitruvius Pollio. ca. 27 BC. *De Architectura.*

Sowa, J. 2000. *Knowledge Representation.* Pacific Grove, CA: Brooks/Cole.

Chapter 4

Structure, Analysis, Design, and Models

The four keywords in the title of this chapter are at the core of systems engineering, software engineering, and architecture. These terms are used in similar but different ways by the engineering and technology communities. For example, in Model Driven Architecture (MDA™), "analysis" is often considered to be accomplished as part of developing "models" of the system. And "design" in MDA is accomplished through model transformations. Also, in both software and systems engineering, the term *structure* is often associated with *architecture* but is usually not associated with the term *model*. However, in mathematical logic, the term *model* has a precise meaning based on relational structures and sentences about those structures. This will be discussed in detail in Chapter 5 ("Architecture Modeling Languages"). Simple matrix representations of models and model transformations will be presented at the engineering level in this current chapter. The matrix representations are based on the concepts of mathematical logic from Chapter 5. The concept of model as a relational structure is the basis for understanding better how the four keywords *structure, analysis, design*, and *model* can be used in a consistent way across software and systems engineering. This better understanding will result in a shift in the focus of system analysis and design from the parameters of a system to parametric models of systems.

Key Concepts

Logical models
Relational structures
Model transformation
Matrix representations

Logical Models

This section introduces the concepts of a logical model without directly using the symbolic language of mathematical logic. Logical models provide the next level of precision after the UML logical diagrams, discussed in Chapter 2 ("Logical and Scientific Approach"). The fundamental concept underlying logical modeling is that truth is to be found not in entities themselves but rather in the relationships and interpretations of entities.

In the predicate calculus introduced in Chapter 5 ("Architecture Modeling Languages"), a logical model is defined as a relational structure and a collection of sentences, which can be interpreted as true in the relational structure. The precise meaning of *sentence*, *interpreted*, and *true* will be given in Chapter 5. In the remainder of this chapter, the focus will be on the relational structure, that is, the relations and relationships on a set of parameters used to model a system.

Herein lies the fundamental difference between the parametric view of a system and a model-based view. In the parametric view, the focus in on the parameters that are used to describe the system. Sentences about the parameters may be in the form of equations or may be in natural language, both of which are typically captured in documents. However, unless this information is captured within a proper database with reference to a logical structure, the knowledge of the system cannot be accurately expressed and managed. Logical models provide a straightforward means of making systematic reference to the relational structure reflected in the knowledge about the system and its parameters.

Using Logical Models in a Document-Based Systems Engineering Process

Logical models can be introduced at any point in a document-based systems engineering process. The ability to introduce the precision of logical modeling at any point in an engineering process is both powerful and attractive. For example, the documents could be taken from any level of decomposition and definition in the Systems Engineering V.

In this section, it will be shown how to introduce logical modeling into a document-based engineering process. This will provide an illustration of the intimate relationships between logical modeling, model transforms, and system design.

In document-based systems engineering, relationships between the system parameters are generally provided in natural language narrative descriptions. Following the procedures of Chapter 2 ("Logical and Scientific Approach"), the UML diagrams of the sentences explicitly expose these relationships graphically and they are easily captured into a database. The relationships can also be depicted using a square matrix, as illustrated in Figure 4.1. Every node in a UML diagram of

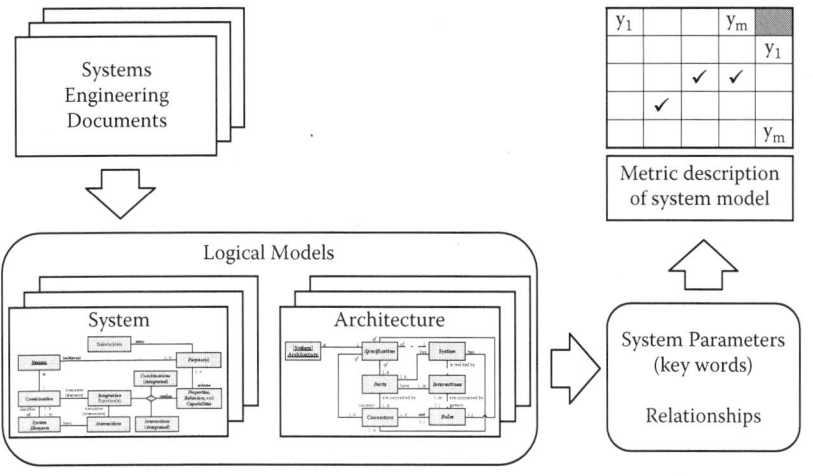

Figure 4.1 **Deriving a system model from systems engineering documents.**

a sentence corresponds to a keyword (parameter) in the sentence, and every line in the diagram corresponds to a relationship between two parameters.

Logical modeling provides a systematic method of creating models from parameters and narratives. This approach can capture far more knowledge about the system than just lists of parameters specified by ranges of values. The approach provides the bridge between document-based and model-based systems engineering.

System Design and Model Transforms

System design relates to how one parametric model of a system is transformed into another parametric model. At a minimum, a model transformation should preserve the relationships between the parameters of the model. However, not all transformations of the parameters will be model transformations.

It is intuitive that model transforms should preserve the relationships between the parameters being transformed. This is well understood in various schools of information theory and computer science. See, for example, the survey article (Walicki and Bialasik 1997) on relation-preserving functions. Relation-preserving functions between relational structures are called *homomorphisms* in the study of universal algebra. The following definitions and notations presented in simple matrix form are an extension of the definition of what is called a *weak homomorphism* of a relational structure. The key difference is that in universal algebra the homomorphisms are *single-valued*; that is, they are functions in the sense of mathematics. For the engineering application of this concept, we must allow for *multi-valued* transformations, as described later. It should be noted that the definition of model transform using matrix notation in the following text reduces to the definition of weak homomorphism of a relational structure whenever the transform is single-valued, that is, whenever it is mathematically a function.

This concept of relation-preserving model transforms can be applied to system design in a very broad sense. It will be seen later in this chapter and in subsequent chapters that this approach provides a common logical basis for system design in both software engineering and systems engineering. See Chapters 16 ("Model Driven Architecture"), 19 ("Systems Design"), and 20 ("Advanced Design Methods").

Model Transformations in MDA™

The software engineering community has more fully developed the concept of model transformations than has the systems engineering community. This will be seen in Chapter 16 ("Model Driven Architecture"), where model transforms are at the heart of system design for Model Driven Architecture (MDA). The concept of relationship-preserving model transforms applies directly to the MDA concept of model transform described in Chapter 16. However, the MDA concept of model transform currently does not rest upon the logical foundation of the definition given in this chapter. Current initiatives at the OMG will begin addressing this question and can be expected to establish the next-generation definition of the term *model transform* as used by the OMG.

Quality Function Deployment (QFD) in Systems Engineering

In systems engineering, the method known as Quality Function Deployment (QFD) is related to the concept of model transforms described in this chapter. In Chapter 19 ("Systems Design"), a QFD procedure will be presented for developing an implementation model (system characteristics) from an essential model (system requirements). However, these concepts of "model" historically have not accounted for model transforms. Similarly, Chapter 20 ("Advanced Design Methods"), which provides an overview of the state-of-the-art methods for the analysis of relationships between system parameters, has not historically addressed the relationships from the viewpoint of logical modeling and model transformations. Chapters 19 and 20 will show how the concept of relationship-preserving model transforms apply to both QFD and the use of surface response equations in system design.

A Matrix Notation for Models and Model Transformations

The concept of representing the relational structure of a model using a matrix was introduced in the section on using logical models in a document-based process. This section will formalize a notation that can be used with the type of matrix introduced in Figure 4.1 and extend the notation to model transformations.

Matrix Notation and Conventions for Models

The relational structure of a model will be represented by a matrix designated by an underlined capital letter, such as $\underline{M}$. The underlying set of parameters will be designated by a capital italics letter, for example, $M = \{y_1, y_2, \ldots y_{m-1}, y_m\}$. A relation on the parameters M will be denoted as a normal capital letter, for example, R = $\{(y_2, y_{m-1}), (y_2, y_m), (y_{m-1}, y_2)\}$. The ordered pairs in R correspond to the tick marks or other entries in the matrix $\underline{M}$.

The square matrix $\underline{M}$ depicted in Figure 4.2 represents an instance of a relational structure for a parametric model related to a system. For example, it could be a model of the problem to be solved or it could be a model of the system requirements. It displays both the names of the parameters and the relationships between the parameters. Natural language, logical variables, and metrics are each suitable for naming the parameters. In Figure 4.2 the matrix representation $\underline{M}$ of the relational structure indicates when a relationship exists between pairs of parameters. Each tick mark in the matrix denotes that a relationship exists, and it associates the parameter in the designated row with the parameter in the designated column.

The convention for reading the matrix $\underline{M}$ is that the row entries relate to column entries (in the sense of mathematical relations and relationships). For example, y_2 is related to y_m in the matrix $\underline{M}$, as depicted in Figure 4.2. It is important to remember that the relationships need not be symmetric. Thus, although y_2 is related to y_m, y_m need not be related to y_2. In the example of the matrix $\underline{M}$ in Figure 4.2, there is no symmetry in the relationship between y_2 and y_m, but there is symmetry in the relationship between y_2 and y_{m-1}.

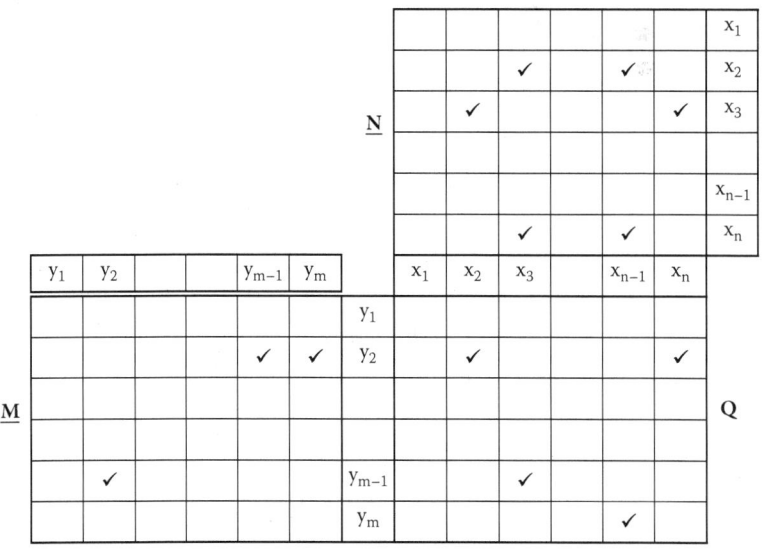

Figure 4.2   Using model transformations in system design.

Matrix Notation and Conventions for Model Transformations

Figure 4.2 also illustrates a model transformation. In the notations in Figure 4.2, the parameters of matrix $\underline{M}$ are transformed by Q into the parameters of the matrix $\underline{N}$. The transformation of the parameters gives rise to a transformation of the matrix $\underline{M}$ into the matrix $\underline{N}$. Not all transformations of the parameters will be model transformations. At a minimum, a model transformation should preserve the relationships between the parameters of the model. Transforms will be denoted by normal capital letters because mathematically they are just relations between two sets, for example, the set M and the set N.

The model $\underline{M}$ has three relationships; that is, in the notation described earlier, $R = \{(y_2, y_{m-1}), (y_2, y_m), (y_{m-1}, y_2)\}$. The model transformation Q acts on values of the three parameters in M: y_2, y_{m-1}, and y_m. The action of Q on y_2 is doubled-valued, so combined with the symmetric relationship between y_2 and y_{m-1}, this maps to two symmetric relationships in the model $\underline{N}$ (a total of four individual relationships). The double-valued action of Q on y_2 and single-valued action on y_m transforms the other relationship that y_2 has with y_m into two relationships in $\underline{N}$. Therefore, Q transforms the three relationships in $\underline{M}$ into the six relationships in $\underline{N}$ depicted in the figure. This relation on N will be designated as RQ.

In general, a model transform is specified only for the system parameters, but it will induce a mapping between the relationships in any model of the parameters. The following notation provides a way of calculating the transformation of relationships in a matrix notation.

First, the matrix Q transforms the parameters $y_1 \ldots y_m$ into another set of system parameters $x_1 \ldots x_n$ by associating one or more of the x_j's with each of the y_i's. We shall use the notation $y_i Q x_j$ to mean that Q has associated y_i with x_j. As a binary mathematical relation, this means that the ordered pair (y_i, x_j) belongs to Q.

To see the implied transformation of the models, we will extend the notation yQx to the notation RQ to show how Q transforms a subset R of $\underline{M}$ into a subset RQ of $\underline{N}$. Recall that a subset R of $\underline{M}$, in general, represents a mathematical relation on the underlying sets of parameters. RQ is defined as follows: for each pair of parameters (y_i, y_k) that belong to the mathematical relation R, if $y_i Q x_j$ and $y_k Q x_l$ then the pair (x_j, x_l) belongs to the mathematical relation RQ in $\underline{N}$.

Model Transformation in a Simple Design Problem

Insight into the meaning of the matrix representation of the model transform in Figure 4.2 can be gained from a simple example. Suppose that a given system is used to perform various tasks and that there is a specified list of enabling functionalities that the system executes to perform the tasks. When the system is used for a stated purpose, a subset of the task list is identified; that is, certain tasks must be performed by the system for the stated purpose to be achieved. The order in which the tasks must be performed is not arbitrary. There is a precedence order to the identified tasks.

The simple system design question is this: what is the precedence order of the system functions? The answer to this question affects resourcing, scheduling, and possible interoperations between the system functions that will be used. It is important to answer the question correctly and to understand that the answer is not independent of the precedence order of the tasks to be performed. The question must not be answered without regard to the precedence order of the tasks to be performed. Model transformations can be used to answer this question about the system functions in a straightforward, repeatable way that lends itself to other systems analyses.

In Figure 4.2, the matrix $\underline{M}$ represents a model of the task list M that the given system must perform, and the set N represents a list of the functionalities that the system can execute. The matrix Q would then represent a model transform of $\underline{M}$ into $\underline{N}$ that associates tasks from the list M to be performed by the system with enabling functions from the list of system functions N, where the relation on N given by RQ is the relational structure for the model $\underline{N}$.

Thus, the relation R, which represents the precedence order for the performance of the tasks, induces a relation RQ, which is the precedence order for the execution of the system functions.

Details of the Matrix Transformation in the Simple Design Problem

The simple design problem can be interpreted into the details of Figure 4.2 as follows: There are exactly three tasks to be performed: y_2, y_{m-1}, and y_m. The checkmarks in the matrix $\underline{M}$ represent the relation R on M, which correspond to the precedence relationships between the tasks. For example, the relationship (y_2, y_m) in R means that the task y_2 must be performed before the task y_m. Note that the symmetry in the relationship between the tasks y_2 and y_{m-1} means that these two tasks must be performed simultaneously.

The transform Q makes four associations between the parameters listed in M and those listed in N, as described in the previous section, "Matrix Notation and Conventions for Model Transformations." The double-valued action of Q on y_2 means that the task y_2 requires the support of two system functions. The single-valued action of Q on y_{m-1} and y_m means that each of these two tasks only requires support from a single system function.

By letting RQ be the relation that is associated with making N into the model $\underline{N}$, the transform Q becomes a relational transform; that is, in the relational structure (N, S) we have set $S = RQ$. The relational transform reflects the precedence order of performing tasks in the precedence order of executing system functions.

Structured Analysis and Design

Systems engineering draws from the principles of structured analysis and design that were originally developed by the software community. These principles can give insight into a model-based approach to architecture and systems engineering. The principles of structured analysis and design are simple but key software engineering concepts that have deep implications for systems engineering, especially from the viewpoint of a model-based approach. The application of logical modeling to structured analysis provides a bridge between document-based and model-based systems engineering that adheres to the principle of definition. And the transformation of models will be seen not only to give a logical foundation for structured design but also to give new insights into structured design itself. The fundamental implications of a model-based approach to structured analysis and design will be introduced later and further elaborated in subsequent chapters, especially Chapter 16 ("Model Driven Architecture").

Structured Design

The principle of structured design can be summarized succinctly as follows:

> Systems should be comprised of modules, each of which is highly cohesive but collectively are loosely coupled.

Cohesion minimizes interactions between elements not in the same module, thus minimizing the number of connections and amount of coupling between modules. Levels of cohesion are generally distinguished by types of association but not strictly ranked in the following order:

Coincidental
Logical
Procedural
Temporal
Sequential
Communication
Functional

Coincidental association is the weakest type of cohesion. Logical association is the first meaningful type. It includes precedence order but is not meant to necessarily imply sequential ordering, which is a stronger type of cohesion. Similarly, temporal association need not imply sequential order. And it may be argued that communication and functional associations are the two strongest types of association because only these two types involve exchanges between entities.

The emphasis on cohesion based on functionality is a common practice when structured design is applied to systems engineering. Although this practice derives from software engineering, we will see in Chapter 16 ("Model Driven Architecture") that functionality is not the basis for cohesion in MDA. Instead, cohesion is achieved by organizing by subject matter. This makes sense because MDA is concerned with software applications for information-intensive systems.

Structured design also considers that the solution should reflect the inherent structure of the problem. Although this consideration may have an intuitive meaning, its true meaning and implications cannot be properly understood without the formalism of models used in the predicate calculus of mathematical logic. The matrix representation of models introduced earlier in this chapter will be used to understand the meaning of the concept that "a solution should reflect the inherent structure of the problem."

Structured Analysis

The principle of structured analysis can be summarized succinctly as follows:

> The specification of the problem should be separated from that of the solution.

Two types of system models are used in structured analysis:

Essential (implementation-free representation)
Implementation (functional, behavior-specific)

In software engineering, the implementation model is a realization of the essential model that exhibits the organization of the hardware, software, and computer code of the software system.

Systems engineering practices for the development of the implementation model will be addressed in Chapter 19 ("Systems Design") and in the Chapter 24 case study ("Long-Range Strike System Design Case Study"). Therefore, this section will focus on the essential model.

The essential model (Yourdon 1989, p. 326), comprises the following:

Environmental model
Behavioral model

The environmental model defines the boundary between the system and the environment.

It provides a short description of the system purpose, a context diagram, and an event list.

The behavioral model identifies what the flows are through the system and establishes controls over the system flows. In this way, a model of what is flowing through the system is developed. It also determines the possible system states and transitions. Detailed examples of this modeling approach are provided in Chapter 16 ("Model Driven Architecture"). The graphical tools most commonly used to model system flow, control of flow, states and transitions, and data are presented in Chapter 7 ("DFD and IDEF0 Graphical Tools"), Chapter 8 ("Rule and State Modeling"), and Chapter 9 ("Data and Information Modeling").

Ed Yourdon is considered the father of structured analysis and design. A more extensive discussion of structured analysis can be found on his Web site http://www.yourdon.com/jesa and his Wiki http://www.yourdon.com/strucanalysis/wiki.

Using Logical Models in Structured Analysis and Design

Logical models can play a key role in structured analysis and design. If the principles of structured analysis and design are introduced at the beginning of the Systems Engineering V, then the specification of the problem will be separate from the specification of the solution. The *essential model* will define the boundary between the system and its environment. It will also provide a short narrative description of the system purpose and a context diagram. Logical models can be derived from the sentences of a proper narrative description supported with the boundary definition and context diagram. In this case, the keywords in this *essential* logical model should include words that will become system requirements. The principle of structured design seeks, among other things, to ensure that the solution should reflect the inherent structure of the problem.

Relation of Model Transformations to Structured Analysis and Design

Recall that the principle of structured design seeks, among other things, to ensure that the solution should reflect the inherent structure of the problem. With the preceding definitions, it is now possible to give this concept a more precise meaning. In this case, the problem would be modeled by $\underline{M}$ and the solution would be modeled by $\underline{N}$. The structure of the problem, as represented by the matrix $\underline{M}$, is "reflected" into the solution $\underline{N}$ by the transformation Q. Specifically, at the level of relationships, R is precisely all the pairs (y_i, y_k) that are related in $\underline{M}$, and the "reflection" of R is represented by the mathematical relation RQ in $\underline{N}$. It is clear that this concept from structured design can be applied at any level of system design.

However, the term "reflected into $\underline{N}$" must be given a more precise meaning because $\underline{N}$ is not just a set; it is a model. Let S be that subset of $\underline{N}$ which contains precisely all the pairs (x_j, x_l) of $\underline{N}$ that are related in the model $\underline{N}$. To say that R is "reflected" into the solution $\underline{N}$ must imply that RQ is a subset of S; that is, Q transforms relationships in $\underline{M}$ into relationships in $\underline{N}$. The transform Q is called weak; that is, it only preserves relationships. A stronger type of transform would be one that preserves relations; that is, each R_mQ is a subset of some S_n in S, where R_m is a relation in R and S_n is a relation in S.

Summary

Models and model transformations as used in this chapter can provide a common basis for the analysis and design of systems across systems engineering and software engineering. In systems engineering, the practice of QFD can be represented as a transformation of one parametric model related to a system into another parametric model. When the transformation is relationship preserving, that is, a model transformation, then properties of the parametric models are also transformed. This is the case, for example, with the transformation of precedence relationships.

The same concept and matrix formalism applies equally to MDA, and structured analysis and design, even though these two software engineering practices have a different basis for cohesion. In MDA, modeling is the basis for analysis and model transforms are the basis for design. In structured analysis and design, if the essential and behavioral models are modeled using the matrix representations presented in this chapter, then model transformations give precise meaning to the structured design principle that the structure of the problem should be reflected in the structure of the solution.

The current state of MDA in software engineering and advanced system design methods in systems engineering provide fertile opportunities for research. As MDA approaches its tenth year of practice, it has attained a level of maturity that could exploit the mathematically founded concept of model transformations presented in this chapter. The OMG is currently in the process of reviewing the original concepts of model transformations that were introduced in MDA. This is an area for future research.

The matrix approach, when practiced with the precision of formal mathematical modeling, has the potential to reduce the representation of models used for architecture and systems engineering to spreadsheets that can be manipulated with macros and wizards. It is hoped that the early artifacts of the model-based approach to architecture and systems engineering presented in this book give the community new concepts and practices that will provide new insights as to the way ahead in the development of model-driven architecture and systems engineering.

References

Walicki, M. and Bialasik, M. 1997. Categories of relational structures. *Lecture Notes in Computer Science; vol. 1376. Selected Papers on the 12th International Workshop on Recent Trends in Algebraic Development Techniques*. London: Springer-Verlag.

Yourdon, E. 1989. *Modern Structured Analysis*. Upper Saddle River, New Jersey: Yourdon Press.

MODELING LANGUAGES, FRAMEWORKS, AND GRAPHICAL TOOLS

2

Chapter 5

Architecture Modeling Languages

What is the best language for modeling system architecture? The answer to this question, of course, depends on the purpose of the model and the languages used in the systems engineering environment. In Chapter 2 ("Logical and Scientific Approach"), a procedure was introduced for modeling sentences using the Unified Modeling Language (UML) and the Systems Modeling Language (SysML). This chapter will review the fundamental notations of UML, but first some key concepts from mathematical logic that are at the heart of both UML and logical models will be reviewed.

Key Concepts

Principle of definition
OMG modeling languages
UML, SysML
Mathematical logic

Mathematical Logic

Recall the principle of definition and how it was applied:

> One needs both a formal definition of a design, for precision, and a prose definition for comprehensibility. (Brooks 1995, p. 234, 6.3)

This principle was applied in Chapters 2 ("Logical and Scientific Approach") and 3 ("Concepts, Standards, and Terminology") to key terms of systems engineering, which were defined in natural language and then logically modeled using UML and SysML.

In software and systems engineering practices, models are frequently considered abstractions that represent a system. Modeling languages bring precision to the process of abstraction. Languages such as UML and SysML are considered technologies.

The concept of a model in mathematical logic is understood using relational structures. The modeling languages of mathematical logic are the

Propositional calculus
Predicate calculus

The term *calculus* is derived from the Latin word for calculation. In logic, calculus refers to the calculation of truth. In the calculus of Newton and Leibnitz, on the other hand, it refers to the calculation of limits. Propositions are declarative statements and are represented by abstract propositional variables. Predicates are statements of relationships between variables and are represented by abstract predicate letters and abstract individual variables. For a more detailed mathematical review of the propositional and predicate calculus and model theory, see the review by Bell and Slomson (1969).

Propositional Calculus

The propositional calculus is a formal language built from propositional variables, logical connectives, and punctuation symbols. The propositional calculus formalizes that part of logic where validity depends only on how propositions are assembled (i.e., with logical connectives) but not on the internal meaning of the propositions.

Propositional Formulae

Propositional variables are denoted by p, q, …. The logical connectives are given in natural language but represented by symbols, as follows:

Not: ¬
And: ∧
Or: ∨

Examples of propositional formulae:

$$p \quad q \quad \neg p \quad p \wedge q \quad p \vee q$$

Interpretation: Truth Values

Propositional formulae are completely interpreted by the truth values of the propositions involved and the Boolean rules of logic. Table 5.1 displays a truth table that

Table 5.1 Truth Table for the Propositional Calculus

p	q	¬p	p ∧ q	p ∨ q
T	T	F	T	T
T	F	F	F	T
F	T	T	F	T
F	F	T	F	F

allows the truth value of any propositional formula to be determined, by repeated application of the Boolean rules. The interpretation of propositions by means of truth values can be very useful. Digital computation ultimately reduces to this type of interpretation. However, this type of logical interpretation is too weak for the modeling of sentences as presented in Chapters 2 ("Logical and Scientific Approach") and 3 ("Concepts, Standards, and Terminology"). The predicate calculus, on the other hand, offers a greater richness of language and precision that yields insights into both models and sentences.

Predicate Calculus

What Is the Predicate Calculus?

The predicate calculus is a formal language built from individual variables (denoted by lowercase letters such as p, q, …) and predicate variables (denoted by capital letters such as P_1, P_2, …), logical connectives (to include equality, =), and punctuation symbols. It formalizes that part of logic where validity depends both on how the predicates are assembled as well as on the internal meaning of the predicates (i.e., their interpretations). Predicate letters are interpreted as mathematical relations.

Examples of Predicates and Interpretation

$P_2(x)$ is a predicate of degree 1 (i.e., 1 variable). The predicate letter P_2 could be interpreted as

$$P_2(\_): \text{``}\_ > 0\text{''} \qquad \text{(This is an open relation.)}$$

where the variable x is interpreted as a number from an unspecified range of real number values. The interpretation of $P_2(x)$ would be: "$x > 0$". It is important to note that both the predicate letter and the predicate variable must be given an interpretation.

$P_1(x, y)$ is a predicate of degree 2. One interpretation of $P_1(x, y)$ could be: "x > y", where the variables x, y are interpreted as numbers from an unspecified range of real number values. If the variable y in $P_1(x, y)$ were to be fixed at the value 0, then the interpretation "$x > y$" of $P_1(x, y)$ would be the same as the interpretation "$x > 0$" of $P_2(x)$. However, this does not mean that $P_1(x, y)$ or even $P_1(x, 0)$ are the same predicate as $P_2(x)$. Each of these predicate letters is just an abstraction until it is given a specific interpretation.

Other formulae can be constructed in the obvious way, using the logical connectives and punctuation marks combined with the Boolean rules of the propositional calculus with the quantifiers of the predicate calculus.

Scope and Interpretation of Predicates

Just as with the propositional calculus, we are interested in asking questions such as: Is $P_2(x)$ true? Is $P_1(x, y)$ true? The answer depends on the values of x and y. If x were selected from the counting numbers (i.e., the values of x could be 1, 2, ...), then $P_2(x)$ would be true. If x were selected from the positive real numbers and y from the negative numbers, then $P_1(x, y)$ would be true. But if x and y were both selected from the positive real numbers, then the truth of $P_1(x, y)$ could only be decided based on specific values (instances) of x, y.

Quantifier Symbols, Models, and Sentences in the Predicate Calculus

Because the truth value of $P_2(x)$ and $P_1(x, y)$ in the preceding examples depends on whether every value (instance) of x, y is considered or only some values (instances) of x, y are considered, the predicate calculus uses what are called *scope operators*, which are quantifiers of the predicate variables. There are two:

Universal quantifier of the variable x: $\forall x$, which is read as "for every x" (i.e., every instance)

Existential quantifier of the variable x: $\exists x$, which is read as "there exists x" (i.e., some instance).

A *sentence* is a fully quantified well-formed formula in the predicate calculus. It is clear from the foregoing discussion why only fully quantified formulae are considered. Formulae that have not been fully quantified are open to possibly unbounded interpretations. This, of course, happens all the time when sentences are given in natural language. Additional information, such as contextual information, must be provided to bound the meaning (i.e., the interpretation) of natural language sentences.

In the model theory of the predicate calculus, *models* are models of sentences:

A *model* is a relational structure for which the interpretation of a (logical) sentence in the predicate calculus becomes valid (true). A *relational structure* is a set *M,* and {R$_a$}, a collection of mathematical relations on the set. We shall use the notation **M** = {*M*, {R$_a$}}.

Example of a Model of a Sentence

The predicate formula

$$S = \forall x \, P_2(x)$$

is true when interpreted in the relational structure **M** = {*M*, { _ > 0}}, where *M* is the counting numbers; that is, every counting number is nonnegative. Because the formula S is fully quantified, it is a sentence. **M** is a model of S because S is true in **M**.

Interpretation: Models and Validity

It should be now clear why the concept of interpretation in the propositional calculus is inadequate for describing and reasoning about models. Sentences are fully quantified well-formed predicate formulae and therefore do not suffer the ambiguity of unbounded interpretations. Models are models of sentences. A model is a relational structure for which the interpretation of a (logical) sentence in the predicate calculus becomes valid. The interpretations of both the predicate letters (as mathematical relations) and their individual variables (via the scope operators) determine whether a predicate or a sentence is true or false.

It is this intimate relationship between (logical) models, sentences, relational structures, and interpretation in the predicate calculus that justifies the modeling approach that was used in Chapters 2 ("Logical and Scientific Approach"), 3 ("Concepts, Standards, and Terminology"), and 4 ("Structure, Analysis, Design, and Models"). The UML provides a widely accepted and broadly practiced technology that allows us to emulate the relationships in the logical modeling of natural language sentences.

Object Management Group

The OMG (Object Management Group) has been an international, open membership, nonprofit computer industry consortium since 1989. Open standards currently managed by the OMG include the following:

UML—Unified Modeling Language, a standardized specification language for object modeling.

SysML—Systems Modeling Language, a general-purpose graphical modeling language for specifying, analyzing, designing, and verifying complex systems.

MOF—Meta Object Facility, a standard for model-driven engineering, is the mandatory modeling foundation for MDA.

MDA—Model Driven Architecture, an approach to using models in software development.

OMG Modeling Languages

Unified Modeling Language (UML)

The OMG UML is the *de facto* standard for software development. It is graphical language that provides semantics and notation for object-oriented problem solving. Why is it important to software development? In the construction of buildings, builders use the blueprints of architects. The more complicated the building, the more critical the communication between architect and builder, and the architect and the customer. There are many excellent books on UML, and the OMG has an open Web site for tutorials on UML: http://www.uml.org/#Links-Tutorials.

UML was originally developed for software engineering, but is moving toward systems engineering. The OMG is on various paths for specifying modeling languages. The Systems Modeling Language (SysML) builds on UML, expanding the scope of UML to systems engineering. The interactions in systems engineering involve the transport of energy, material, and information. The interactions in software engineering are focused on the processing and the transport of information. This is reflected in how software objects interact by sending messages to each other.

UML has many different notations, but the key types of UML diagrams needed for modeling systems are

Use case	Collaboration
Class	Statechart
Package	Activity
Object	Component
Sequence	Deployment

This chapter will focus on the five types of diagrams on the left.

Use Case Diagrams

These diagrams show the interactions between a system and the users or other external systems. The emphasis is on what the system does rather than how it does it. The viewpoint is that of an external observer. Use case diagrams should not be confused with scenarios that are examples of interactions, but the collection of scenarios for a use case may suggest a suite of test cases. Use cases are part of requirements engineering. Any aspect of system behavior that provides a measurable result should be captured in a use case. They should be a means of communication.

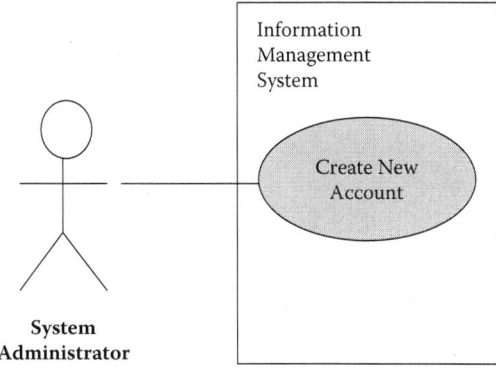

Figure 5.1   Use case diagram.

Figure 5.1 illustrates a simple use case in which a system administrator uses an information management system to create a new account. The graphical elements of the diagram are

Use cases (ovals)
Actors (stick figures)
Communication (lines)
Boundary (rectangle)

Class Diagrams

A class diagram provides system definition by showing its classes and their relationships. A class integrates attributes and behaviors and becomes a system object. In class diagrams, the emphasis is on *what* interacts rather than what happens in the interaction. The viewpoint is internal to the system.

There are three types of class relationships in a class diagram:

Association: A relationship between instances (objects) of two classes, a reference to an object in the form of an attribute
Aggregation: A stronger form of association in which one class owns but may share a collection of objects of another class
Generalization: An inheritance relationship indicating that a class is a type of (i.e., inherits properties from) superclass

Figure 5.2 illustrates a simple class diagram in which a system administrator uses an account and an account entry in an information management system to create a new account. The graphical elements of the diagram are

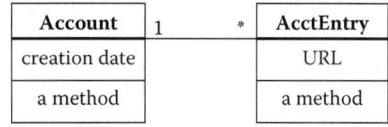

Figure 5.2   Class diagram.

Class (rectangle)
Name
Attributes
Methods
Association (line)
Aggregation (diamond, line)
Generalization (triangle)
Multiplicity (0, 1, 1.. n, 1.. *, *)

A special type of class called an abstract class was used in Chapters 2 ("Logical and Scientific Approach") and 3 ("Concepts, Standards, and Terminology") for the logical modeling of sentences. This type of a class is a partial declaration of a class, omitting behavior that may be realized in another class. Abstract classes can also be used in generalization to create a generic reusable class for which all the behaviors cannot be implemented. The graphic of an abstract class diagram may omit attributes and behaviors.

Package Diagrams

A package diagram is a collection of groupings of classes (called *packages*) that provide structure or organization to the system model. The emphasis is on organization of and dependencies between the packages. A dependency between two classes is a declaration that one class needs to know about another class to use its objects. The dependency relationship is weaker than association. Packages can be defined on the basis of logical relationships, subject matter, or responsibilities.

Figure 5.3 illustrates a simple package diagram in which a system administrator uses two components of an information management system (account management and security) to create a new account. The previously defined classes of account and account entry have been grouped into the UML package named account management. The graphical elements of the diagram are

Package (tabbed rectangle)
Name
Classes Contained
Dependency (dashed arrow)
Hierarchy (nested; or ::)

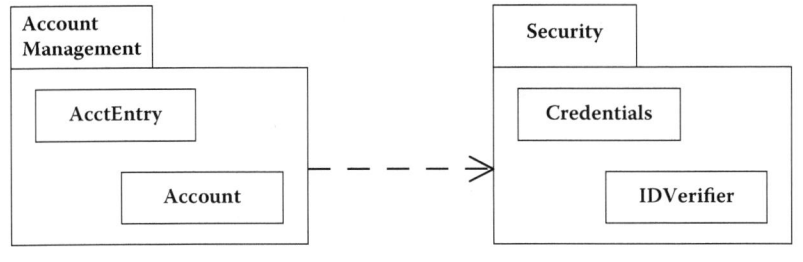

Figure 5.3 **Package diagram.**

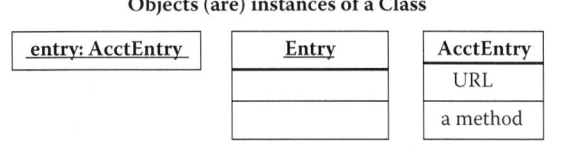

Figure 5.4 **Object diagram.**

Object Diagrams

Objects are instances of classes, and classes are blueprints for objects. The notation is similar to that for classes: object names are underlined, and colons are used in association with the instantiated class.

Figure 5.4 illustrates a simple object diagram that illustrates two instances (i.e., objects) of a previously defined class (account entry) in an information management system.

Sequence Diagrams

Sequence diagrams are a type of interaction diagram. Unlike the static views presented by class and object diagrams, interaction diagrams are dynamic views that show how objects collaborate by means of interactions. A sequence diagram is an interaction diagram that shows the details of how operations are carried out, what messages are sent, and in what order. The emphasis is on the order of interactions between parts of the system, which should not be confused with timing.

Figure 5.5 is a simple sequence diagram that illustrates four interactions between two objects, which are instances of (previously defined) classes representing the system administrator and the information management system. The graphical elements of the diagram are

Participants (rectangles)
Time (dashed line, down)

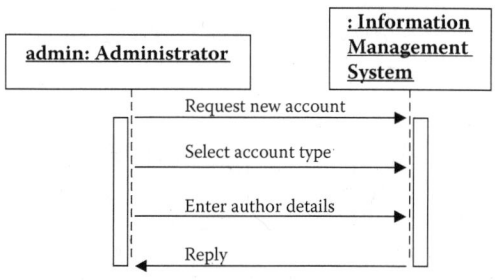

Figure 5.5   Sequence diagram.

Activation (thin rectangles)
Events (annotated arrow)

Participants are the actors that interact with the system, the system itself, and (at the next level of detail) the parts of the system. Participants are usually represented as objects.

Systems Modeling Language (SysML)

SysML is a visual (graphical) modeling language that extends UML. It provides semantics and notation for systems engineering SysML to support the specification, analysis, design, verification, and validation of systems that include hardware, software, data, parametrics, personnel, procedures, and facilities. SysML supports interchange of models and data via XMI and AP233. It is being implemented in tools provided by a variety of vendors (IBM, Telelogic, No Magic, and Sparx to name a few) and is part of the OMG UPDM specification (the UML Profile for the DoDAF and MODAF architecture frameworks).

Figure 5.6 illustrates the relationship between UML and SysML. Because SysML is focused on the semantics and notation for systems engineering, a large portion of the UML 2.0 standard is not used, although a large part of SysML directly comes from UML 2.0. The remaining portion of SysML is various stereotypes and extensions to UML 2.0 specialized for systems engineering.

The high-level relation between UML and SysML diagram types can be seen in Figure 5.7. At the highest level, SysML uses three primary types of diagrams: behavior, structure, and requirements. Of these, only the requirements diagram is completely new. At the next level down, SysML does make specialized modifications to UML activity diagrams, and it has made significant modifications to UML package diagrams to create what are called *block diagrams*. Also, a completely new type of diagram called the *parametric diagram* has been introduced, which allows for the storage and management of various types of metric data, which might, for example, be used for system simulation. Parametric diagrams have a key role

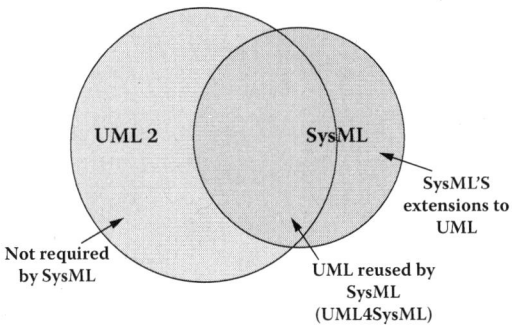

Figure 5.6   Relation between UML and SysML.

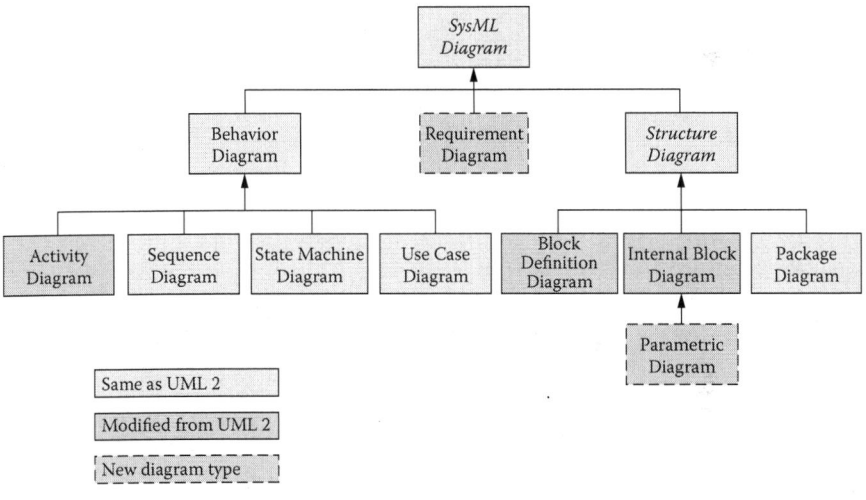

Figure 5.7   SysML diagram types.

in system modeling and will be discussed in detail in Chapter 6 ("The Systems Modeling Language [SysML]").

Altogether, SysML builds on four major types of diagrams: structure, behavior, requirements, and parametrics, as illustrated in Figure 5.8.

Summary

Both formal and natural language definitions of a system are needed for precision and comprehension. Logic interpretation is the basis for the concept of *model*. Interpretation in the propositional calculus depends only on how propositions are

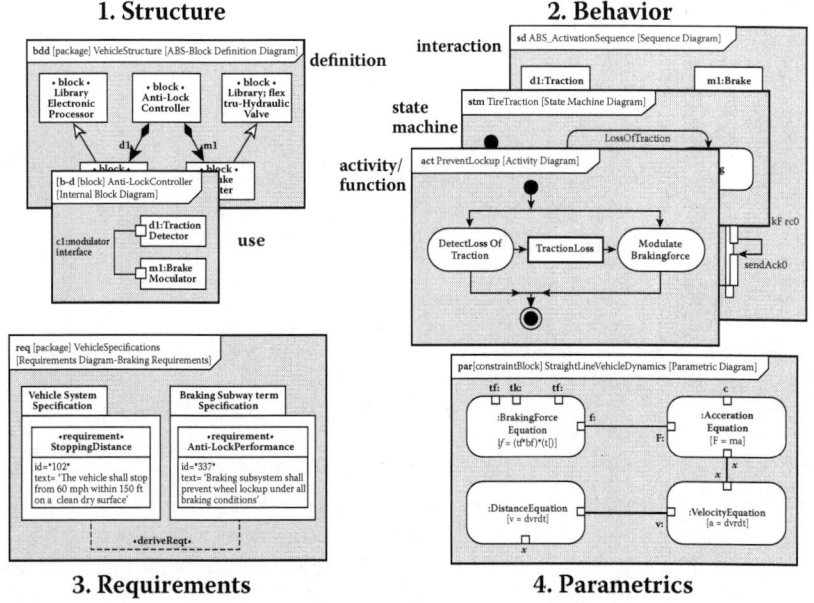

Figure 5.8 The four pillars of SysML.

assembled, and is therefore too weak to support the logical modeling of sentences. The predicate calculus supports the logical modeling of sentences because interpretation depends on both how predicates are assembled and the logical interpretation of the predicates. The concept of a model in logic is understood using logical sentences and relational structures in the predicate calculus, which can be supported by UML and SysML.

References

Bell, J.L. and Slomson, A.B. 1969. *Models and Ultraproducts: An Introduction*. Amsterdam: North-Holland Publishing Company.

Brooks, F.P. 1995. *The Mythical Man Month*. Boston: Addison Wesley Longman

Friedenthal, S., Moore, A., and Steiner, R. 2006. OMG Systems Modeling Language (OMG SysML™) Tutorial. http://www.omgsysml.org/SysML-Tutorial-Baseline-to-INCOSE-060524-low_res.pdf

OMG. Introduction to OMG's Unified Modeling Language™ (UML®). http://www.omg.org/gettingstarted/what_is uml.htm

Chapter 6

Applications of SysML to Modeling and Simulation

Russell Peak

Chapter 5 ("Architecture Modeling Languages") introduced modeling languages including the Unified Modeling Language (UML). A brief overview of the System Modeling Language (SysML) and its relationship to UML was also given. This chapter focuses on how SysML applies to modeling and engineering analysis. Several resources are emerging that describe how to use SysML for systems engineering, including overview papers such as Hause (2006) and in-depth books such as Friedenthal, Moore, and Steiner (2008); Holt and Perry (2008); Weilkiens (2008).

Key Concepts

Object relationships
SysML parametrics
Composable objects
Analysis building blocks

OMG SysML™ is a modeling language for specifying, analyzing, designing, and verifying complex systems. It is a general-purpose graphical modeling language with computer-sensible semantics. The target audience for this chapter is end users wanting to learn about SysML parametrics in general and its applications to engineering design and analysis in particular. We include background information on the development of SysML parametrics that may also be useful for other stakeholders (e.g., vendors and researchers). We walk through models of simple objects that progressively introduce SysML parametrics concepts. To enhance understanding by comparison and contrast, we present corresponding models based on composable objects (COBs). The COB knowledge representation has provided a conceptual foundation for SysML parametrics, including executability and validation. We end with sample analysis building blocks (ABBs) from mechanics of materials showing how SysML captures engineering knowledge in a reusable form.

The object and constraint graph concepts embodied in SysML parametrics and COBs provide modular analysis capabilities based on multidirectional constraints. These concepts and capabilities provide a semantically rich way to organize and reuse the complex relations and properties that characterize simulation-based design (SBD) models. Representing relations as noncausal constraints, which generally accept any valid combination of inputs and outputs, enhances modeling flexibility and expressiveness. We envision SysML becoming a unifying representation of domain-specific engineering analysis models that include fine-grain associativity with other domain-level and system-level models, ultimately providing fundamental capabilities for next-generation systems life-cycle management.

Introduction to SysML and COBs

This section introduces SysML parametrics technology in general and its application to engineering analysis and design in particular. Composable object (COB) technology is included as one possible means to execute SysML parametric models. This section overviews these technologies and assumes the reader is familiar with basic object-oriented concepts and engineering analysis.

OMG SysML and Parametrics

The official Object Management Group (OMG) Web site describes SysML as follows (OMG 2008):

> The OMG Systems Modeling Language (OMG SysML™) is a general-purpose graphical modeling language for specifying, analyzing, designing, and verifying complex systems that may include hardware, software, information, personnel, procedures, and facilities. In particular, the language provides graphical representations with a semantic foundation for modeling system requirements, behavior, structure, and integration with a broad range of engineering analysis. SysML represents a subset of UML2 with extensions needed to satisfy the requirements of the UML™ for Systems Engineering RFP. SysML uses the OMG XML Metadata Interchange (XMI®) to exchange modeling data between tools, and is also intended to be compatible with the evolving ISO 10303-233 systems engineering data interchange standard.

The UML for Systems Engineering request for proposal (RFP) was developed jointly by the OMG and the International Council on Systems Engineering (INCOSE) and issued by the OMG in March 2003. The RFP specified the requirements for extending UML to support the needs of the systems engineering community. The SysML specification was developed in response to these requirements by a diverse group of tool vendors, end users, academia, and government representatives. The OMG announced the adoption of OMG SysML on July 6, 2006.

Unless otherwise noted, the diagrams and description in this chapter conform to the SysML Final Adopted Specification as well as changes proposed by the SysML Finalization Task Force released on February 23, 2007 (OMG 2007). The latest specification and additional information on OMG SysML publications, tutorials, vendor tools, and links to other items mentioned in this subsection can be found via http://www.omgsysml.org/, including Burkhart (2006) and Friedenthal, Moore, and Steiner (2006).

Motivation for SysML Parametrics

SysML is organized around four main concept groups and their corresponding diagrams: structure, behavior, requirements, and parametrics. These concepts and their relationship to UML were introduced in Chapter 5 ("Architecture Modeling Languages"). Parametrics is not part of UML and were introduced as a new modeling capability as part of SysML. A primary focus for SysML parametrics is to support engineering analysis of critical system parameters including evaluation of performance, reliability, and physical characteristics. Engineering analysis is a critical aspect of systems engineering. However, there has been a fundamental gap in previous modeling languages including UML, Integration Definition (IDEF), and behavior diagrams. As a result, nonstandardized engineering analysis models are often disjoint from the system architectural models that specify the behavioral and structural aspects of a system. The lack of integration and synchronization between the system architectural models and the engineering analysis models is aggravated as the complexity, diversity, and number of engineering analysis models increase. Parametrics provides a mechanism to address this gap and integrate engineering analysis models with system requirements and design models for behavior and structure. In addition, parametrics is a constraints representation that can be used more generally to capture other types of knowledge beyond engineering analysis. Peak et al. (2007b) provide further motivation related to parametrics and engineering analysis in particular.

Key SysML parametrics concepts include the following:

- A *constraint* is analogous to an equation (e.g., $F = m^*a$, which is useful in many engineering problems). A *constraint block* is a definition of this equation in a way that it can be reused. A *constraint property* is a particular usage of a generic constraint block (e.g., to support engineering analysis of a specific design).
- A *parameter* is a variable of an equation (a constraint) such as F, m, and a in the aforementioned example.
- A *value property* is any quantifiable characteristic of the environment, system, or its components that is subject to analysis such as the mass of a system or

its components. The basic properties of a system architectural model and its components are typically represented in SysML models as value properties.

Value properties are bound to the parameters of a constraint. This binding is the mechanism to keep the generic equations, such as $F = m * a$, linked to the value properties that are used to specify the system and its components. The complete terminology includes additional terms and concepts, but the aforementioned ones are the essential elements to keep in mind as you begin to work with SysML parametrics.

The Composable Object (COB) Knowledge Representation

SysML parametrics is based in part on a theory called composable objects (COBs). COBs were developed at the Georgia Institute of Technology (GIT) originally with a focus on representing and integrating design models with diverse analysis models. Design and analysis information is typically represented by a collection of inter-related models of varying discipline and fidelity. Thus, a method for capturing diverse multi-fidelity models and their fine-grained relations was needed. It was also desirable for this method to be independent of the specific computer-aided design/computer-aided engineering (CAD/CAE) tools used to create, manage, and compute these models.

The COB representation is based on object and constraint graph concepts to gain their modularity and multidirectional capabilities. Object techniques provide a semantically rich way to organize and reuse the complex relations and properties that naturally underlie engineering models. Representing relations as constraints makes COBs flexible because constraints can generally accept any combination of input/output (I/O) information flows. This multidirectionality enables design sizing and design verification using the same COB-based analysis model. Engineers perform such activities throughout the design process, with the former being characteristic of early design stages and vice versa.

The COB representation includes several modeling languages (Peak et al. 2004). It has lexical formulations that are computer interpretable, as well as graphical forms that aid human comprehension (Figure 6.1). Two lexical languages, COB Structure (COS) and COB Instance (COI), are the master forms that are both computer interpretable and human friendly. Other forms depict subsets of COS and COI models. For example, the graphical constraint schematic notation (Figure 6.2) emphasizes object structure and relations among object variables and has strong electrical schematic analogies. The structure-level languages define concepts as templates at the schema level, whereas the instance level defines specific objects that populate one or more of these templates.

Over the past few years, the GIT has worked together with other SysML developers to embody COB concepts within SysML—especially regarding its block and parametric constructs (GIT 2007c; Peak 2002a; Peak et al. 2005a; Peak, Tamburini,

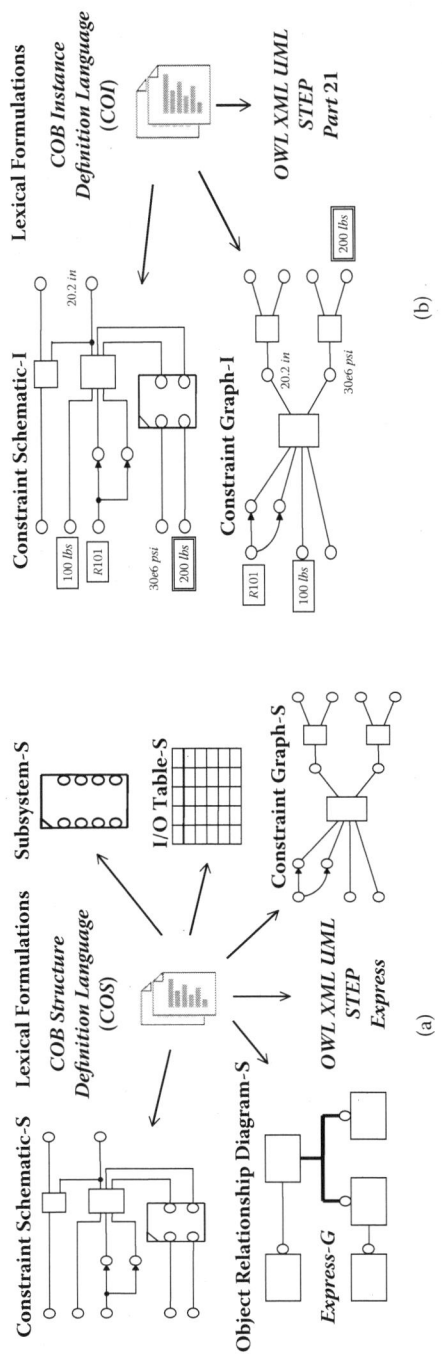

Figure 6.1 **Classical lexical and graphical formulations of the COB representation: (a) COB Structure languages, and (b) COB Instance languages.**

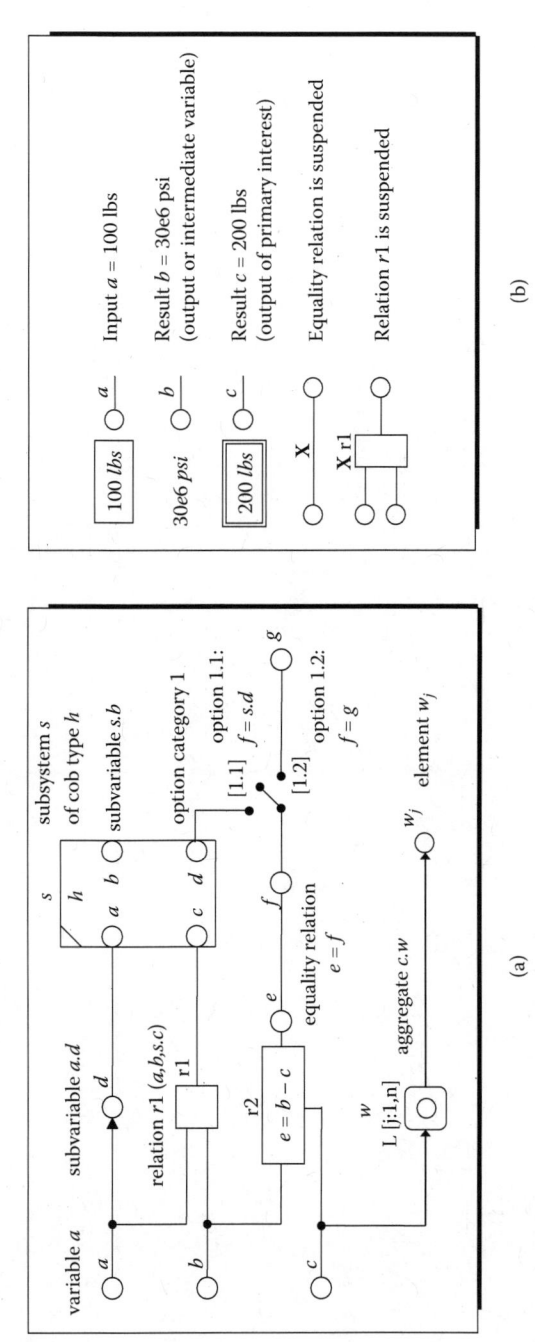

Figure 6.2   Basic COB constraint schematic notation: (a) Structure notation (-S), and (b) Instance notation (-I).

and Paredis 2005b). This approach has benefited both SysML and COBs; the former by providing conceptual formalisms and a broad variety of examples, and the latter by leveraging a richer set of SysML/UML2-based constructs that integrates with other system specification and design features, and by providing graphical modeling capabilities via multiple commercial tool vendors.

Introductory Concepts and Tutorial Examples

This section introduces SysML and COB concepts using tutorial-style examples. COBs are leveraged here as a preexisting body of work that illustrates typical parametrics concepts. The intent is to help the reader understand better such concepts as used in SysML parametrics. Over time, we intend to convert these examples to fully use SysML as their content representation (e.g., for all graphical modeling views), while continuing to employ COB constraint management algorithms for parametrics execution. See GIT (2007c); Peak et al. (1999); Peak (2002a); Peak et al. (2005a); Peak, Tamburini, and Paredis (2005b); Wilson (2000); and Wilson, Peak, and Fulton (2001) for further information and examples.

Triangles and Prisms

SysML and COB concepts and graphical methods will be used to model two geometrical objects. Classical COB diagrams will be developed first for these objects and then compared with the related SysML diagrams.

Representation Using Classical COB Formulations

Figure 6.3 provides the main classical COB formulations for a primitive object (a right triangle) and its usage in a compound object (a prism). The upper-left portion of the figure shows two forms traditionally used to convey knowledge about such objects: equations that hold in the object and a corresponding graphical view depicting variables in these equations. In COB terms, we call these *relations-S* and a *shape schematic-S*, respectively. The *-S* denotes that the representational construct describes a structural facet of the object—a class or schema-level facet in object-oriented terms—which can be thought of as a kind of template.

Representation of the first object as a COB right_triangle template is shown in the left-hand side of Figure 6.3, including (a) the *shape schematic-S* form, (b) the *relations-S* form, and (c) the *constraint schematic-S* form, which depicts its equation-based relations and variables as a kind of graph. *Subsystems*, (d), are encapsulated views that represent this object when it is used as a building block in other COBs. The last entry in the template is (e), the *lexical COB structure (COS)* form, which is

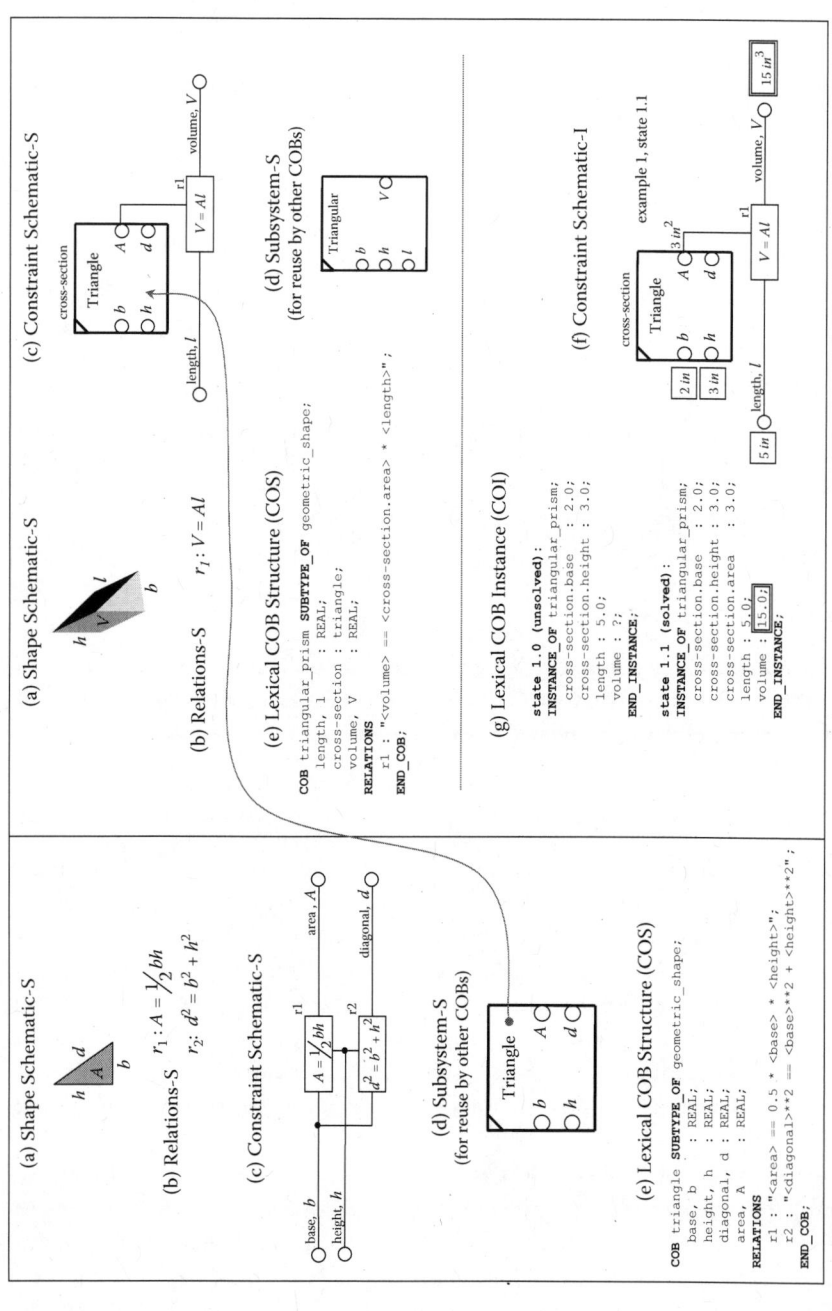

Figure 6.3   Triangles and prisms tutorial: classical COB formulations.

executable. It is also the master template from which structural forms (b)–(d) can be derived.

Structural forms of a COB triangular_prism template are similarly shown in the right-hand side of Figure 6.3. Besides basic variables such as volume and length, this template has one compound variable, cross-section, which is a usage of the right_triangle template defined previously.

The equation relations in all the COB structural forms are noncausal; that is, they are undirected and support any valid input/output combination at the instance level. For example, relation r1 in right_triangle can have variables base and height as inputs to produce area as the output, or area and height can be inputs to produce base as the output.

A COB instance can be loosely thought of as a copy of a COB template that has values filled in and input/output directions specified. In Figure 6.3 for the prism, (f) is the *constraint schematic-I* form of a triangular_prism instance after it has been solved (example 1). And (g) shows the corresponding *lexical COB instance* (*COI*) form for state 1.1, as well as the form for state 1.0 (before solving). Variables cross-section.base, cross-section.height, and length are provided as inputs in these states, and the primary target to be solved for is volume. As seen in state 1.1, note that cross-section.area was produced as an intermediate (ancillary) output along the way.

Representation Using SysML

The same objects can be represented using similar concepts in SysML, as seen in Figure 6.4 SysML diagrams. In brief, COBs are represented using SysML *blocks*, primitive variables are called *value properties*, and SysML *parametric diagrams* essentially replace COB constraint schematics. SysML represents relations themselves as a special type of SysML block known as a *constraint block*, which elevates their semantics and makes them explicitly reusable (versus their simple string form in COS). Usages of constraint blocks are known as *constraint properties* and are shown as rounded-corner rectangles (e.g., r1 and r2 in RightTriangle).

Figure 6.4a is a SysML *block definition diagram* (*bdd*) that shows a block called RightTriangle that includes value properties named area, base, diagonal, and length. Block definition diagrams present structural relationships among blocks based on concepts including *part-of* and *is-a* relationships. This type of diagram is based on the UML2 *class diagram* and essentially replaces the classical COB *object relationship diagram-S* (Figure 6.1). The value properties represent quantifiable characteristics that can include units and dimension such as units of meters and dimension of length. This is accomplished in SysML by typing the value property by a *value type*. In the bdd shown in Figure 6.4a, the value types are shown as LengthMeasure, AreaMeasure, and VolumeMeasure. These represent abstract value types that users can employ to make their model libraries usable with a wide variety of unit systems. Figure 6.5 defines these value types and sample concrete

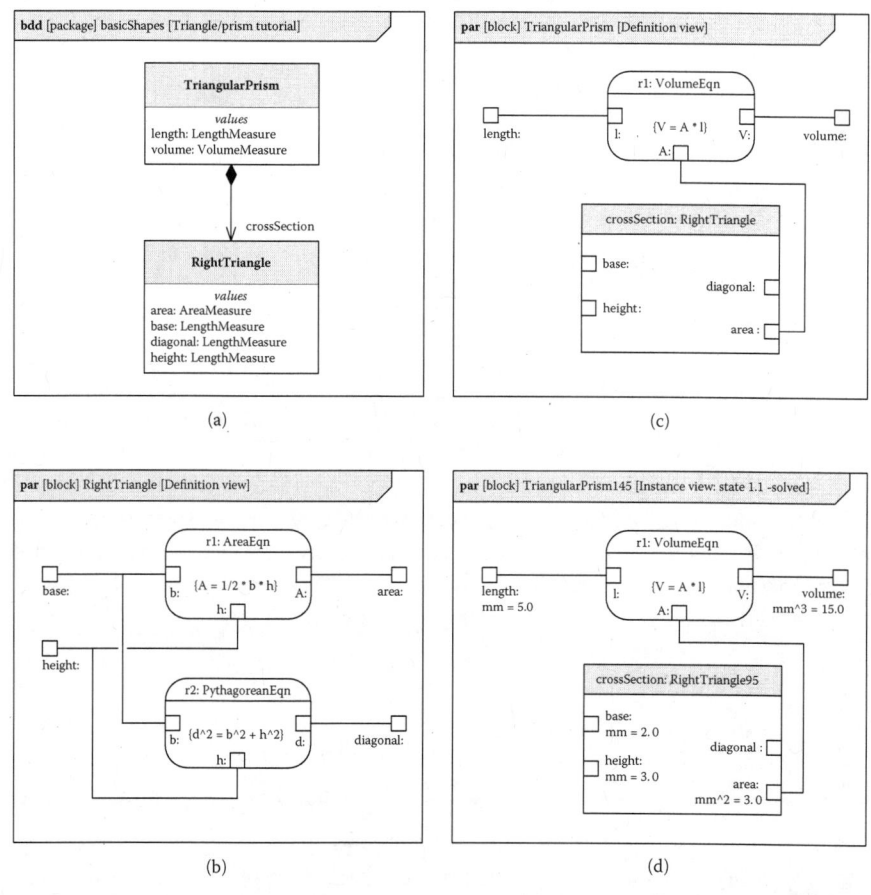

Figure 6.4 Triangles and prisms tutorial—SysML diagrams: (a) Triangle/ prism tutorial block definition diagram, (b) RightTriangle parametric diagram, (c) TriangularPrism parametric diagram, and (d) TriangularPrism sample instance.

specializations that are used throughout this chapter. It utilizes units and dimensions based on Appendix C.4 in the SysML specification (OMG 2008).

SysML parametrics is introduced in Figure 6.4b, which is a *parametric diagram* (*par*) that shows how these value properties are constrained by area equation and Pythagorean equation usages (i.e., constraint properties). This is accomplished by binding the parameters of the equations to the value properties of the RightTriangle block, which are depicted by the small sqaures in the figure. This binding is achieved using *binding connectors,* which are graphically represented as solid lines that "wire together" value properties and parameters.

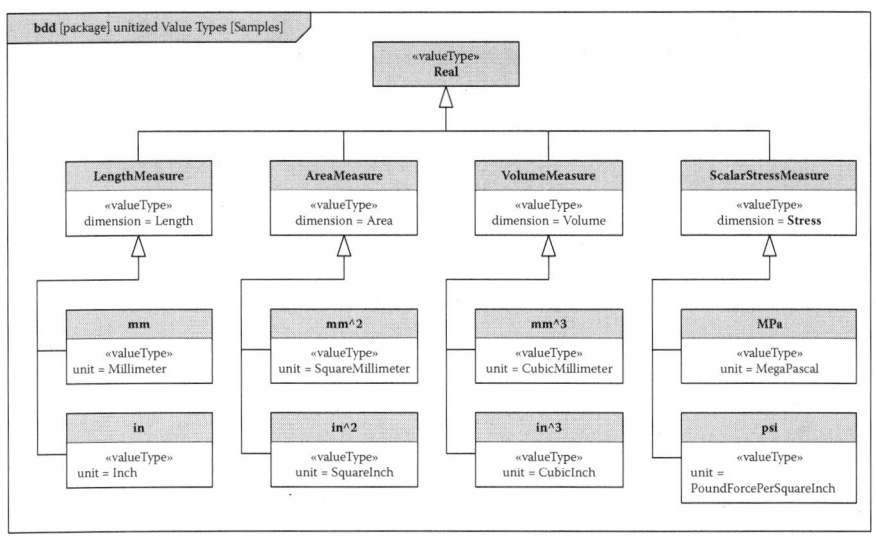

Figure 6.5   Sample SysML value types with specified dimensions and units.

Note that the parametric diagrams in this chapter use an electrical schematic-like branching convention for connecting a value property to more than one constraint parameter (e.g., height connecting to both r1.h and r2.h in Figure 6.4b). This branching approach reduces the graphical space needed to connect to a value property and typically results in a cleaner looking diagram. The convention for unambiguously interpreting such branching is that every connection between the value property and the indicated parameters is implemented as a distinct binding connector. Thus, two binding connectors attach to height in the right triangle parametric diagram given in Figure 6.4b. This convention applies similarly for the case in which a parameter connects directly to two or more other parameters.

Figure 6.6 contains a sample SysML tool screenshot, which shows *binding connectors* being used to graphically wire together value properties and parameters. These models have been implemented on the basis of earlier drafts of the SysML specification.

Figure 6.4a also defines a TriangularPrism block, which is composed of a RightTriangle and has two value properties. The particular usage of the RightTriangle for TriangularPrism is referred to by its part property name called crossSection. Figure 6.4c is a parametric diagram that further defines the TriangularPrism block by showing how the length, volume, and crossSection.area value properties are constrained by the r1 volume equation.

Figure 6.4d shows a particular instance of a triangular prism parametric. This is an instance based on the diagram in Figure 6.4c that includes specific values for

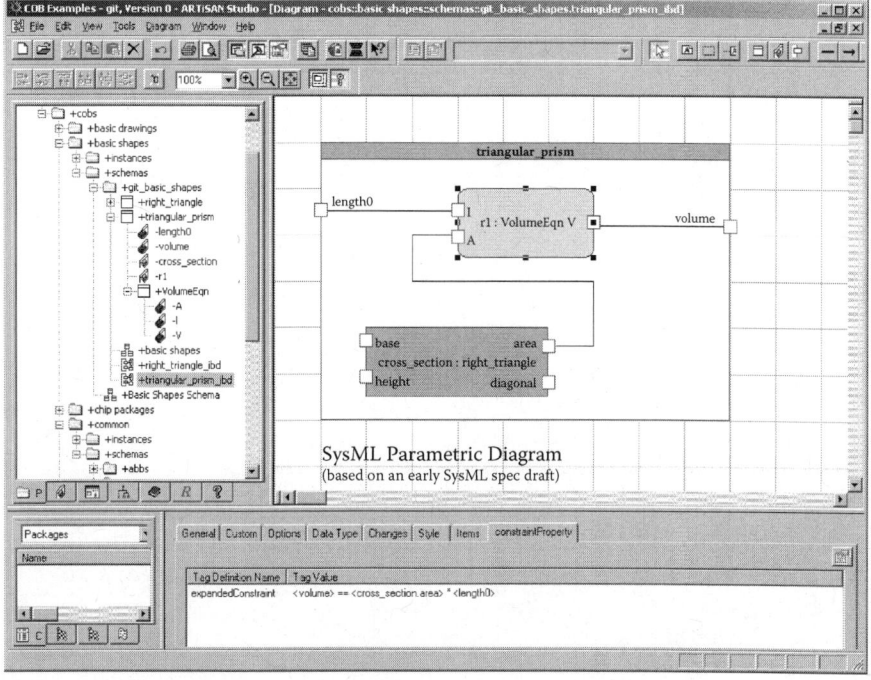

Figure 6.6   Sample parametrics implementation in a SysML tool.

the value properties. The term *instance* is loosely used in its generic form. Briefly, we employ an instantiation by specialization approach. The instance of the triangular prism represents the same COB instance shown in Figure 6.3 and uses a similar color coding scheme to depict the current inputs and outputs. (This causality color scheme is a GIT user/tool convention.) This example utilizes value types based on millimeters. SysML syntax places what looks like the units to the left of the value because these units are actually concrete value types. For example, mm^2 is a specialization of AreaMeasure as defined in Figure 6.5.

There is an important distinction between a constraint block, which defines a constraint, and a constraint property, which is a usage of a constraint block. A constraint block definition can represent a generic equation, such as the volume, area, and Pythagorean equations. Constraint blocks may be reused to constrain value properties of many different contexts such as the example shown in Figure 6.4. The colon notation distinguishes the usage from the definition. In the right triangle parametric diagram shown in Figure 6.4b, for example, the heading r1: AreaEqn indicates that r1 is a specific usage of the generic constraint block definition called AreaEqn.

In general, we envision that analysis libraries representing concepts and reusable equations for specific analysis domains such as fluid dynamics will be implemented as SysML blocks and constraint blocks. The remaining examples in this chapter demonstrate one way to capture such knowledge. Peak et al. (2007b) show how these analysis libraries can be readily used to bind the value properties of the system design or architectural models to various types of analysis.

Implementation and Execution

SysML is intended to capture actual parametric models—not just documentation—which can be solved by external tools. Here, we describe one way to achieve this executable intent. As seen in Figure 6.6, vendors are now supporting SysML in their modeling tools, often by leveraging their existing UML2 capabilities. Underneath the graphical diagrams and model tree browsers, these tools typically capture model structure in a semantically rich manner. This enables ready execution of SysML constructs based on UML2 such as sequence diagrams. However, because parametric constructs are new, current tools typically do not support parametrics execution.

XaiTools™ (Wilson, Peak, and Fulton 2001; Peak 2000) is a toolkit that implements practically all COB concepts except the graphical forms. Originally developed at GIT for design–analysis integration, it imports and exports COB models in various lexical formats, including COS and COI files (Figure 6.3). It includes constraint graph management algorithms that determine when to send what relations and values to which solvers, and how to pass the subsequent results to the rest of the graph. In this way XaiTools enables links with design tools and effectively provides an object-oriented, constraint-based front end to traditional CAE tools, including math tools such as Mathematica and finite element analysis tools such as Abaqus and Ansys (Figure 6.7).

Leveraging these capabilities, we implemented the architecture shown in Figure 6.8 to demonstrate how one can author COBs as parametric models in a

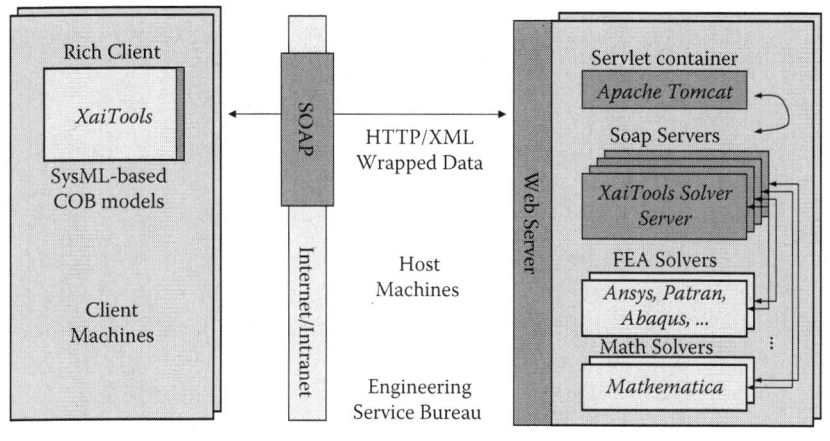

Figure 6.7   CAE solver access via XaiTools engineering Web services.

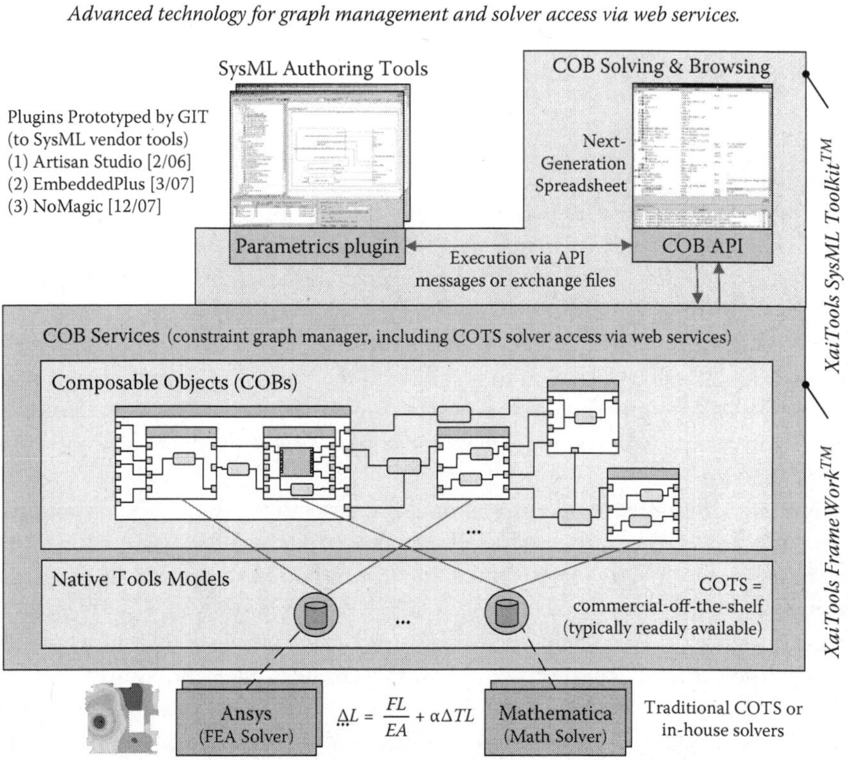

Advanced technology for graph management and solver access via web services.

Figure 6.8 Tool architecture enabling executable SysML parametrics.

SysML tool and then execute them using commercial solving tools. COS and COI models are generated from the SysML model structure and exported as text files. XaiTools then loads these models and executes them as usual. Figure 6.9 shows the triangular prism instance described earlier in before and after solving states. This XaiTools COB browser application can be considered an object-oriented noncausal spreadsheet that captures and manages the execution of parametric models.

More about Composable Object (COB) Technology

Work began at GIT in 1992 on the technology that eventually led to what we today call COBs. See Peak (2000) for further background, including early engineering analysis applications to electronics. COBs were originally known as *constrained objects* because of their constraint graph basis. Since 2005, they have been called *composable objects* to better reflect their interactive model construction nature, which vendor support for SysML parametrics is beginning to make practical. (The word *composable* here is inspired by the composable simulation work that Professor Chris Paredis and colleagues initiated at CMU [Paredis et al. 2001]).

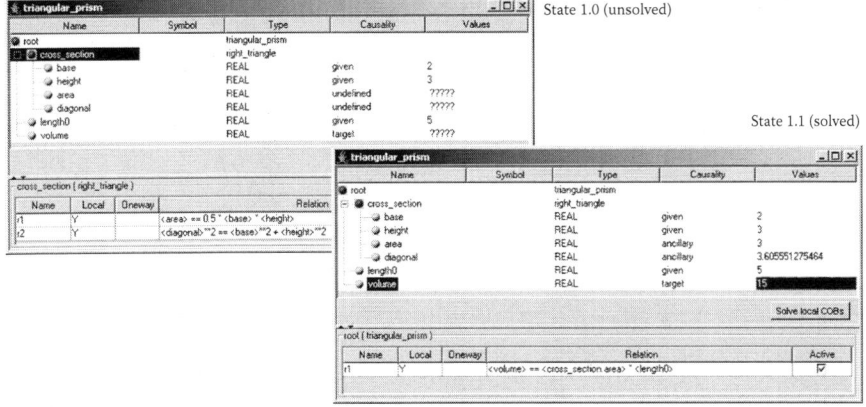

Figure 6.9 TriangularPrism instance from Figure 6.4 in XaiTools COB browser: an object-oriented noncausal spreadsheet.

Analytical Spring Systems

Next, we walk through models of several analytical objects from basic physics similar to examples by Cowan et al. (2003). The primary purpose of these tutorial examples is to introduce additional SysML and COB modeling concepts. It also introduces thought processes regarding how one can go about capturing such knowledge in the form of reusable SysML-based libraries.

Representation Using Classical COB Formulations

Figures 6.10a and 6.10b show traditional forms of an idealized linear spring object. Its shape schematic identifies variables and their idealized geometric context, and relations define algebraic equations among these variables. As with the triangle/prism tutorial, Figures 6.10c–e provide other structural formulations that define linear_spring as a COB template.

Figure 6.11 shows a linear_spring instance in two main states. Throughout state 1, spring_constant, undeformed length, and force are the inputs, and total_ elongation is the desired output. The COI form, (b), represents state 1.0 as this COB instance exists before being solved. State 1.1 shows the instance after solution, where one can see that length was also computed as an ancillary result. Variables end and start have no values in state 1.1 because sufficient information was not provided to determine them. State 5 shows this same linear_spring instance later after the user changed the length variable to be an input, provided it a length value (the desired length after deformation), and changed the spring_constant variable to be the target output.

Considering the engineering semantics behind such scenarios, one sees that state 1 typifies a kind of design verification scenario; the natural inputs (i.e., key

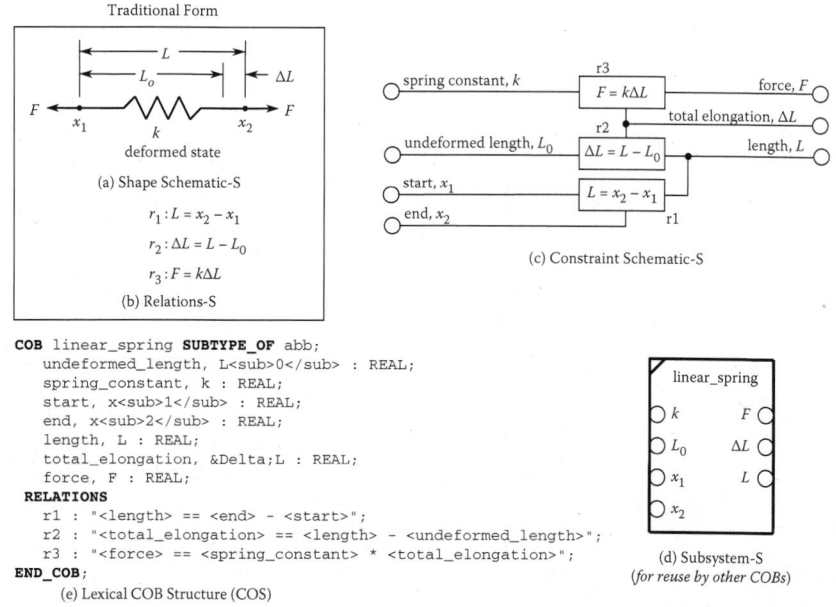

Figure 6.10 Analytical springs tutorial: linear_spring classical COB structural formulations.

design-oriented properties and a load) are provided as inputs, and a natural output (i.e., total elongation, as a response to the load) is the requested output. Hence, the analytical spring design is being checked to ensure that it gives the desired response.

Along these lines, state 5 is a kind of simple design synthesis (sizing) scenario. It reverses the situation by making one natural output an input and setting one natural input as the target output. It effectively asks, What spring_constant (a design-oriented variable) do I need to achieve the desired deformed length (a response)? This COB capability to change input and output directions with the same object instance can be applied to more complex situations. It is a multidirectional capability in that there are generally many possible I/O combinations for a given constraint graph structure. It enables a form of simulation-based design (SBD) by driving design from simulation and vice versa.

Next, we look at how systems can be composed from basic elements such as linear_spring. Given an analytical system containing two idealized springs as in Figure 6.12 (a1), in traditional modeling approaches one might start by drawing their freebody diagrams, (a2). From there, one could list the analytical spring formulae and boundary conditions, as in Figure 6.12b, and solve the resulting set of equations for target outputs (e.g., elongations at the individual and system levels). One could use symbolic math tools such as Maple or Mathematica to aid this

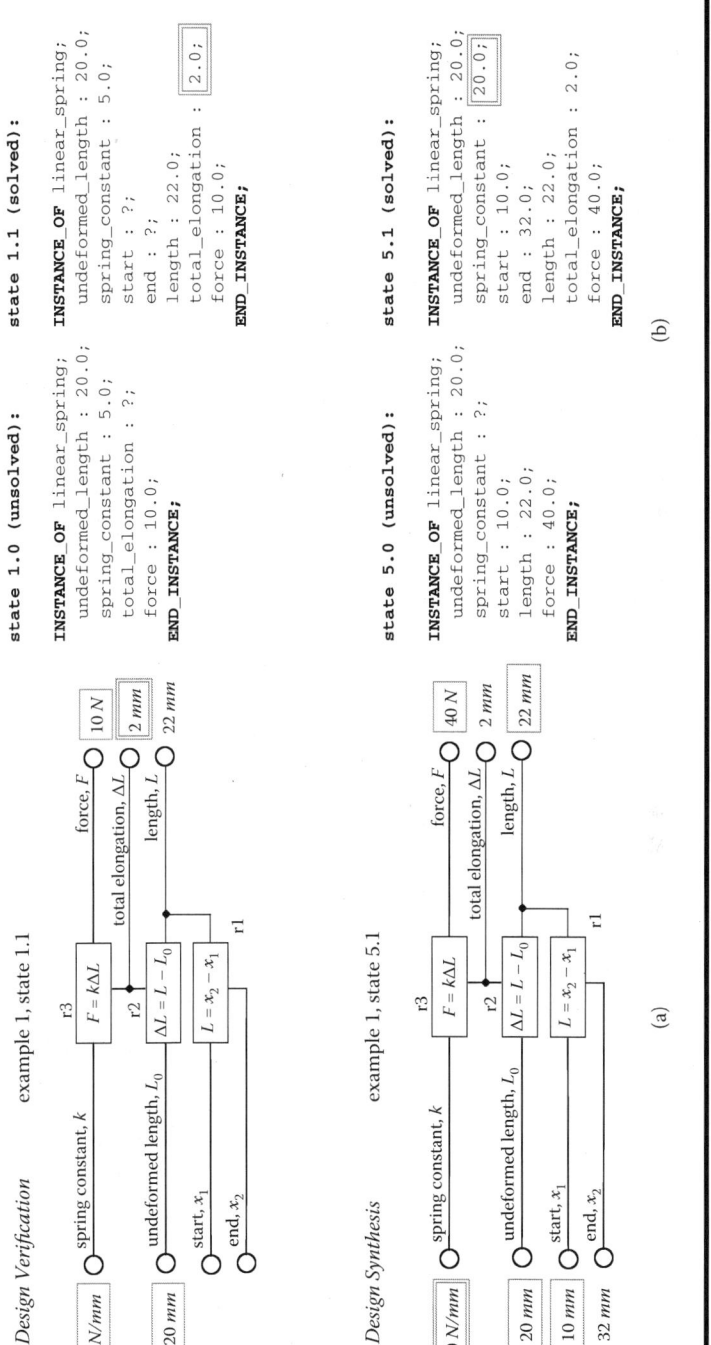

Figure 6.11 Multidirectional (noncausal) capabilities of a COB linear_spring instance: (a) Constraint schematic-I, and (b) lexical COB instance (COI).

Traditional Forms

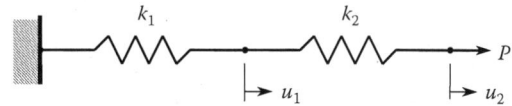

(a1) Shape Schematic-S: Analytical Assembly

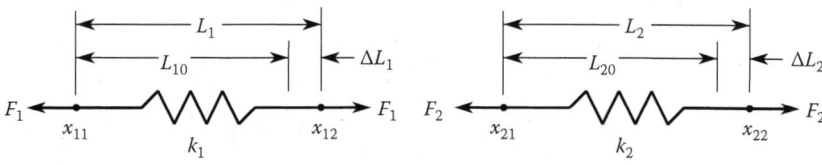

(a2) Shape Schematic-S : Freebody Diagrams

Kinematic Relations

$r_{11}: L_1 = x_{12} - x_{11}$ $bc_1: x_{11} = 0$
$r_{12}: \Delta L_1 = L_1 - L_{10}$ $bc_2: x_{12} = x_{21}$ *Boundary Conditions*
$r_{13}: F_1 = K_1 \Delta L_1$ $bc_3: F_1 = F_2$
Constitutive Relations $r_{21}: L_2 = x_{22} - x_{21}$ $bc_4: F_2 = P$
$r_{22}: \Delta L_2 = L_2 - L_{20}$ $bc_5: u_1 = \Delta L_1$
$r_{23}: F_2 = K_2 \Delta L_2$ $bc_6: u_2 = \Delta L_2 + u_1$

(b) Relations-S

Figure 6.12   Analytical springs tutorial: two_spring_system traditional math and diagram forms.

process and change I/O combinations. However, one would essentially end up having a list of equations with only implied engineering semantics. For example, one could not query relation r11 in the list and directly know that it is part of the first spring. Furthermore, adding and deleting equations to change input/output directions for a large system of equations could become unwieldy.

Next, we redo the definition of this system using similar COB formulations seen in Figure 6.3. When one considers the constraint graph illustrated in Figure 6.13d for this spring system, one recognizes that the shaded portions are essentially duplications of the same kind of relations (e.g., r11 versus r21). Traditionally, one would have to manually replicate and adjust these similar relations via a tedious and error-prone process. COBs address these semantic and usability issues by grouping relations and variables according to their engineering meaning and capturing them within explicit reusable entities.

For example, by applying object-oriented thinking, the shaded regions in Figure 6.13d can be represented by two linear_spring subsystems, as is done in Figure 6.13c. There is no need to specify these relations in the corresponding system-level COS form, Figure 6.13e, as they are already included in the linear_spring entity per its COS definition (Figure 6.10). System-level relations such as boundary conditions are the only relations that need to be specified here. With this

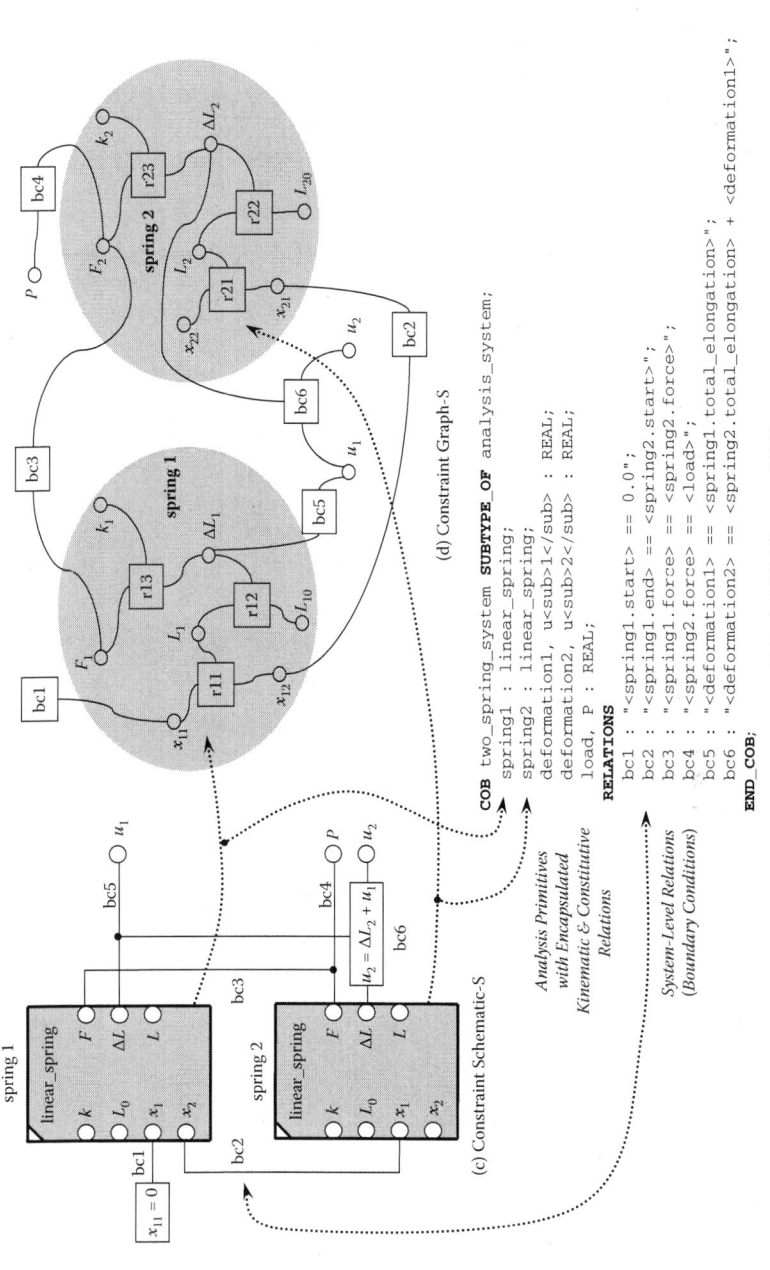

```
COB two_spring_system SUBTYPE_OF analysis_system;
  spring1 : linear_spring;
  spring2 : linear_spring;
  deformation1, u<sub>1</sub> : REAL;
  deformation2, u<sub>2</sub> : REAL;
  load, P : REAL;
  RELATIONS
  bc1 : "<spring1.start> == 0.0";
  bc2 : "<spring1.end> == <spring2.start>";
  bc3 : "<spring1.force> == <spring2.force>";
  bc4 : "<spring2.force> == <load>";
  bc5 : "<deformation1> == <spring1.total_elongation>";
  bc6 : "<deformation2> == <spring2.total_elongation> + <deformation1>";
END_COB;
```

(d) Constraint Graph-S

(e) Lexical COB Structure (COS)

Analysis Primitives with Encapsulated Kinematic & Constitutive Relations

System-Level Relations (Boundary Conditions)

(c) Constraint Schematic-S

Figure 6.13   Analytical springs tutorial: two_spring_system classical COB formulations.

definition completed, the constraint graph can now be seen as another COB form that is derivable from the lexical form. A *constraint graph-S* such as (d) is effectively the fully decomposed form of its corresponding constraint schematic-S (c); it is flattened such that no subsystem encapsulations are present. COBs can also have elements such as options, conditional relations, and variable length aggregates, in which case they have not one but a family of corresponding constraint graphs.

Representation Using SysML

One can similarly render SysML models for this analytical springs tutorial. Figure 6.14a is a block definition diagram showing that TwoSpringSystem is composed of two part properties named spring1 and spring2 that utilize LinearSpring. Figures 6.14b and

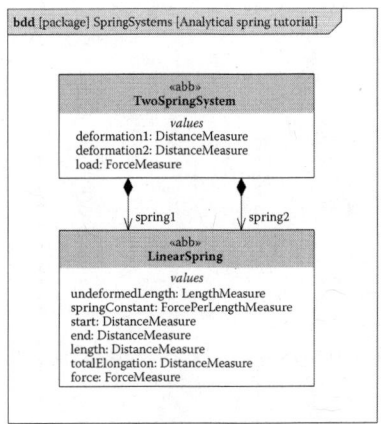

(a) Analytical springs tutorial block definition diagram.

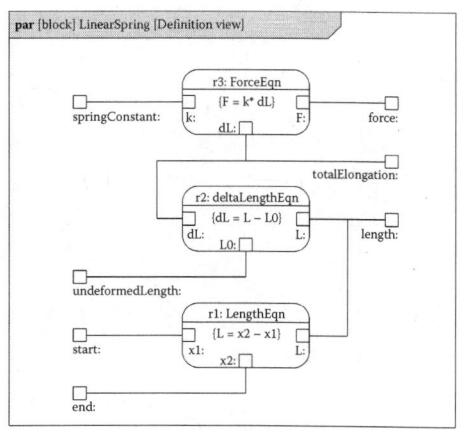

(b) LinearSpring parametric diagram.

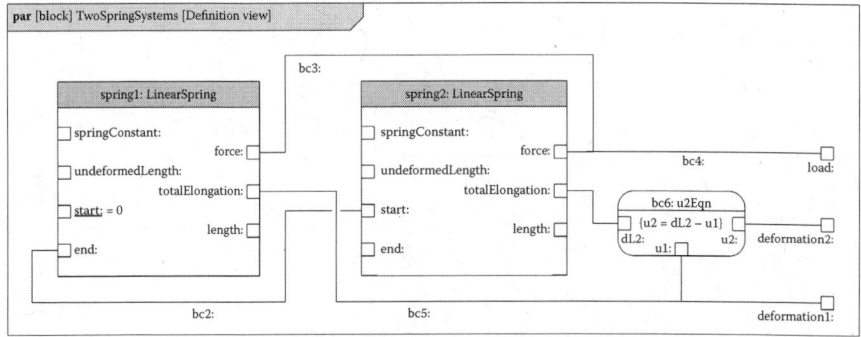

(c) TwoSpringSystem parametric diagram.

Figure 6.14 Analytical springs tutorial: TwoSpringSystem SysML diagrams: (a) block definition diagram, (b) LinearSpring parametric diagram, and (c) TwoSpringSystem parametric diagram.

6.14c show SysML parametric diagrams for LinearSpring and TwoSpringSystem, respectively. Note again the richer semantics enabled by SysML value types. DistanceMeasure values are intended to represent position distance in space relative to an origin and can take on any real number, whereas LengthMeasure values represent shape measures of physical things and thus they must be nonzero positives.

Figure 6.15 represents an instance of TwoSpringSystem with state 1.0 being the inputs (givens) for a kind of system design verification scenario. The user provides input values for the intrinsic characteristics of the analytical springs and a system load, and also indicates the primary target properties to solve for (deformation1 and deformation2) as denoted by the question marks. State 1.1 shows the results after solving, where one sees that the total system deformation (u2) has been determined to be 3.49 mm. At this point, the user could change one of the input values and re-solve (e.g., increasing spring1.springConstant to reduce the total deformation). Or, one might change I/O directions and calculate the required spring constant to achieve the target total deformation.

Implementation and Execution

Figure 6.16a is lexical COB instance corresponding to Figure 6.15 state 1.0 as generated from a parametrics model in a SysML tool (using the approach outlined in Figure 6.8). CXI is a relatively newer COB lexical form that is essentially an XML-based form of COI plus new capabilities (e.g., capturing causality more completely). XaiTools reads in this CXI model and displays it in a COB browser before and after solving (Figure 6.16b). These figures thus exemplify the complementary relationship between SysML parametrics (for model authoring and management) and COBs (for model execution).

Analysis Building Block Examples

SysML can be used to represent analytical engineering concepts as analysis building blocks (ABBs) (Peak 2000; Peak and Wilson 2001). This section highlights sample ABBs representing mechanics of materials domain knowledge.

Figure 6.17a is the SysML parametric diagram for a material behavior ABB known as OneDLinearElasticModel. This analytical model includes material properties such as Young's modulus and relations such as stress–strain–temperature constitutive equations. It is typically reused in building continuum primitives such as the two shown in the same figure: ExtensionalRod (Figure 6.17b) and TorsionalRod (Figure 6.17c). These continuum primitives in turn can be used to build other analysis models, including product-specific analysis models as seen in Peak et al. (2007b).

Note that traditional relations such as Equation 6.1 (for total deformation in an extensional rod) need not be explicitly included in parametric models. Rather,

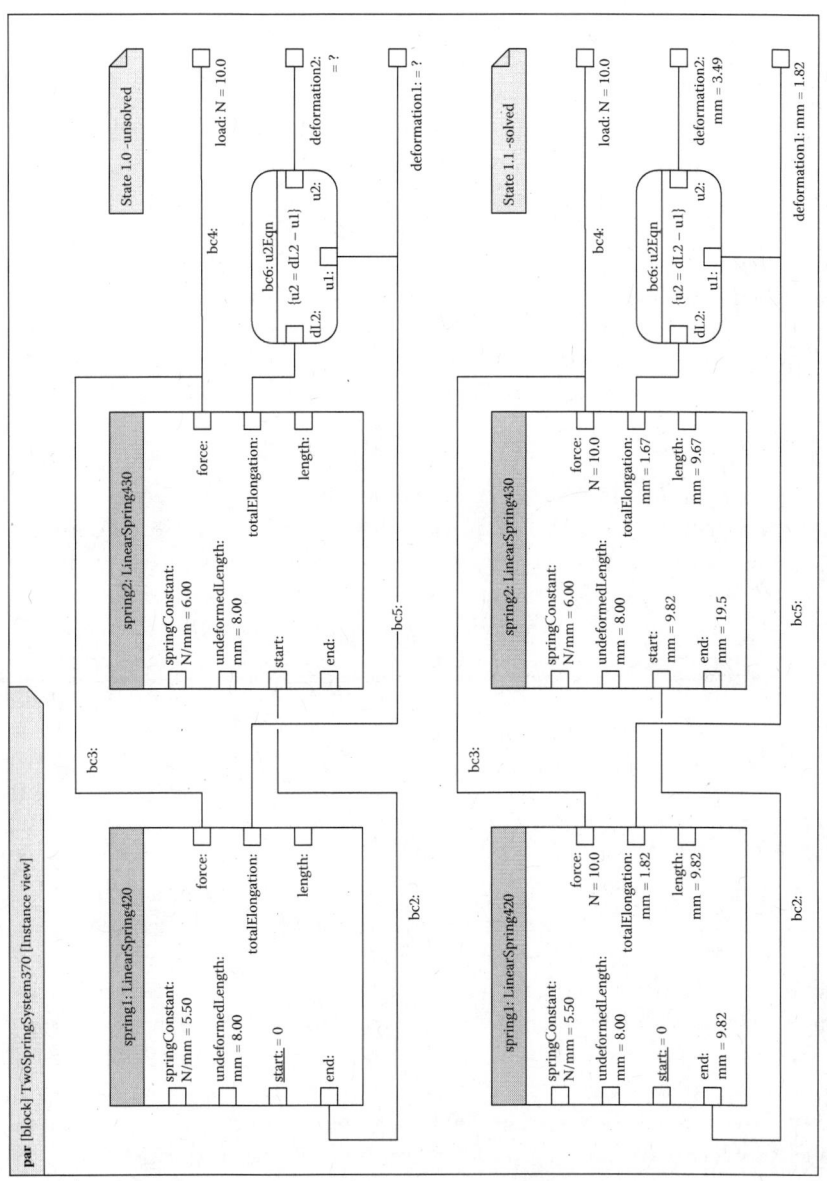

Figure 6.15   TwoSpringSystem parametric diagram: sample instance.

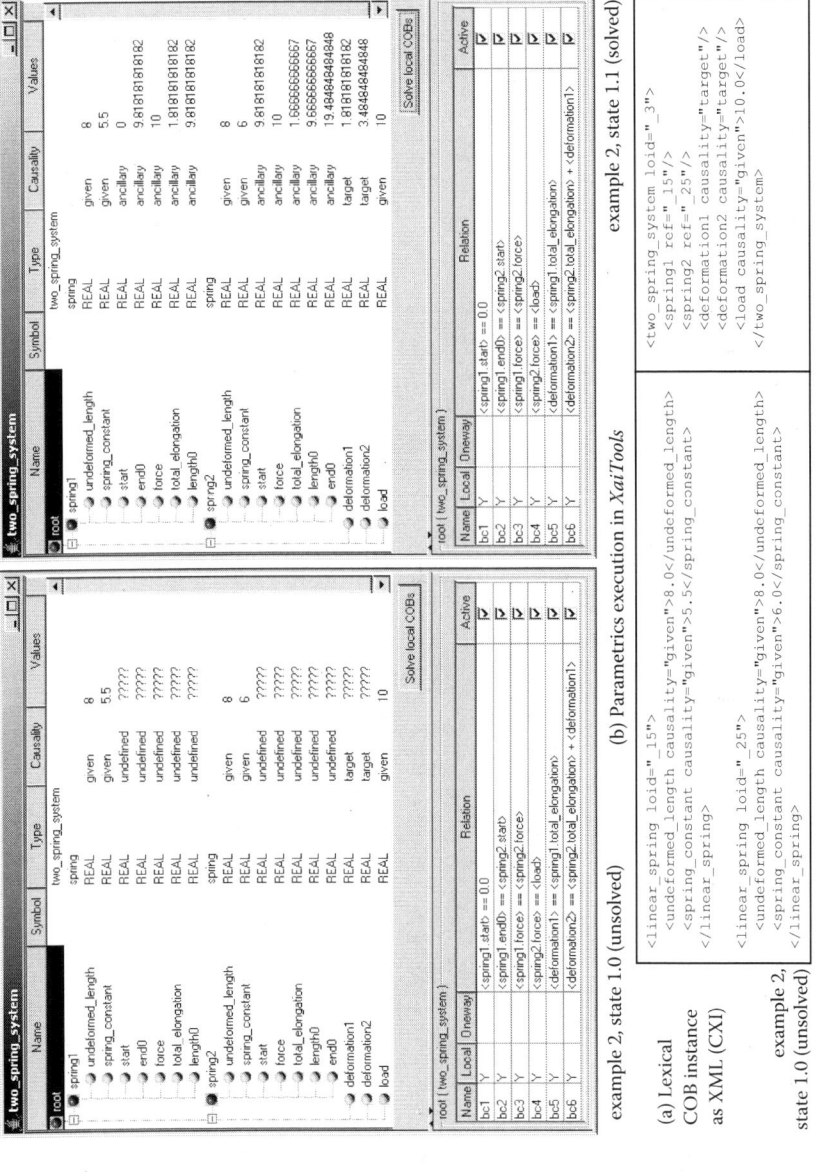

Figure 6.16 COB instance generated from a SysML TwoSpringSystem model (Figure 6.15) and its execution in XaiTools.

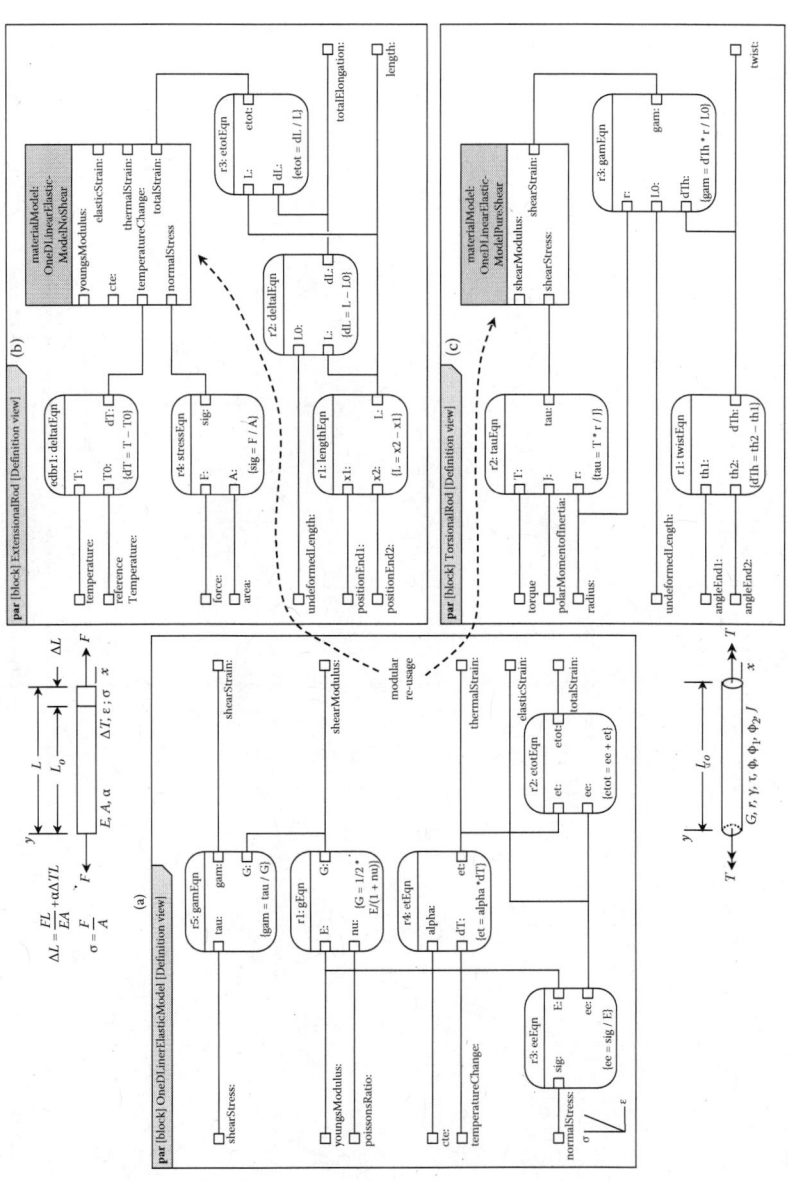

Figure 6.17 Example mechanics of materials analysis building blocks (ABBs)—SysML parametric diagrams: (a) OneDLinear ElasticModel, (b) ExtensionalRod, and (c) TorsionalRod.

these types of relations are derivable from the fundamental relations present in their corresponding ABBs; thus, the effective behavior of Equation 6.1 is automatically included in ExtensionalRod.

$$\Delta L = \frac{FL}{EA} + \alpha\Delta TL \tag{6.1}$$

System Examples

This section highlights a few recent applications of SysML parametrics technology to system-level problems and corresponding simulations. The reader is referred to the references in the figure captions for further information.

The first example is a road scanning system that consists of a squadron of unmanned aerial vehicles (UAVs), fuel used in the UAVs, and crews to deploy and maintain the UAVs. Figure 6.18 informally overviews the problem including multiple equations that relate various system/subsystem parameters. Figure 6.19 shows various views of a model that implements this problem in a SysML authoring tool.

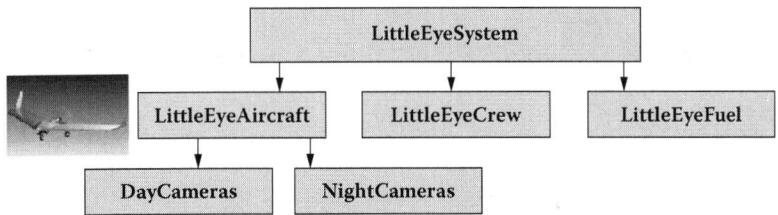

NumberMilesScannedPer24Hours = NumberAvailablePlanes × MilesScannedPerHour × 24

NumberAvailableSystems = MINIMUM(NumberAvailablePlanes, NumberAvailableCrews, NumberAvailableFuelLoads)

NumberAvailablePlanes = 0.5*(NumberAvailablePlanesByDay + NumberAvailablePlanesByNight)*DutyCyclePlane

DutyCyclePlane = (1-DutyCycleTurnaround) × (1-DutyCycleCameraRefit) × (1-DutyCycleMaintenance)

NumberAvailablePlanesByDay = MINIMUM(NumberPlanes, NumberDayCameras)

NumberAvailablePlanesByNight = MINIMUM(NumberPlanes, NumberNightCameras)

NumberAvailableCrews = NumberCrews × CrewTimeOn

NumberAvailableFuelLoads = FuelSupplyPerDay/DailyFuelLoadPerPlane

Information Provided
MilesScannedPerHour = 40 mph
CrewTimeOn = 0.42 (120 hours on over 12 day period)
DutyCycleTurnaround = 0.23 (30 hours turnaround per 100 flight minutes)
DutyCycleCameraRefit = 0.02 (15 minutes refit per 12 hour period)
DutyCycleMaintenance = 0.09 (3 hours maintenance per 30 flight hours)
DailyFuelLoadPerPlane = 50 gallons

Figure 6.18 **Problem overview: road scanner system using LittleEye UAVs. (From Zwemer, D.A. and Bajaj, M. 2008. SysML parametrics and progress towards multi-solvers and next-generation object-oriented spreadsheets.** *Frontiers in Design & Simulation Workshop,* **Georgia Tech PSLM Center, Atlanta.)**

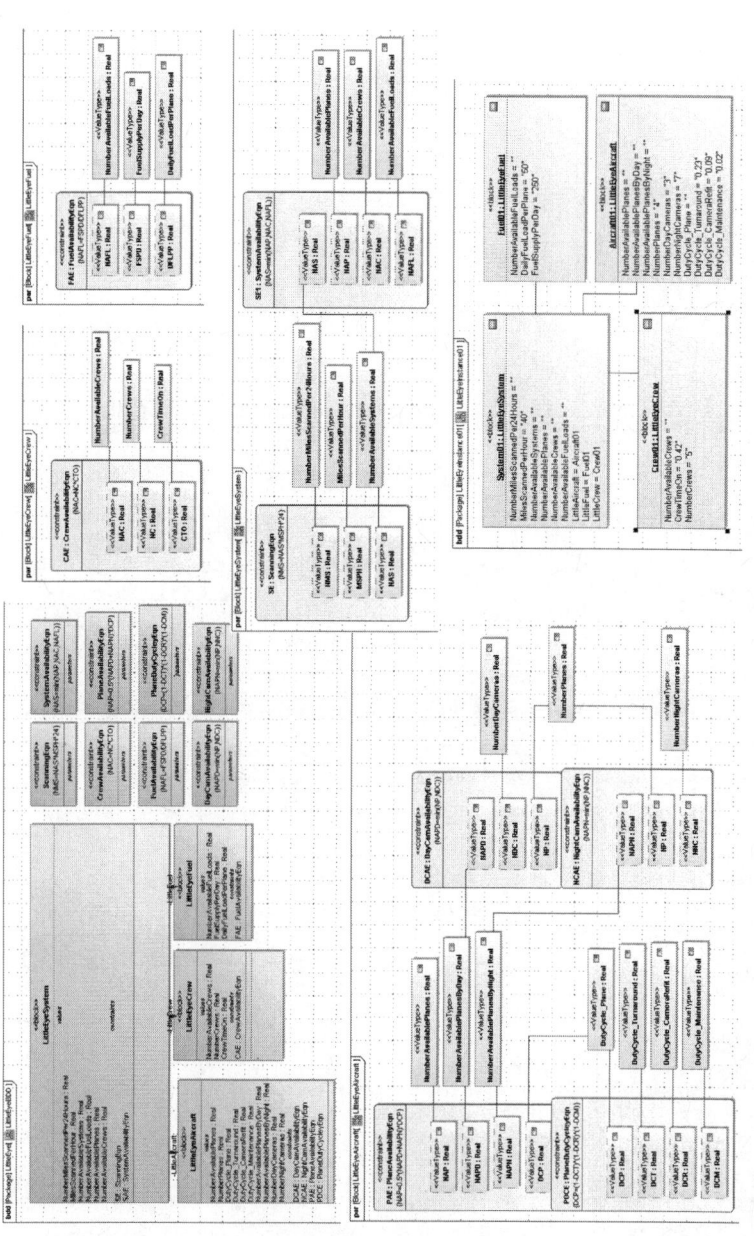

Figure 6.19 LittleEye SysML model: various diagram views. (From Zwemer, D.A. and Bajaj, M. 2008. SysML parametrics and progress towards multi-solvers and next-generation object-oriented spreadsheets. *Frontiers in Design & Simulation Workshop,* Georgia Tech PSLM Center, Atlanta.)

Instance 1 - Before Solving

Name	Symbol	Type	Causality	Values
Root		LittleEyeSystem		
LittleAircraft		LittleEyeAircraft		
DutyCycle_CameraReft		REAL	given	0.09
DutyCycle_Maintenance		REAL	given	0.02
DutyCycle_Plane		REAL	undefined	?????
DutyCycle_Turnaround		REAL	given	0.23
NumberAvailablePlanes		REAL	undefined	?????
NumberAvailablePlanesByDay		REAL	undefined	?????
NumberAvailablePlanesByNight		REAL	undefined	?????
NumberDayCameras		REAL	given	3
NumberNightCameras		REAL	given	7
NumberPlanes		REAL	given	4
LittleCrew		LittleEyeCrew		
CrewTimeOn		REAL	given	0.42
NumberAvailableCrews		REAL	undefined	?????
NumberCrews		REAL	given	5
LittleFuel		LittleEyeFuel		
DailyFuelLoadPerPlane		REAL	given	50
FuelSupplyPerDay		REAL	given	250
NumberAvailableFuelLoads		REAL	undefined	?????
MilesScannedPerHour		REAL	given	40
NumberAvailableCrews		REAL	target	?????
NumberAvailableFuelLoads		REAL	target	?????
NumberAvailablePlanes		REAL	target	?????
NumberAvailableSystems		REAL	undefined	?????
NumberMilesScannedPer24Hours		REAL	target	?????

Instance 1 - After Solving

Name	Symbol	Type	Causality	Values
Root		LittleEyeSystem		
LittleAircraft		LittleEyeAircraft		
DutyCycle_CameraReft		REAL	given	0.09
DutyCycle_Maintenance		REAL	given	0.02
DutyCycle_Plane		REAL	ancillary	0.686686
DutyCycle_Turnaround		REAL	given	0.23
NumberAvailablePlanes		REAL	ancillary	2.403401
NumberAvailablePlanesByDay		REAL	ancillary	3
NumberAvailablePlanesByNight		REAL	ancillary	4
NumberDayCameras		REAL	given	3
NumberNightCameras		REAL	given	7
NumberPlanes		REAL	given	4
LittleCrew		LittleEyeCrew		
CrewTimeOn		REAL	given	0.42
NumberAvailableCrews		REAL	ancillary	2.1
NumberCrews		REAL	given	5
LittleFuel		LittleEyeFuel		
DailyFuelLoadPerPlane		REAL	given	50
FuelSupplyPerDay		REAL	given	250
NumberAvailableFuelLoads		REAL	ancillary	5
MilesScannedPerHour		REAL	given	40
NumberAvailableCrews		REAL	target	2.1
NumberAvailableFuelLoads		REAL	target	5
NumberAvailablePlanes		REAL	target	2.403401
NumberAvailableSystems		REAL	ancillary	2.1
NumberMilesScannedPer24Hours		REAL	target	2,016

Figure 6.20 Solving LittleEye SysML parametrics: ParaMagic browser views. (From Zwemer, D.A. and Bajaj, M. 2008. SysML parametrics and progress towards multi-solvers and next-generation object-oriented spreadsheets. *Frontiers in Design & Simulation Workshop***, Georgia Tech PSLM Center, Atlanta.)**

Figure 6.20 illustrates solving the parametrics in this SysML model to answer one of the overall systems questions, How many miles of road can a given system configuration scan in a 24-hour day? This tool is called ParaMagic™, which is a commercialized version of the COB-based XaiTools toolkit described earlier (Figures 6.7–6.9). The user can readily change various parameters such as available fuel, UAV scanning capability, etc.; re-solve the model; and see the impact on miles of road scanned per day, and thus address system-level questions such as the following: If I had a UAV with 20 percent better scanning capability, how much more road could I scan each day?

Figure 6.21 shows a similar informal problem overview in a totally different domain: financial systems. In this case, the model deals with a three-year financial projection for a small business. Figure 6.22 shows views from a SysML model implementation of this system, and Figure 6.23 shows before and after execution of an example instance. With this model, you can answer questions such as: Given these starting values and projected sales each year, what will be my cash-on-hand at the end of Year 3? or in reverse, If I want this amount of cash-on-hand at the end of Year 3, what do my sales in Year 2 need to be?

Figure 6.24 overviews a more extensive test bed underway at Georgia Tech. This excavator domain test bed uses SysML to interconnect several types of models and tools. Figure 6.25 is a top-level view of the system context, and Figure 6.26 highlights an associated hydraulics model that is autogenerated from the SysML model and executed in a third-party solver for continuous system dynamics. The primary test bed demo scenario involves a trade study that considers the impacts of increasing both the dig cycle capacity (an excavator system measure of effectiveness) and the factory production rate (a manufacturing system measure of effectiveness) on

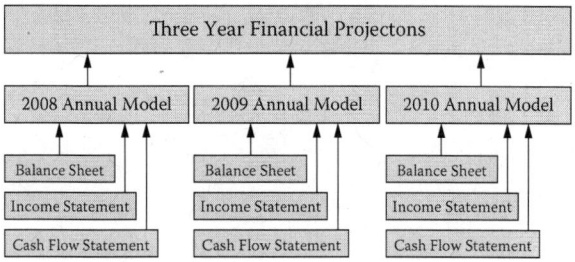

Key Questions for the Model to Answer
- Given projected sales, expenses and financing, what is the financial position of the company at the end of 3 years?
- Given the desired financial position at the end of 3 years, what are the required sales, expenses and financing?
- Etc.

Figure 6.21 Problem overview: financial projections system. (From Zwemer, D.A. and Bajaj, M. 2008. SysML parametrics and progress towards multi-solvers and next-generation object-oriented spreadsheets. *Frontiers in Design & Simulation Workshop*, Georgia Tech PSLM Center, Atlanta.)

the excavator design itself as well as the manufacturing systems. See Peak et al. (2008) for numerous other SysML views and simulation aspects for this test bed.

Finally, we also refer the reader to Peak et al. (2007b), which describes SysML models for a simulation template tutorial (Figure 6.27). It demonstrates how SysML supports a form of simulation-based design (SBD) including executable parametrics based on COB technology. It overviews concepts from the multi-representation architecture (MRA) for simulation templates that enables advanced CAD–CAE interoperability. Employing an object-oriented approach, the MRA defines natural partitions of engineering knowledge that occur between traditional design and analysis models. This benchmark tutorial—for a flap linkage parts family—demonstrates how SysML, COBs, and MRA together support capabilities important to engineering design and analysis. Many of these capabilities are relevant to broader simulation and knowledge representation domains, including the following:

■ Diversity of design information sources, analysis behaviors, analysis fidelities, solution methods, and solution tools
■ Modular, reusable analytical building blocks and fine-grained intermodel associativity

Additionally, this flap linkage tutorial introduces additional SysML and COB modeling concepts beyond this chapter, including packages, building block libraries, and requirements–verification–simulation interrelationships. Experiences to date reinforce our belief that SysML holds great promise as a unifying language for a diversity of models—from top-level system models to discipline-specific leaf-level models.

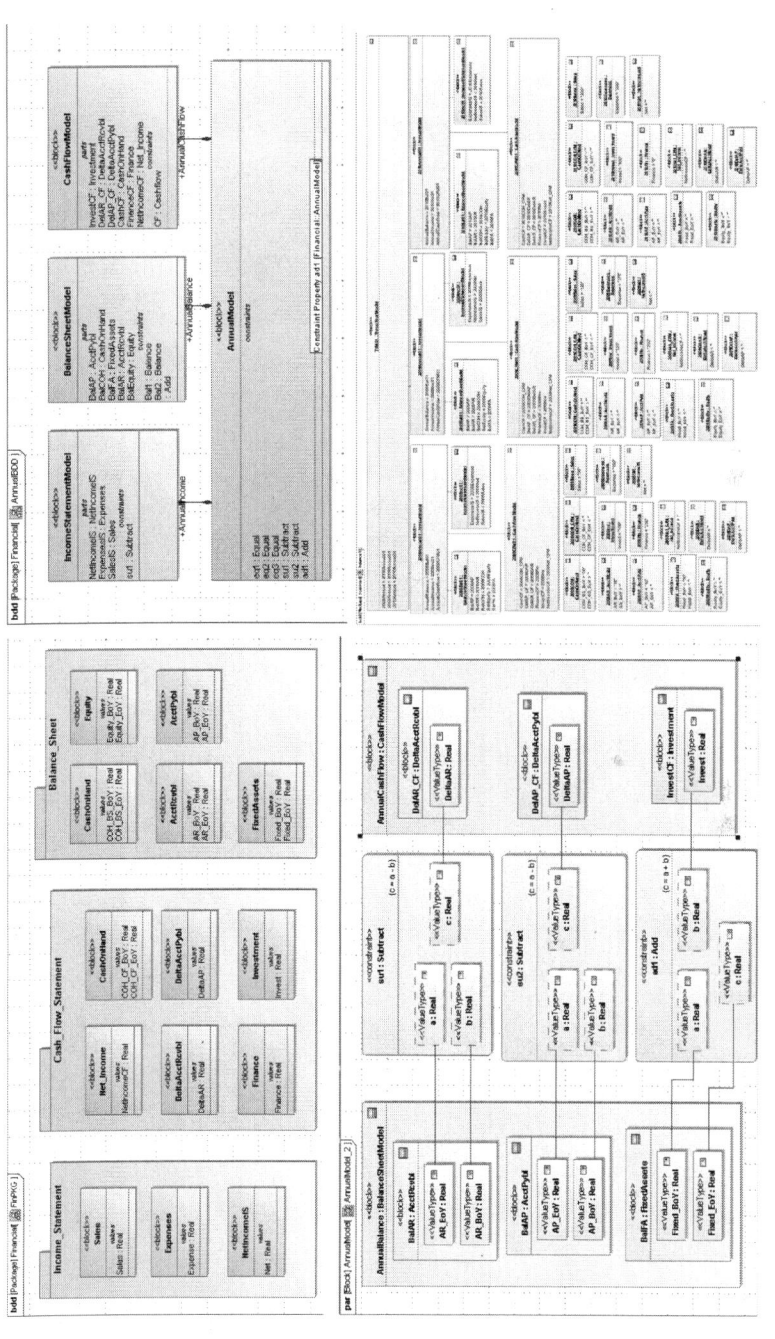

Figure 6.22 Financial projections SysML model: various diagram views. (From Zwemer, D.A. and Bajaj, M. 2008. SysML parametrics and progress towards multi-solvers and next-generation object-oriented spreadsheets. *Frontiers in Design & Simulation Workshop*, Georgia Tech PSLM Center, Atlanta.)

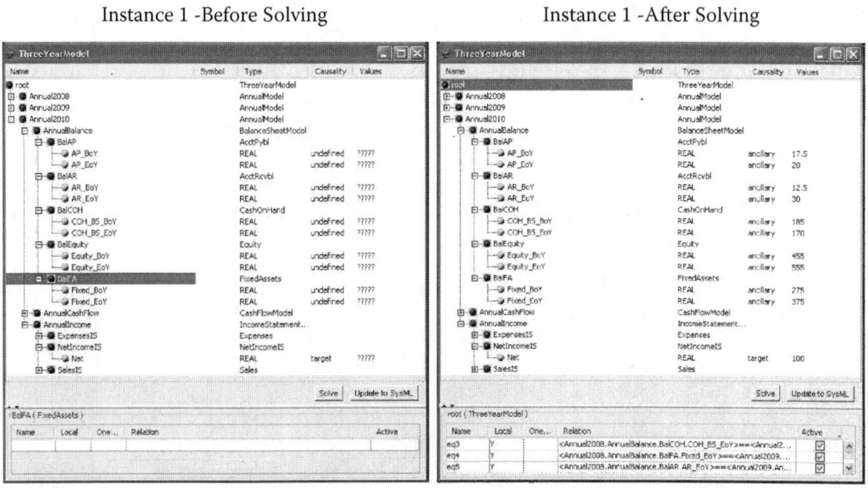

Instance 1 -Before Solving   Instance 1 -After Solving

Figure 6.23 Solving financial projections SysML parametrics: ParaMagic browser views. (From Zwemer, D.A. and Bajaj, M. 2008. SysML parametrics and progress towards multi-solvers and next-generation object-oriented spreadsheets. *Frontiers in Design & Simulation Workshop*, Georgia Tech PSLM Center, Atlanta.)

Discussion

Normally, end users will not need to create from scratch primitive templates such as those mentioned earlier—triangle, prism, linear spring, one-dimensional (1D) material model, extensional rod, and torsional rod. Such commonly used entities should be predefined and available in ready-to-use SysML libraries, along with frequently occurring ABB systems. This approach will empower users to concentrate on composing higher-level models to represent and solve their engineering analysis problems. However, it will take time and effort to create such libraries. Various vendor-specific engineering model libraries are available. See, for example, the following Web sites:

http://www.mathworks.com/applications/controldesign/
http://www.modelica.org/library/
http://www.wolfram.com/products/applications/mechsystems/

However, SysML offers a way to wrap existing libraries in a vendor-neutral form and to create new libraries from fundamental principles. Burkhart, Paredis, and colleagues (GIT 2007c) are pursuing a multi-aspect library approach for fluid power components. Peak (2002b, 2003) provides an initial characterization of such libraries, estimating that there are, for example, on the order of several hundred ABB concepts in the structural analysis domain.

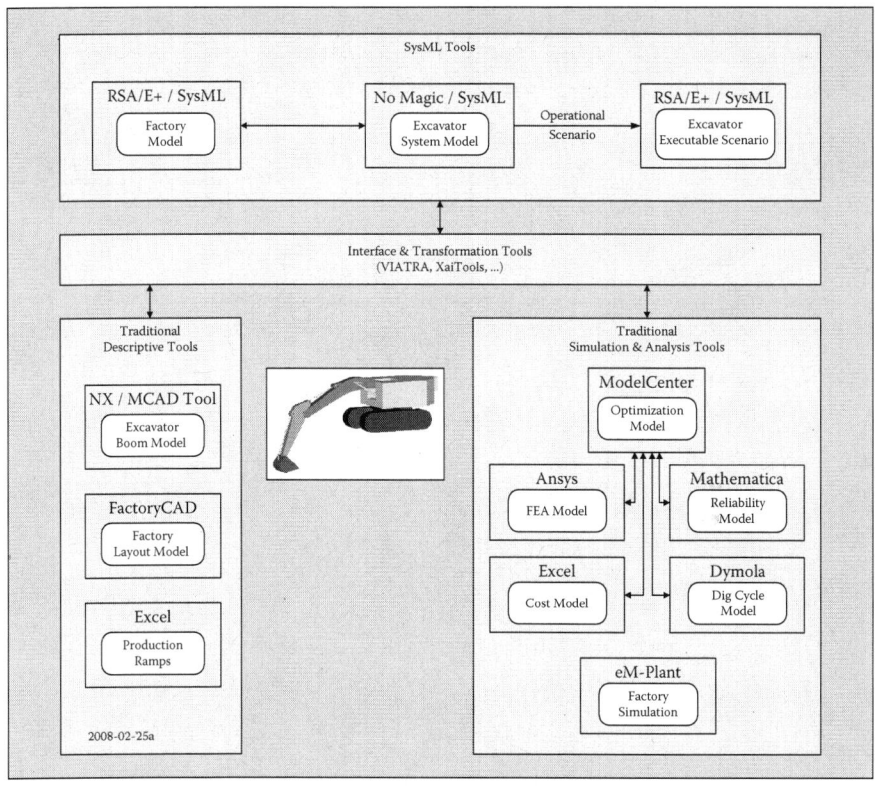

Figure 6.24 **Excavator modeling and simulation test bed: tool categories view. (From Peak, R.S., Burkhart, R.M., Friedenthal, S., Paredis, C.J.J., and McGinnis, L.M. 2008. Integrating design with simulation & analysis using SysML—Mechatronics/interoperability team status report. Presentation to INCOSE MBSE Challenge Team, Utrecht, Holland.)**

The aforementioned examples demonstrate how SysML parametrics has largely subsumed COB graphical modeling capabilities. As one becomes more familiar using a SysML modeling tool, other SysML capabilities beyond those in COB formulations become evident (e.g., stereotypes, value types, and automated consistency checks), and interactively composing diverse models becomes more of a reality.

These examples also show COB formulation capabilities that are not yet readily available in SysML itself or associated tools. Compare, for example, Figure 6.10c with Figure 6.14b and note that constraint blocks could use a way to show their relations in a localized usage manner (and also offer the option to use symbols)—for example, representing $F = k\Delta L$ using a generic $a = b*c$ relation for $r3$. Shape schematics and flattened constraint graphs such as Figures 6.12 and 6.13 are other COB views that may be worth considering for future versions of SysML.

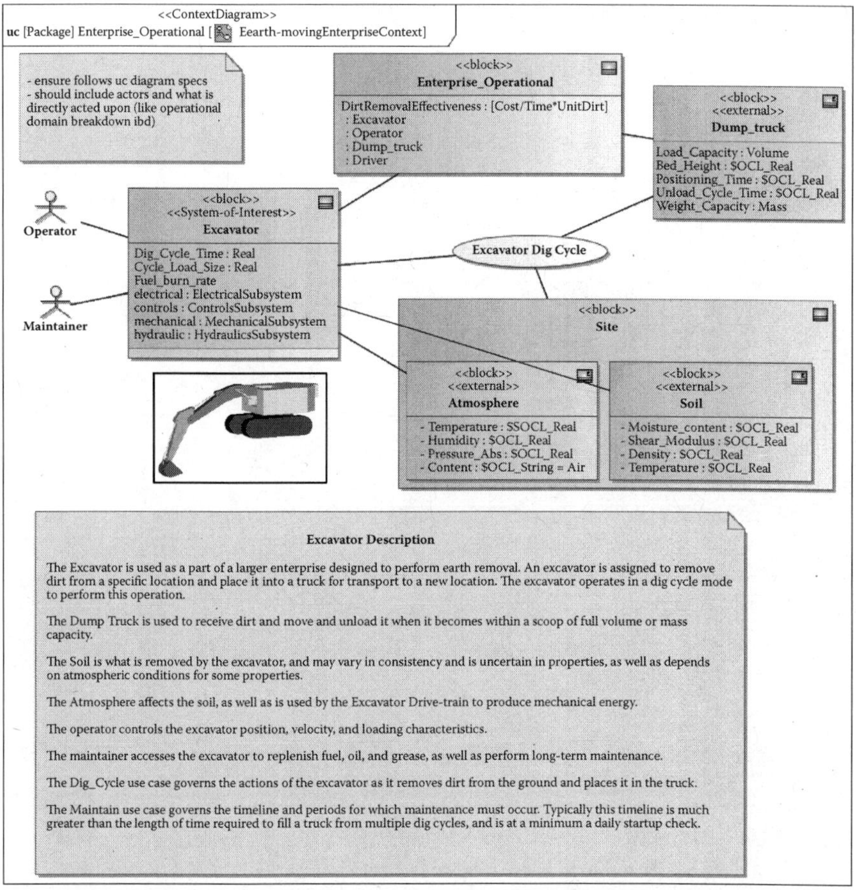

Figure 6.25 Excavator operational domain: top-level context diagram. (From Peak, R.S., Burkhart, R.M., Friedenthal, S., Paredis, C.J.J., and McGinnis, L.M. 2008. Integrating design with simulation & analysis using SysML—Mechatronics/ interoperability team status report. Presentation to INCOSE MBSE Challenge Team, Utrecht, Holland.)

Instances, causality, and human-friendly lexical forms (e.g., XML in Figure 6.16) are other areas that may warrant further attention, as seen by comparing Figure 6.3g–f with the triangular prism sample instance in Figure 6.4d. Frequent interaction with instances (and associated instance libraries) may be one of the key differences between traditional UML2 usage for software development versus envisioned SysML usage for interactive system design, modeling, and simulation.

Finally, as users begin to get familiar with SysML, it is helpful to keep in mind that SysML is in its infancy, relatively speaking, as a v1.0 technology. Additional capability needs are bound to be identified as the user experience base grows over

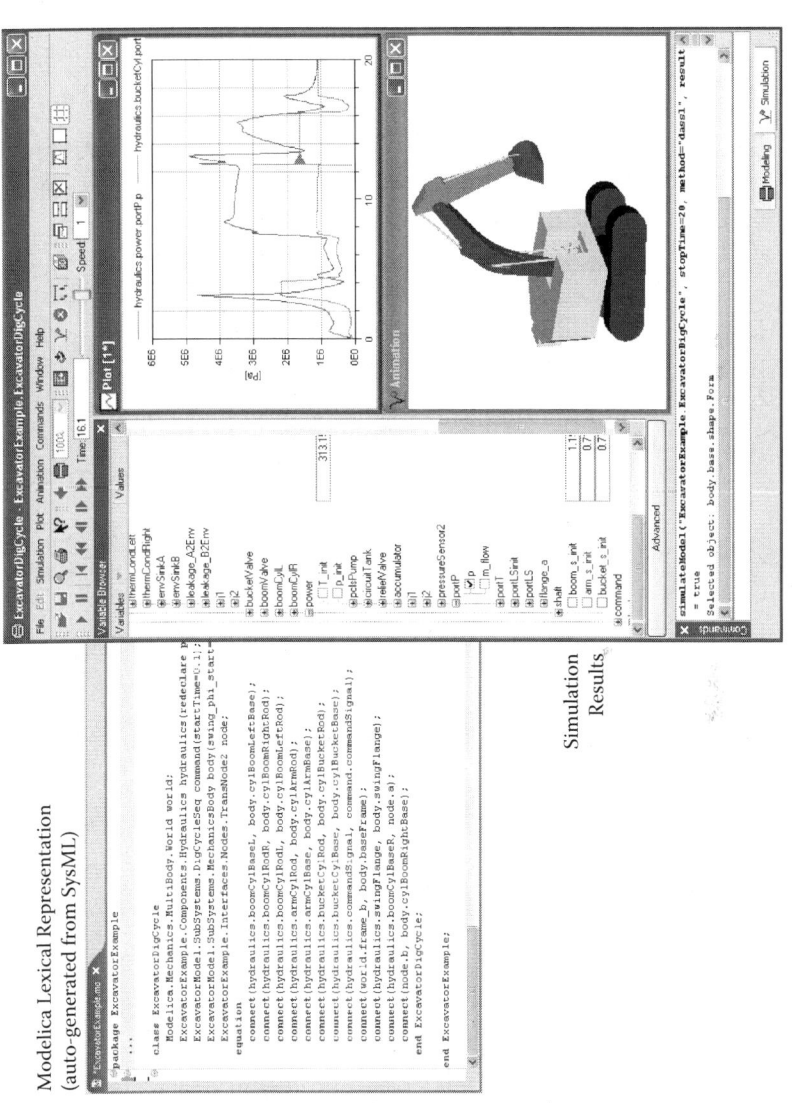

Figure 6.26 Excavator hydraulics system simulation in Dymola. (From Johnson, T.A., Paredis, C.J.J., and Burkhart, R.M. 2008. Integrating models and simulations of continuous dynamics into SysML. *Proc. 6th Intl. Modelica Conf.*)

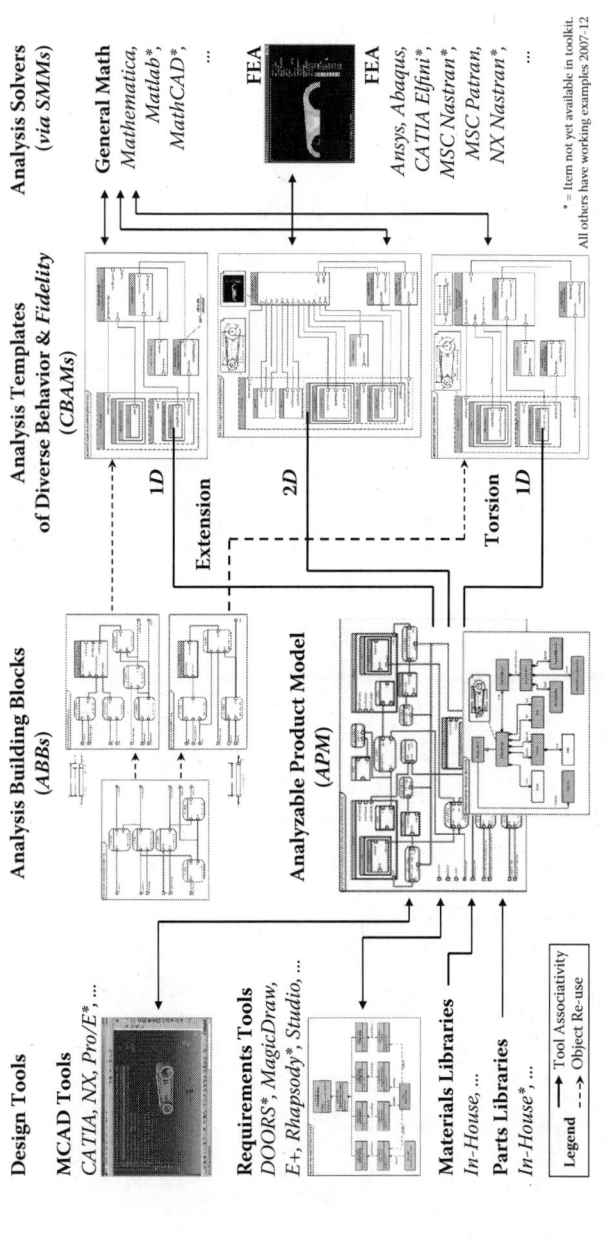

Figure 6.27 SysML-based panorama for high diversity CAD–CAE interoperability: flap linkage tutorial. (From Peak, R.S., Burkhart, R.M., Friedenthal, S., Wilson, M.W., Bajaj, M., and Kim, I. 2007b. Simulation-based design using SysML—Part 2: Celebrating diversity by example. *INCOSE Int. Symposium,* San Diego.)

time. Some may require v1.x specification updates, and others may not be addressed until v2.0 or beyond. However, as evidenced by this chapter, experience to date indicates there is good value to be gained from utilizing SysML v1.0.

Summary and Conclusions

This chapter progressively introduced SysML parametrics concepts through models of simple objects to demonstrate how engineering analysis models can be integrated with system architecture/design models. Sample ABBs demonstrate how SysML captures engineering knowledge from mechanics of materials as a representative domain.

This chapter also presented an overview of the COB knowledge representation that (a) provided motivation, semantics, and examples during the development of SysML parametrics; and (b) continues to provide means for executing SysML parametrics. We presented classical COB formulations alongside SysML diagrams for the same models. This approach has hopefully given the reader a better understanding of SysML parametrics and how to use it to create executable parametric models. It has also highlighted the following mutual benefits.

Benefits from COB technology for SysML parametrics:

- Provides a conceptual foundation:
  - Declarative knowledge representation (noncausal)
  - Combination of object and constraint graph techniques
- Contributes engineering analysis test cases:
  - Variety: domains, CAD tools, behaviors, fidelities, CAE tools, etc.
- Exercises and verifies numerous constructs systematically
- Enables architectures and methods for parametrics executability

Benefits from SysML technology for COBs:

- Provides richer modeling constructs:
  - Reusable relations, stereotyping, extended object constructs, etc.
- Integrates with other specification and design features of SysML
- Broadens and enhances support by commercial tools
- Increases modeling effectiveness (leveraging capabilities supported by numerous vendor tools):
  - Tool-aided graphical view creation (e.g., all the SysML diagram types)
  - Automated consistency between views

Finally, we highlighted several emerging examples that apply this technology to system-level problems. The range of applications, from UAVs to financial modeling to excavators, demonstrates the generality of the approach, which can be viewed as a next-generation, object-oriented, spreadsheet-like capability in some respects.

These and other examples indicate that COBs and SysML together provide these advantages over traditional engineering analysis representations:

- Richer semantics and greater expressivity
- Capture of knowledge in a modular, reusable manner
- Greater solution control

As new developments in model-based architecture and systems engineering continue to unfold, SysML is poised to become a unifying representation for both domain-specific engineering analysis models and system-level models, and thus to enhance multidisciplinary communication and speed system development. Richer knowledge management becomes feasible, including capturing fine-grain associativity among diverse models and leveraging this associativity for model execution. If SysML does indeed attain a central role in model-based systems engineering practices, then SysML and COB-like representations might ultimately provide fundamental advances for next-generation systems life-cycle management.

Acknowledgments

This chapter is based on Peak et al. (2007a); Zwemer and Bajaj (2008); and Peak et al. (2008), and the contributions of the corresponding authors/co-authors are gratefully acknowledged. The work described was funded in part by NASA (GIT 2007a), NIST (GIT 2007b), Deere (Johnson, Paredis, and Burkhart 2008), and Lockheed (Peak et al. 2008). SysML modeling work by Andy Scott and Bennett Wilson at GIT has contributed to these examples. While at Artisan Software, Alan Moore played a key role in the development of SysML parametrics, including aiding the incorporation of COB concepts into the specification, as well as their prototype implementation in Artisan Studio. While on the research faculty at Georgia Tech (2005–2006), Diego Tamburini implemented some of the aforementioned examples in SysML tool prototypes. He also developed the executable interface for SysML parametrics between Artisan Studio and XaiTools. Similarly, EmbeddedPlus and NoMagic Inc. supported GIT efforts to develop interfaces to their tools for COB-based solving. Finally, our colleagues in the Georgia Tech SysML Forum have been a steady source of helpful feedback.

References

Some references are available online at www.pslm.gatech.edu and http://eislab.gatech.edu/.

Burkhart, R.M. 2006. Modeling system structure and dynamics with SysML blocks. *Frontiers in Design & Simulation Workshop*, Georgia Tech PSLM Center, Atlanta.

Cowan, F.S., Usselman, M., Llewellyn, D., and Gravitt, A. 2003. Utilizing constraint graphs in high school physics. *Proc. ASEE Annual Conf. & Expo*, http://www.cetl.gatech.edu/services/step/constraint.pdf

Friedenthal, S., Moore, A., and Steiner, F. 2006. OMG Systems Modeling Language (OMG SysML) tutorial. *INCOSE Int. Symposium*, Orlando.

Friedenthal, S., Moore, A., and Steiner, F. 2008. *A Practical Guide to SysML: Systems Modeling Language*. San Francisco: Morgan Kaufmann.

GIT. 2007a. The composable object (COB) knowledge representation: Enabling advanced collaborative engineering environments (CEEs). [Project Web Page]. Georgia Tech, http://eislab.gatech.edu/projects/nasa-ngcobs/

GIT. 2007b. Capturing design process information and rationale to support knowledge-based design-analysis integration. [Project Web Page]. Georgia Tech, http://eislab.gatech.edu/projects/nist-dai/

GIT. 2007c. SysML focus area. Product & Systems Lifecycle Management Center, Georgia Tech, http://www.pslm.gatech.edu/topics/sysml/

Hause, M. 2006. The SysML modelling language. *15th European Systems Engineering Conf.*

Holt, J. and Perry, S. 2008. SysML for Systems Engineering: The Emperor's New Modelling Language: INSPEC/IEE.

Johnson, T.A., Paredis, C.J.J., and Burkhart, R.M. 2008. Integrating models and simulations of continuous dynamics into SysML. *Proc. 6th Int. Modelica Conf.*

OMG. 2007. OMG Systems Modeling Language (OMG SysML™) specification. SysML 1.0 Proposed Available Specification (PAS), OMG document [ptc/2007-02-03] dated 2007-03-23. http://www.omg.org/

OMG. 2008. OMG Systems Modeling Language (OMG SysML™). http://www.omgsysml.org/

Paredis, C.J.J., Diaz-Calderon, A., Sinha, R., and Khosla, P.K. 2001. Composable models for simulation-based design. *Engineering with Computers* 17:112–128.

Peak, R.S., Scholand, A.J., Tamburini, D.T., and Fulton, R.E. 1999. Towards the ubiquitization of engineering analysis to support product design. *International Journal Computer Applications Technology* 12(1) (Invited Paper for Special Issue: Advanced Product Data Mgt. Supporting Product Life-Cycle Activities):1–15.

Peak, R.S. 2000. X-analysis integration (XAI) technology. EIS Lab Report EL002-2000A, Georgia Tech. Atlanta. [X = Models throughout the product lifecycle including design (DAI), manufacturing (MAI), and sustainment (SAI).]

Peak, R.S. and Wilson, M.W. 2001. Enhancing engineering design and analysis interoperability—Part 2: A high diversity example. *1st MIT Conf. Computational Fluid & Structural Mechanics*, Boston.

Peak, R.S. 2002a. Part 1: Overview of the constrained object (COB) engineering knowledge representation. Response to UML for Systems Engineering Request for Information (SE DSIG RFI 1) OMG document ad/2002-01-17. http://syseng.omg.org/

Peak, R.S. 2002b. Techniques and tools for product-specific analysis templates towards enhanced CAD-CAE interoperability for simulation-based design and related topics. *Int. Conf. Electronic Packaging*, JIEP/ IMAPS Japan, IEEE CPMT, Tokyo.

Peak, R.S. 2003. Characterizing fine-grained associativity gaps: A preliminary study of CAD-E model interoperability. ASME DETC, Chicago.

Peak, R.S., Lubell, J., Srinivasan, V., and Waterbury, S.C. 2004. STEP, XML, and UML: Complementary technologies. Product Lifecycle Management (PLM) Special Issue, *J. Computing & Information Science in Engineering* 4 (4):379–390.

Peak, R.S., Friedenthal, S., Moore, A., Burkhart, R.M., Waterbury, S.C., Bajaj, M., and Kim, I. 2005a. Experiences using SysML parametrics to represent constrained object-based analysis templates. *NASA-ESA Workshop on Product Data Exchange (PDE)*, Atlanta, http://eislab.gatech.edu/pubs/conferences/2005-pde-peak/

Peak, R.S., Tamburini, D.T., and Paredis, C.J.J. 2005b. GIT update: Representing executable physics-based CAE models in SysML. Presentation to *OMG SE DSIG*, Burlingame, CA. http://eislab.gatech.edu/pubs/seminars-etc/2005-12-omg-se-dsig-peak/

Peak, R.S., Burkhart, R.M., Friedenthal, S., Wilson, M.W., Bajaj, M., and Kim, I. 2007a. Simulation-based design using SysML—Part 1: A parametrics primer. *INCOSE Int. Symposium*, San Diego.

Peak, R.S., Burkhart, R.M., Friedenthal, S., Wilson, M.W., Bajaj, M., and Kim, I. 2007b. Simulation-based design using SysML—Part 2: Celebrating diversity by example. *INCOSE Int. Symposium*, San Diego.

Peak, R.S., Burkhart, R.M., Friedenthal, S., Paredis, C.J.J., and McGinnis, L.M. 2008. Integrating design with simulation & analysis using SysML—Mechatronics/interoperability team status report. Presentation to INCOSE MBSE Challenge Team, Utrecht, Holland.

Weilkiens, T. 2008. *Systems Engineering with SysML/UML*. San Francisco: Morgan Kaufmann.

Wilson, M.W. 2000. The constrained object representation for engineering analysis integration. Masters thesis, Georgia Tech, Atlanta.

Wilson, M.W., Peak, R.S., and Fulton, R.E. 2001. Enhancing engineering design and analysis interoperability—Part 1: constrained objects. *1st MIT Conf. Computational Fluid and Structural Mechanics*, Boston.

Zwemer, D.A. and Bajaj, M. 2008. SysML parametrics and progress towards multi-solvers and next-generation object-oriented spreadsheets. *Frontiers in Design & Simulation Workshop*, Georgia Tech PSLM Center, Atlanta.

Chapter 7

DFD and IDEF0
Graphical Tools

Developing models of system flow provides a description of what a system must do before the architect or systems engineer performs functional decomposition. Data flow diagrams (DFDs) and IDEF0 are two popular graphical tools that support system functional definition and decomposition. DFD is more focused on the data and its flow, whereas IDEF0 is more focused on system functions and their decomposition. Graphical tools can be attractive to the practitioner because relationships between entities are captured visually by the structure of the graph created.

Key Concepts

Environmental Model
Behavioral Model
System flow
Functional decomposition

Graphical Tools

DFD and IDEF0 are commonly used graphical tools for modeling the Behavioral Model. Each also has provisions for modeling the context diagram of the Environmental Model. DFD supports activity and functional modeling by first focusing on what data is flowing through the system. Activities and functions are therefore defined by the need to transform data from one form into another. The points of transformation identified in this way become the nodes of the DFD graph. Data is the only entity associated with the nodes in DFD, and the nodal semantics express only whether data is flowing into a node or out of it or is being stored by the node. IDEF0 is graphically similar in form, although the semantics of graphical representation of the nodes differ somewhat from the DFD semantics; in some ways, IDEF0 is richer. Practitioners of IDEF0 also tend to focus more on the nodes than on the data flow. Both of these graphical tools are used in the practice of functional modeling of information-intensive systems, but the differences in their nodal semantics preclude full exchange of information between a DFD and an IDEF0 graph.

Data Flow Diagrams

A DFD is a graph with a defined syntax and semantics that shows the flow of data from external sources to external destinations, and identifies the transformations of data and the stores of data. There are many variants and extensions of the data flow model, but the original work was done by Edward Yourdon (Yourdon 1989). His approach is still popular with practitioners of DFD. More extensive details on DFD and open discussions may be found on his wiki: http://www.yourdon.com/strucanalysis/wiki. The description of DFD in the following text follows the Yourdon approach but also shows some of the relationships between DFD and UML, how DFD can be used to support systems engineering definition and decomposition, and how the logical elements of a DFD diagram might be realized in the Implementation Model.

DFDs have four types of elements: flows, processes, stores, and terminators. Figure 7.1 is an example of the symbology typically used for the first three types of elements. Flows are represented graphically as directed arrows. Processes may be represented graphically as circles, ovals, rectangles with rounded edges, or even rectangles. A store is typically represented graphically as two parallel lines, as in Figure 7.1, or as a variation on that construct. Terminators are most typically represented by rectangles. There are many minor variations of symbology across the practitioners of DFD, but the differences between shapes and styles are generally cosmetic. It is obviously important to use the same shapes and styles consistently to represent the four types of elements when modeling a system using DFDs.

The DFD element "flow" is used to describe the movement, or transport, of data from one part of the system to another (data in motion). Flow occurs between

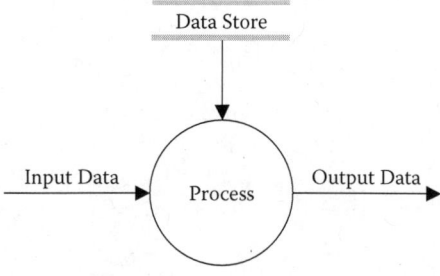

Figure 7.1 Typical symbols for DFD elements.

pairs of processes, stores, or terminators. A flow is named by a noun or noun phrase that describes what moves along the flow. Thus, a flow is described by what data is flowing, the pair of elements between which the data flows, and the direction of the flow. Generally, flows carry only one type of data, but it is possible to consolidate several elementary flows into one flow. This is called a *convergence of flow* (or a *join*). The converged flow should be given a name distinct from the individual flows being joined. Flows can also diverge (branch). As with the converged flow, the integrity of the names of the flows should be preserved. The appropriate names of the diverged flows should be given to the individual flows on the branches.

A process is a function or transformation of the data that flows through the system. The process shows a part of the system that transforms inputs into outputs. Processes can also be thought of as operations on data, acting on inputs to produce outputs. Communication of the data between processes is not modeled in any detail in DFDs, although in an Implementation Model, communication can be represented by trivial processes that simply pass or route data but do not change its content. The name of the process describes what it does in a brief verb–noun phrase.

A DFD store models data at rest (in contrast to data in motion, as in the flows). The name of the store is the plural of the name of the data carried by flows in and out of the store. The store can act as a database, which may be a derived system requirement. An example of this would be a records store in an accounting system. Stores may also be created directly in response to user requirements.

A DFD terminator represents an external entity with which the system communicates. The names of terminators are nouns or noun phrases. Terminators are outside the system being modeled; the flows between terminators and the various processes represent the interfaces between the system and the environment. As defined in Chapter 3 ("Concepts, Standards, and Terminology"), the system processes that interface with the environment will be implemented by the boundary elements of the system. Terminators are external to the system and in the system flow; they simply represent sources and sinks. Neither the internal details of the terminators nor the relationships between terminators are shown in a DFD.

From the viewpoint of structured analysis, DFDs can be used to create the context diagram of the Environmental Model and diagrams of the system flows within the Behavioral Model. DFD diagrams also have a hierarchy that follows the pattern of definition and decomposition.

The highest-level diagram in DFD is the context diagram, which describes the system and its flows in the context of its environment. The context diagram contains a single unnumbered process, as illustrated in Figure 7.2. Unlike the DFD in Figure 7.1, the context diagram is not permitted to contain any stores. The perspective of the context diagram is from the system outward toward the environment. This is similar to a UML use case, which indicates the interactions of a system with its environment, but from the DFD viewpoint of system flows; that is, the interactions are expressed with the details of a flow between the system and the terminators in its environment. A statement of system purpose and definition of the

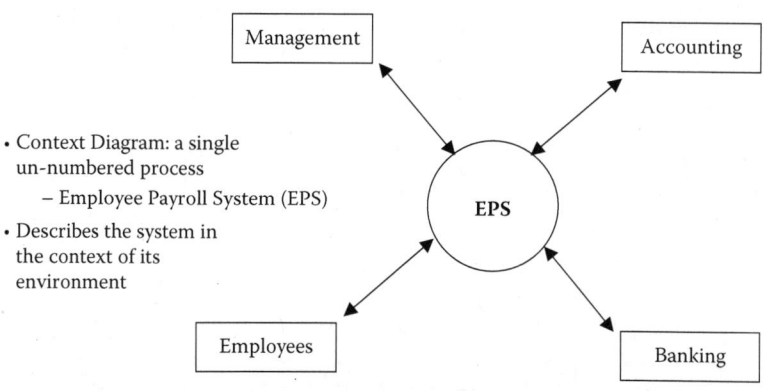

Figure 7.2   Context diagram (Environmental Model).

system boundary from the Environmental Model are also appropriate for the context diagram.

When the (single) process identified in the context diagram is decomposed into the main system processes, stores, and terminators, the result is a new DFD that is referred to as a *0-diagram*. The perspective is from the system boundary inward toward the system. The terminators in the context diagram must appear in the 0-diagram. The flows from the terminators in the context diagram are labeled in the same way in the 0-diagram. The terminators in the 0-diagram will be at the system interfaces with the environment from the context diagram. In the Implementation Model, these interfaces should be implemented as boundary elements. Figure 7.3 depicts a 0-diagram.

The processes and stores in the 0-diagram can be arrived at in one or more ways. One way is for user requirements to directly imply the need for certain

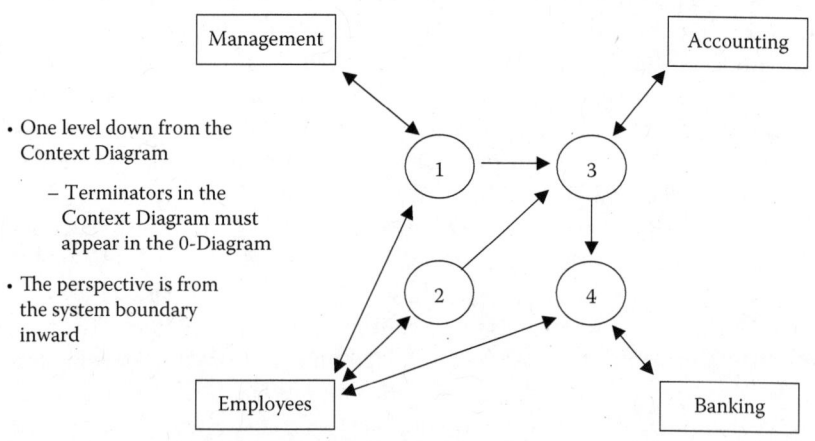

Figure 7.3   0-diagram (Behavioral Model).

processes and stores. There may also be derived requirements that further identify processes and stores. Among the derived requirements can be internal associations between the external flows; that is, a process is identified that transforms one or more inputs (into the single process of the context diagram) from the external flows into one or more of the output flows in the context diagram. However, because relationships between the terminators in the context diagram are not exposed, it is not completely possible to adhere to the principles of structured design. Without the relationships between the terminators in the context diagram, the structure of the problem is not completely exposed and therefore cannot be reflected into the structure of the solution (e.g., the relationships between the processes and stores of the 0-diagram).

If any of the processes in the 0-diagram can be decomposed further into lower-level processes and stores, then a new DFD can be created following the same approach that was used to create the 0-diagram. The resulting DFD is called a *1-diagram*. All the terminators from the 0-diagram (inherited from the context diagram) should be maintained in the 1-diagram, but it is not permitted to add new terminators. The data structures and processes of the 1-diagram must be consistent both within the diagram and in relationship to the 0-diagram.

Successive decompositions of the DFDs can be performed as required, following the same approach used in deriving the 1-diagram from the 0-diagram. The resulting DFDs are referred to as *N-diagrams*. Each process in the 0-diagram is decomposed into a lower-level DFD until the lowest-level processes are reached. These unpartitioned processes are called *primitive processes*. They correspond to the leaves in the functional decomposition tree.

In the Yourdon approach to DFDs, the following guidelines are recommended. Begin with the 0-diagram. Identify the main system processes, terminators, and stores. Construct the context diagram (but do not include stores). Include all terminators in the 0-diagram and the context diagram; maintain terminators in lower-level diagrams. Do not introduce new terminators in lower-level diagrams. Make sure that each DFD is internally consistent and also consistent with any associated DFDs. Choose meaningful names for processes, flows, stores, and terminators. Number the processes. Avoid overly complex DFDs.

Strengths and Weaknesses of DFDs

The graphical notation is straightforward and facilitates traceability of problems through inheritance of the terminators from the context diagram. DFDs are hierarchical and support functional definition and decomposition in a way that emphasizes components and interfaces. They suppress the internal details of the processes to focus on the flow of the system as a whole.

However, DFDs do not show flow of control and temporal aspects (time) of the system, such as delays and durations. They do show precedence relations (through information dependencies) between processes, but not timing.

IDEF0 Standard

The basic IDEF0 concepts are conciseness and communication through graphical representation of activity modeling. An IDEF0 diagram is a graph with a defined syntax and semantics that is simple. The rules for IDEF0 include purpose and viewpoint, syntax for graphics, and rules for uniqueness of names and labels on a diagram. The prescriptive arrangements of each box in the graph and its related arrows and text to specify meanings convey a coherent sense of visual meaning.

IDEF0 is hierarchical and supports functional definition and decomposition. It has a hierarchy of diagrams with major functions at the top level of description and subfunctions at the successive levels of decomposition. The separation of organization from function is also included in the purpose of the IDEF0 model to avoid organizational viewpoints. IDEF0 also provides a methodology that is based on a step-by-step procedure for modeling, review, and interview tasks.

IDEF0 is standard for Integration Definition for Function Modeling, maintained by the U.S. National Institute of Standards and Technology (NIST). It is available in Publication 183: Integration Definition for Function Modeling (IDEF0), 21 December 1993. This publication may be found on the NIST public Web site, http://www.nist.gov/itl/div897/pubs/by-num.htm.

An IDEF0 model uses two basic graphical elements: a box and arrows. A box is a rectangle containing a name and number that is used to represent a function or an activity. An arrow is a directed line, composed of one or more arrow segments, that models an open channel or conduit conveying data or objects from source (no arrowhead) to use (with arrowhead). There are four classes of arrows:

Input
Output
Control
Mechanism and call (communication)

Arrow segments are labeled with nouns or noun phrases to express meanings of the data or objects flowing through the system. The box name, on the other hand, is a verb or verb phrase that is descriptive of the function or activity that the box represents. The construct of input, control, output, and mechanism is called the ICOMs of the function.

Each side of the box has a standard meaning in terms of relationships between the function that the box represents and data or objects that the arrow represents, as shown in Figure 7.4. The left-hand side of the box is always an input to the function that the box represents. The right-hand side is always an output. The top of the box is reserved for control over the function, as represented by an arrow point toward the box. The bottom of the box is reserved for the mechanisms that enable the function, and the communications from the function to other functions.

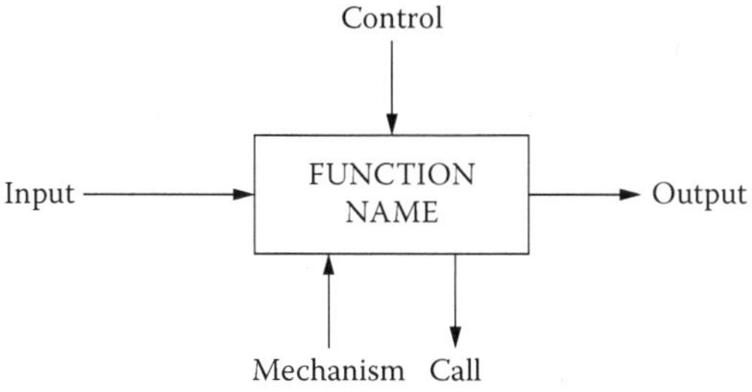

Control

Input — FUNCTION
NAME — Output

Mechanism Call

Figure 7.4 IDEF0 basic notation.

IDEF0 notation also allows for joins and branches of arrows. An arrow may branch or join, indicating that the same kind of data or object may be needed or produced by more than one function. Labels on branching arrow segments provide detailing of the arrow content, just as lower-level diagrams provide detailing of parent boxes. All or part of the contents of an arrow may follow a branch. A forking arrow may denote the "unbundling" of meanings that had been combined under a more general label. The joining of two arrow segments may denote "bundling," that is, the combining of separate meanings into a more general category.

IDEF0 also has page semantics. The first page only contains one box. The page number is A0, and the box number is 0. The arrows are labeled to indicate the inputs of the environment into the system and the outputs of the system to the environment. Textual definitions of the viewpoint and purpose of the model are also provided. An example of an A0 page is provided in Figure 7.5.

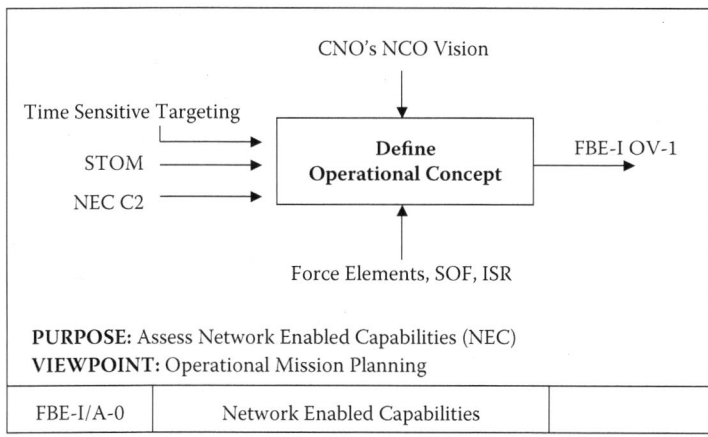

Figure 7.5 IDEF0 top-level context diagram (A0).

The A0 page is the context diagram, the root of the tree of the functional decomposition. In this sense, IDEF0 is just like DFD. Both have a context diagram with a single function (process in the case of DFD) that represents the system at the highest level in its environment. However, IDEF0 does not permit any other boxes in the A0 page that could serve as the terminators that are used in the DFD context diagram. And the DFD context diagram does not have any symbology for the control of a function or the mechanisms that enable the function.

Each page of an IDEF0 model under the context diagram is the decomposition of an individual function or activity from the previous page. On each page one level down from the previous page, a function represented by a box on the given page is decomposed into a set of boxes laid out diagonally from the upper left to the lower right on the next IDEF0 page, as illustrated in Figure 7.6. Ideally, after the A0 page, each page should contain three to six new boxes that represent the functional decomposition of an individual box from the previous page. An example of a second-level decomposition would be annotated as the pages

A0 → A1, A2, A3 ...
A1 → A11, A12, A13 ...
A11 → A111, A112, A113 ...

This simple example shows the indexing scheme for the page numbers that decomposes A0 into lower-level functions, and then decomposes the first function from page A1 into lower-level functions. The page numbers are used to track the decomposition. The single highest-level function on the context diagram page A0 might be decomposed into functions numbered 1, 2, 3, ..., each of which will have its own page. The first function is described on page A1, the second on page A2, the third on page A3, and so forth. In this scheme, each function represents the transformation of one or more inputs at the A0 level into one or more outputs at that same level. The pages A1, A2, A3 are a functional partitioning of A0. If any of these pages contains only one function, then that page is at the leaf level of the functional decomposition.

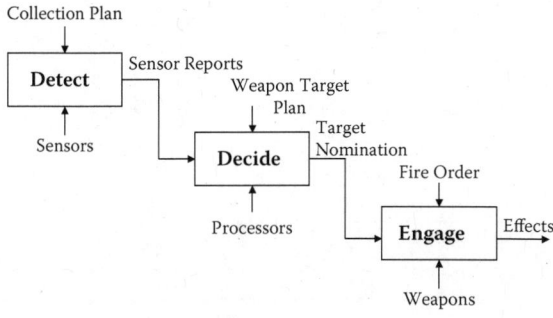

Figure 7.6 IDEF0 diagram example.

A box may perform various parts of its function under different circumstances, using different combinations of its input and controls and producing different outputs. These different performances are called the different activations of the box.

Strengths and Weaknesses of IDEF0

IDEF0 is a comprehensive and expressive graphical modeling tool, capable of representing a wide variety of business, manufacturing, and other types of enterprise operations to any level of detail. It provides a simple and coherent graphical syntax and semantics, providing for precise expression and promoting consistency of usage and interpretation. This enhances communication between systems analysts, developers, and users through ease of learning and its emphasis on hierarchical exposition of detail. As a long-standing standard from the U.S. National Institute of Standards, it is well tested and proven through many years of use in government and private industry. It is also supported by a variety of computer graphics tools and implemented by numerous commercial products that specifically support development and analysis of IDEF0 diagrams and models.

However, despite the representation of flow between functions, IDEF0 is focused more on functional decomposition than system flow. It also does not emphasize the storage of the data and objects flowing through the system. Also, the IDEF0 context diagram does not expose where the system is receiving its inputs from or where it is sending its outputs to.

Summary

DFD and IDEF0 can each be used to support aspects of the essential model in structured analysis. Both can be used to model functions and data flow. Both support functional definition and decomposition. However, IDEF0 and DFD cannot be used in conjunction with each other because they do not contain the same information. IDEF0 includes flow of control and, when appropriate, mechanisms. DFD models data stores explicitly, but IDEF0 has no graphical representation of data stores.

Other graphical tools will be needed to model the rules for system control, the system states and transitions, and to model the data itself. There is no one graphical tool that can be used to model all four aspects of the Behavioral Model.

References

U.S. National Institute of Standards and Technology (NIST). 21 December 1993. Publication 183: Integration Definition for Function Modeling (IDEF0). http://www.nist.gov/itl/div897/pubs/by-num.htm.

Yourdon, E. 1989. *Modern Structured Analysis*. Englewood Cliffs, NJ: Prentice Hall.

Chapter 8

Rule and State Modeling

In addition to modeling the system flow and system data, the Behavioral Model must also include models of how the system flow is controlled and give models of the system states and its transition between states. A model of the system behavior is captured by system state transition diagrams. As indicated in Chapter 7 ("DFD and IDEF0 Graphical Tools"), there is no single graphical tool that can be used to describe the Behavioral Model in structured analysis. Therefore, this chapter will continue with the presentation of a variety of commonly used graphical modeling tools for the Behavioral Model. Chapter 7 dealt with graphical tools for modeling only one of the four aspects of the Behavioral Model: system flow. This chapter will focus on two of the other four aspects: using rules for the control of flow, and system states and their transitions. The design of the control of system flow is accomplished by writing rules that implement the control structure over the system flow. The system behavior is determined by the integration of the rules into the system activity model and state transition diagrams.

Key Concepts

The structure of control
Using rules for control of flow
System states and events
State transitions

Control Structures

The control structure over the system flow can be built from three elementary types of structure: sequential, decisional, and loops. Null actions are permitted within these three types. The three elementary structures can be used in combination to develop highly complex process logical representations of the system-level control structure. When building up a system-level control structure, attention must be paid to the scope of the constructs. Similar to functional decomposition, nested

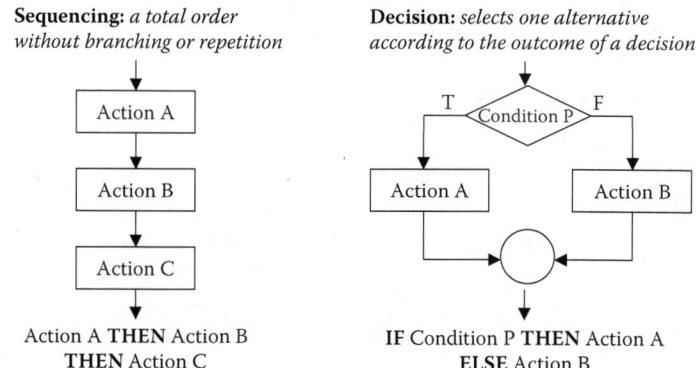

Figure 8.1 Control structures for sequences and decisions.

constructs bound the scope of control and should be displayed when the control structure is shown in graphical form. Indentation of phrases should be used to show subordination of constructs when represented in natural language.

Figure 8.1 illustrates the elementary sequential and decisional control structures. In sequencing, a total ordering is established over a collection of instructions (i.e., without branching, iteration, or repetition). In the left-hand side of Figure 8.1, whenever a given action or process has finished executing, the next action in the sequence begins execution. On the right-hand side of the figure is an elementary decisional control structure. In this construct, a condition is placed before one or more actions (or processes) and, based on the outcome of the condition, a decision is made to select which one will be executed. If the actions (or processes) can be executed simultaneously, then more than one can be selected for execution, depending on the outcome of the condition.

There are two types of loops in the elementary control structures: iteration and repetition. They are illustrated in Figure 8.2. Iteration allows an action or process

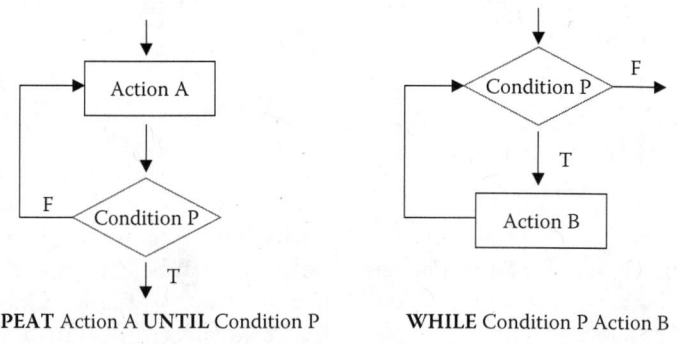

Figure 8.2 Loops in control structures.

to repeat itself until a set of conditions is satisfied. In this case, the precedence order of the structure is as follows: first, the condition, and next, the action or process to be executed. Thus, while the condition is valid, the execution will be continued iteratively. This would be the case for a process that requires repeated execution to converge, for example, a tracking filter in a radar system. In repetition, on the other hand, the precedence between the condition and the execution is reversed. The condition is evaluated after the execution of the action or process. The condition could be something as simple as a counting condition that is set for the purpose of executing the action or process a specified number of times.

Rule Modeling

Control structures are implemented through the specification of rules. Three types of rule models will be reviewed in this section: one in natural language, one in graphical form, and one in tabular form. The structured English rule model can be used for all three types of elementary control structures. The graphical and tabular rule models presented in this section are suitable for decision making, that is, as decisional control structures. Details of the DoDAF and MODAF with regard to rule modeling will not be addressed in this chapter. However, in DoDAF, the Operational Rule Model (OV-6a) is used to model the rules for control of flow in the operational views.

The rules used to specify the control of system flow must establish relationships among system elements and their interactions. Rules must eventually be formulated precisely using algorithms. Modeling can bring precision to the specification of rules that natural language generally cannot. Imprecise formulation of rules corrupts the structure of control in the system flow. In large-scale systems, it is more likely that the corruption of the structure of control leads to system faults than execution failures of the processes embodied within the system.

Rule models are usually not created as independent models but instead are embedded in the system activity model (e.g., IDEF0) or in the system dynamics model (state transition diagram).

Structured English

Structured English is based on a theorem in Turing machines that guarantees that any flowchartable logic can be represented using only the three elementary types of control structures, that is, sequential, decisional, and loops (Boehm and Jacopini 1966). Structured English expresses process logic by using the nested pattern, If–then–else–so.

An action is expressed in structured English by an imperative sentence or command. A verb or verb phrase associated with a process symbol in a graphical representation (e.g., a circle in DFD or a box in IDEF0) must be expressed in one of these sentence forms before it can be modeled in structured English. The following simple example illustrates the method (Solvberg and Kung 1993). Consider the following three sentences in natural language English:

> Vegetables that are both leafy and crispy should be fried, while those that are crispy but not leafy should be boiled. Prior to cooking, all vegetables that are not crispy should be chopped. Then those that are green and hard should be boiled, while those that are hard but not green are steamed; those that are not hard are grilled.

In structured English, these sentences would be written as follows:

```
if crispy
        if leafy then fry
        otherwise boil
otherwise chop
        if hard
                if green then boil otherwise steam
        otherwise grill
```

Note: "otherwise" has been substituted for "else" for readability.

The use of phrasing and indentation in structured English exposes the underlying logic. And because structured English expresses process logic by using the nested pattern of If–then–else–so, it can be used to model process logic constructed from all three elementary types of control structures.

Decision Trees

A decision tree represents a decisional control structure in the graphical form of a tree. Recall that a tree has a single root, but multiple branches and leaves. The root represents the first decision. Each node (i.e., a point at which branching occurs) represents a decision point. The possible outcomes of a decision are shown as branches. The leaves of the tree represent the actions.

A path from the root to a leaf represents a decision rule. Using a decision tree is appropriate when each rule results in a single action. Figure 8.3 is a graphical representation of the culinary example that was previously modeled in structured English. On the far left is the root decision, Is the vegetable crispy? After one or two branches of further decision making, the path arrives at the leaves of the tree, which are the actions. The leaves are represented on the far right in the graph. There are five leaves. Now, a closer examination of the leaves reveals that some of the actions may be described in two parts. Logically, this is not a problem. The three leaves that have two actions are actually in a sequential control structure. In this example, these three leaves share the common action "chop." Thus, "chop,

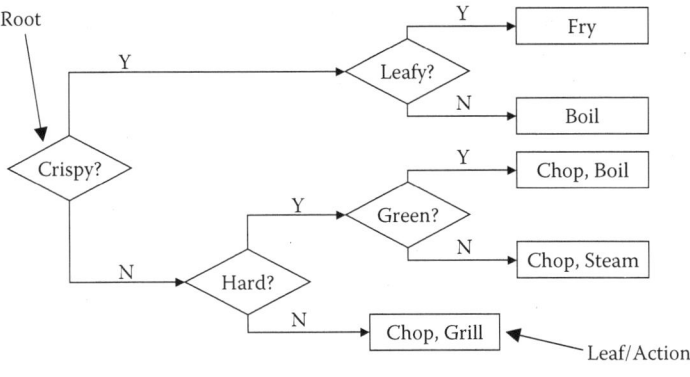

Figure 8.3   Decision tree example.

boil" is the sequence of actions "chop then boil," which is evident in the structured English model. "Chop, steam" and "chop, grill" follow the same sequential control structure. There are actually five actions at the atomic level of description (i.e., at a level where there is no intersection between the actions). Only four of these actions are what will be called *end states* in the next section. In general then, decisional control structures can have embedded sequential control structures and the sequential structure can be exposed in the graphical representation of a decision tree.

Decision Tables

A decision table represents a decisional control structure in tabular form. As indicated by the discussion of decision trees, decision tables can also have embedded control structures, but the precedence order is not evident from the table without further annotation.

A decision table has four parts that are used to model the rules, as illustrated in Figure 8.4. The lower half of the table is associated with the actions to be taken. The upper half of the table is associated with the conditions related to the rules under which the actions will be taken. The upper-left-hand quadrant of the table displays the names of the decision variables used in the conditions. The upper-right-hand quadrant of the table is a matrix of entries for the values of each variable associated with a rule (i.e., a column). The set of values in a given rule column is the list of the conditions (i.e., combinations of values of the decision variables) under which the rule will result in the actions indicated in the portion of the column in the lower half of the table. Only the actions to be taken are marked. The names of the actions are in the lower-left quadrant of the table. Thus, the action matrix organizes the actions to be taken by the rules (columns).

It is important to understand that Boolean-type decision variables can actually take on three values, not just two. In this case, the conditions that define a

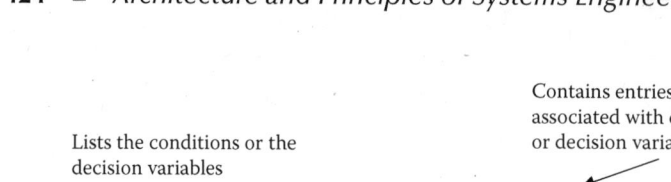

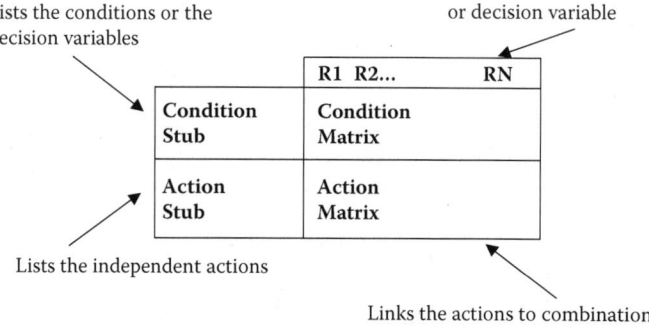

Figure 8.4   The four parts of a decision table.

rule are based on the designated decision variables taking on the values Y (= yes) or N (= no). However, not all decision variables in a rule need to be assigned one of these two values. Recall that in Figure 8.3 each path from the root to a leaf defines a rule. The tree has five leaves and therefore has five rules. Consider the first rule to be the uppermost branch in the tree. There are exactly two decision variables: "crispy," "leafy." The rule would be as follows: if crispy = Y and leafy = Y then take the action fry (the vegetable). But what about the other two decision variables: "hard," "green"? How do they relate to the decision? The answer is that they don't. The vegetable could be "hard" or it could not be "hard". Either way, it will be fried if the two designated variables ("crispy", "leafy") take on the value Y (= yes). The same can be said of the decision variable "green." The decision as to whether to fry a given vegetable is independent of the decision variables "hard," "green." This is obvious from looking at the decision tree in Figure 8.3. These two decision variables are on different branches of the tree than the two variables that the rule depends on (crispy, leafy). In the case of any decision variable from which the rule is independent, a symbol such as a dash is placed in the appropriate row and column. This indicates that the decision rule is independent (or indifferent to) of the values of that variable. This leads to what is called a *limited entry* table, because not all decision variables must be assigned a value in the condition matrix.

Decision tables may be constructed using the following procedure. First, identify all the decision variables and the possible values for each; use the names of the variable to fill in the condition stub. Next, determine the number of rules and then generate the combinations of values of the decision variables associated with each rule (column); use these values to systematically fill out the condition matrix. Now, identify the independent actions to be taken and fill out the action stub. Finally, identify the action or combination of actions to be taken for each rule, that

• Limited Entry Decision Table

	R1	R2	R3	R4	R5
Crispy	Y	Y	N	N	N
Hard	-	-	Y	Y	N
Leafy	N	N	-	-	-
Green	-	-	Y	N	-
Fry	X				
Chop			X	X	X
Boil		X	X		
Steam				X	
Grill					X

Y : Yes	- : Indifferent	R1, R2,... rules (if... then...)
N : No	X : Action Marker	

Figure 8.5   Rule specification example.

is, combination of conditions (decision variable values) in each column. In this way, fill in the action matrix one column at a time.

Validate the decision table by formulating questions for users, which can be used to resolve omissions, inconsistencies, or ambiguities in the process logic represented by the table. Revise and complete the decision table based on the users' answers to the questions.

The culinary example modeled first in structured English and then again in a decision tree is displayed as a decision table in Figure 8.5. The three (logical) leaves of "chop, boil"; "chop, steam"; and "chop, grill" can be found in the row for "chop," which has action markers in the columns with rules R_3, R_4, and R_5. Unlike the decision tree model, where the sequential structure of control could have been exposed to produce independent leaves, in the tabular form it cannot expose the individual independent leaves without further annotations. The structured English model, for example, would be needed here to resolve the precedence order of the actions.

Decision tables are based on the assumption that the decision variables take on a limited (small) number of discrete values or that the condition matrix has a limited number of decisional entries. That is, the condition matrix of the limited entry decision table may be a large array, but it is sparsely populated. However, decision tables are well suited for specifying actions to take under complex combinations of conditions (vice complex conditions).

Decision Tables or Decision Trees?

It is appropriate to use a table as a working tool to generate the decision part of a logic process; use a tree to present the results when the tree is not too large to visualize. Decision tables are suitable when there are many decision variables, or if the decision rules result in combinations of actions. Decision trees can be easily visualized when there are only a few decision variables, and the decision rules result in only one action per rule.

State Modeling

State modeling has roots in the physical sciences and plays a key role in the analysis of system behavior in engineering. In physics, the canonical variables used to describe the state of a system are position and energy. Transitions between states in physics refer to the relationship between changes in position and changes in energy. See, for example, Jauch (1968), for a mathematical discussion of states in the so-called phase space of classical and quantum mechanics in physics. The mathematical concept of states applies also to engineered systems. A detailed example of states for a simple electric circuit, when viewed as a system, can be found in Wymore (1967).

The changes in energy and position can be easily observed as an object falls to the Earth under the influence of gravity. The object gains kinetic energy as it changes position. An attractive feature of state models used in physics is that only the current state of a system need be known to predict future states of the system. The prior history of the states of a system in physics is not necessary for the prediction of future states if the current state and the state equations of the system are known. In systems engineering, the state transition diagram plays the role of the state equations used in physics.

States in Engineering and Physics

There are many examples of states of systems. One of the simplest is a simple light switch. The state of the light switch is described by one of two values: "on" or "off." The state space is the set of all values for the state of the system. In the case of the light switch, the state space consists of two discrete values. The values "on" and "off" are sometimes referred to as the states of the light switch, but strictly speaking, they are the values of the state of the light switch. In this idealized model, details such as the time to switch between states are neglected.

Another simple but more interesting example is the autopilot on an automobile. The (value of the) state of the autopilot could be "off" or "on," as with the light switch. However, when the autopilot is on, there are three further state values: cruise, accelerate, and decelerate. The automobile autopilot therefore has four possible state values.

How do the light switch and the autopilot relate to the example of the falling object described earlier in the physics of the dynamics of objects (bodies with mass)? The answer is that they are not a subject of the dynamics of objects. So, what are they? They are control systems. This is why their state spaces have a different character compared to the falling object.

In the case of the automobile autopilot, the system is controlling the physical dynamics of another system, that is, the automobile. When the automobile is cruising, its kinetic energy is constant (although it is constantly changing position). In fact, the driver of the car would have no sense of motion if it were not for the visual sense of motion and the occasional turn or bump in the road. When the automobile is decelerating, it is losing kinetic energy and the driver will feel that change in the dynamic state. During deceleration, the (value of the) dynamic state of the automobile is continuously changing from the instant that the deceleration begins until the automobile either stops or enters a new state of cruise. In the mathematical description of deceleration, the dynamic state of the automobile will transition through an infinite number of values (different positions with different levels of kinetic energy) between the initial state of cruise and the final state of (slower) cruise or a full stop.

So, when the architect or systems engineer models the state of a system, it is important to ask what is being modeled. The subject of the term *deceleration* is associated with different models of the state space, depending on whether it is associated with the state of control for an autopilot or with the dynamical state of the automobile itself. This is an example of using a term without specifying its scope. The term state can be interpreted in many ways that are consistent with its original usage in physics.

For the purpose of modeling system dynamics in systems engineering, the following definition is commonly accepted. The state of a system at time t_1 is the information required such that the system response $y(t)$, for any given future time $t > t_1$, is uniquely determined from this information and any given system input. The state space of the system is the set of all possible values that the state can have. In these definitions of state and state space, the information required in the definition of state is assumed to be an organization of the values in the state space. In discrete state models, the state space is a discrete set.

States and Events in Chess

In this section and the next, the focus will be on discrete models of states and events. Insights into questions and subtleties regarding the term *state* can be gained by considering the game of chess. It is meaningful to model the states of the chessboard upon which the game is played, and it is also meaningful to model the states of the game of chess itself. From the viewpoint of models, the game is at a higher level of description than the chessboard and the pieces; that is, models of the game relate to statements about the relationships of the pieces on the chessboard, which are a lower-level model.

The chess system includes a chess set, which comprises an 8 × 8 board and 32 pieces (16 white and 16 black, of various roles), rules of play, and two players. The chessboard has 64 squares, which are the possible positions of the pieces. (*Note*: there are 4 pieces that are not allowed to move to all 64 positions, as they are restricted to remain on the color of square that they start from.) One player is designated as White and the other as Black. At any time, the state of the chessboard comprises the remaining pieces, their positions on the board, as well as the information of whether White or Black has the next move. Given this information, the two players can continue any game from that point on. They do not need to know how the system arrived at the current state. This sense of state is strongly aligned with that of physics, in that there is no reference to time. There are a finite number of states that precede the current state, but the precedence is in the order of play, without reference to time. There are, of course, timed chess games, but time would be additional information in the description of the state.

The values of a state of the chessboard have a standard notation in chess. Each piece has a two- or three-letter designator based on its type and original starting position. For example, the White King can be designated as WK and the Pawn that starts the game directly in front of WK would be WKP. If the chessboard is numbered from the perspective of White, then WK would start the game in the position K1 and WKP would start in the position K2. At the end of the board directly opposite K1 is the position K8. The values of a state could then be organized into 64 pairs of alphanumeric values with a 65-th single alphanumeric that indicates which player has the next move, for example, a two-valued parameter M with values: W, B. The 64 pairs could use the board position as the first alphanumeric value and the designation of the piece at that position as the second value. Thus, the pair representing the starting state at the position K1 would be (K1, WK), that is, the White King is at the prescribed starting position. Another pair would be (K2, WKP). Moving one position further up the K1 column would be the pair, (K3, O) to indicate that (initially) there is no piece in that position, the capital letter "O" being used to indicate that there is no piece at that position. The vector of those 65 data now becomes the information (organized data with meaning) that is associated with a single state of the chessboard.

Clearly, the chessboard and the pieces have a very large state space. The state space represents all possible legitimate arrangements of the 32 pieces over the 64 possible positions. The chess system has events, which in this case are caused by the actions of the players, governed by the rules of play, which cause changes in the state of the system. Players move pieces on the board, capture each other's pieces, etc., in accordance with the rules of play. This narrative description of the chessboard could be used to develop a lower-level model of the chess system.

At a higher level, the game can be modeled in a very simple way that abstracts the chessboard, the pieces, and the state of play. The higher-level model has two elements: "next move" and "game status". In this model, as with the chessboard model, the game is initialized by the start of play (i.e., White moves from the original arrangement of the pieces). As with the chessboard, "next move" has two values: W (White move) or B (Black move). At the end of each move, the "game status" has two values: C (continuation [of play]) or E (end of game). There is a rule that relates E and C: if not E, then C. Also, note that "next move" and "game status" are also governed by a rule: The "next move" is not permitted (i.e., is a null state) unless the "game status" takes on the value C. The state of the game in this model can be represented by the discrete two-dimensional vector, (M, G) where M takes on the values W and B, and G takes on the values C and E. There are precisely four (values of the) states of the game in this model.

This model of the game relates to model of the chessboard in two ways. One way is obviously through the shared variable M (i.e., "next move"). The other way is through the variable G ("game status"). The value "game over" must be further resolved into one of two values: checkmate or stalemate. The first of these values is based on a decision rule as to whether the state of the chessboard has certain properties of the arrangements of the pieces. The second of these values is triggered by certain types of repetitions of states of the chessboard.

State Transitions and Events: Concepts and Terminology

An event is a primitive concept related to state transitions. It should be thought of as occurring instantaneously and causing transitions of the system from one state to another. An event may be identified with

A specific action taken
A natural occurrence governed by the laws of science
The result of a set of conditions being satisfied

A discrete event system has a discrete set of events associated with its states. For example, in the model of the light switch there is only one event—the flipping of the switch to the other position. And the events in the chess game are the set of allowable moves of the pieces on the chessboard, which are the result of (physical) actions taken by the two players.

A discrete event dynamic system is a discrete-state, event-driven system; that is, its state evolution depends entirely on the occurrence of possibly asynchronous discrete events over time. The states of many systems are typically driven by both time and events. Every event E occurs at some instance of time and defines a distinct process through which the instance of time when E occurs is determined. State transitions are the result of combining these asynchronous and concurrent event processes. These processes need not be independent of each other.

State Diagrams

A state diagram relates events and states. Recall that a change of state caused by an event is called a *transition*. A state diagram is a graph whose nodes are states and whose directed arcs are transitions labeled by event names. A state is typically drawn as a box, or a rounded box, or an oval. As with the cautions given for data flow diagrams in Chapter 7 ("DFD and IDEF0 Graphical Tools"), the architect and systems engineer must be mindful of dealing with variations in graphical notation.

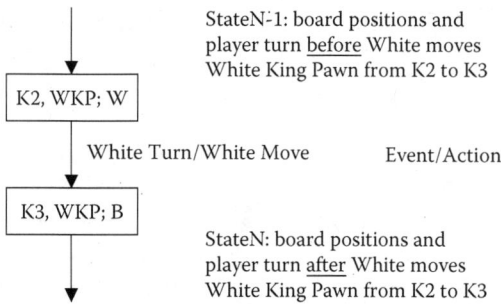

State in this example is information about the positions and pieces on the chess board and which player has the next turn.

Figure 8.6   Relation of events, actions, and states.

In a state diagram, the name of the state is inscribed in the node symbol. An activity (of observable duration) can be associated with a state. In this case, the activity starts when the system enters that state and ends when it exits the state. The execution of the activity can also be annotated in the node symbol for the state (see Figure 8.6).

A transition is represented as an arrow from a state to the next state. The label on the arrow is the name of the event causing the transition. If multiple arrows leave a state, then each must correspond to a different event. Actions (instantaneous) are shown as labels adjacent to the event labels; the two are separated by a slash or a horizontal line. Rules may be shown inside brackets; sometimes they are called *guard functions*. The rule must be evaluated for the transition to occur. Figure 8.7 illustrates the graphical notations for a state transition diagram.

Systems can also perform actions during a transition between states. When an event occurs that is associated with a transition from a given state, the system will go to the state indicated by the transition arc and will simultaneously perform the indicated action. Transitions are sometimes loops: the arc starts and ends on the same state. This means that the event causes an action, but cannot a change of state. An event can also cause a change of state but cannot produce an action.

State Transition Diagrams

A state transition diagram has a single initial state (or start state). The initial state is denoted by an incoming arrow not connected to any other state but typically anchored with a large dot.

A state transition diagram may have one or more final states. A final state has no outgoing transitions. The final states are mutually exclusive.

There are two approaches to state transition diagram construction. One is to define the state space and then identify state transitions. The other is to start with

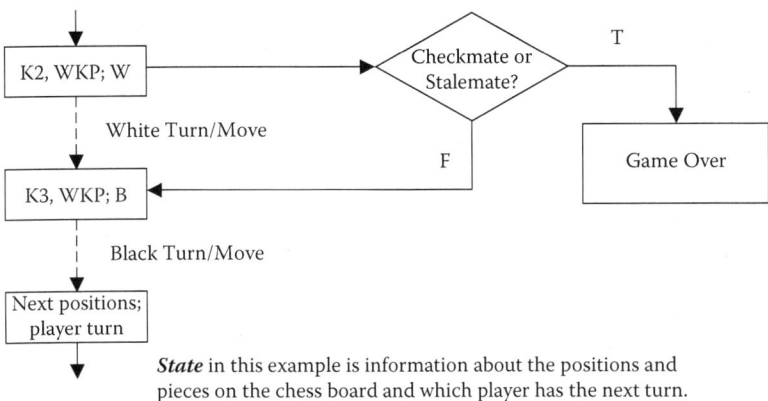

State in this example is information about the positions and pieces on the chess board and which player has the next turn.

Figure 8.7 Example of state transition diagram.

an initial state and then define transitions to next states, then transitions to following states, etc., until final states are reached.

Some basic consistency checks should be applied when constructing a state space:

Is the state space complete?
Are all states reachable?
Do all nonfinal states have exit transitions?
For each state, are all conditions accounted for?

State Transition Diagram Summary

State transition diagrams describe the behavior of a system; they can be thought of as a trajectory in the discrete finite state space caused by a sequence of events. When fully annotated, they contain much useful behavioral information not contained in the other models. There are many other models that describe behavior, for example, operational sequence diagrams, event traces, and others.

References

Boehm, C. and Jacopini, G. 1966. Flow diagrams, Turing machines and languages with two formation rules. *Commun. of the ACM*, vol. 9, no. 5: 366–371.

Jauch, J. 1968. *Foundations of Quantum Mechanics*. Reading, MA: Addison-Wesley.

Solvberg, A. and Kung, K.C. 1993. *Information Systems Engineering*. Berlin: Springer-Verlag.

Wymore, A.W. 1967. *A Mathematical Theory of Systems Engineering*. New York: John Wiley & Sons.

Chapter 9

Data and Information Modeling

Chapters 7 ("DFD and IDEF0 Graphical Tools") and 8 ("Rule and State Modeling") dealt with graphical tools for modeling three of the four aspects of the Behavioral Model: system flow, control of flow, and state transitions. This chapter will focus on the fourth aspect of the Behavioral Model, modeling the data that is flowing through the system.

Key Concepts

Data relationships
Semantic modeling
Instances and attributes of entities
Primary and foreign keys
Views and domains

This chapter will continue with the presentation of common graphical modeling tools by reviewing two of the more widely used graphical tools in data modeling: (1) entity-relationship (E-R) diagrams and (2) a companion to the IDEF0 activity modeling tool, denoted as IDEF1x, that is used for data modeling. The data structures underlying the Behavioral Model must be modeled independently of the system processes.

Entity-Relationship (E-R) Diagrams

The E-R diagram is a data modeling approach introduced by Peter Chen in 1976. E-R is a high-level data modeling language that integrates the concepts of semantic modeling and object-oriented modeling. Semantic modeling is used by linguists to represent the meaning of words and by artificial intelligence researchers for knowledge representation. The core concepts and terminology of E-R have much in common with the core concepts and terminology of the Unified Modeling Language (UML), which have already been used in several of the previous chapters of this book. This is to be expected, as E-R, UML, and other data modeling approaches have their roots in mathematical logic.

E-R Concepts and Terminology

Entities are conceptual data units or data objects. The term *class*, especially as used in UML, is, in effect, identical. *Attributes* are the characteristics or properties of entities. *Relationships* are structural associations that exist between entities. The more precise definition of relationship given in Chapter 3 ("Concepts, Standards, and Terminology") can also be used. An *occurrence* or an *instance* of an entity is the result of an entity having some or all its values assigned to the attributes. (This is similar to UML; i.e., objects are instances of classes.)

In the graphical notation for the three basic elements of E-R, entities are depicted by rectangles, attributes are depicted by circles, and relationships are represented by diamonds. In each case, the name of the entity, attribute, or relationship is annotated inside the node. And in the graphical notation, the elements are connected by lines. Figure 9.1 depicts the graphical notation. Although having a similar appearance and form as UML, there is a minor difference. For example, in the UML abstract class diagrams introduced in Chapter 2 ("Logical and Scientific Approach") for the logical modeling of sentences, the relationships were the lines connecting the (abstract) classes, and text that identified the relationship was placed directly on the line between two classes. In E-R, the corresponding diagram would have two lines, one from each of the

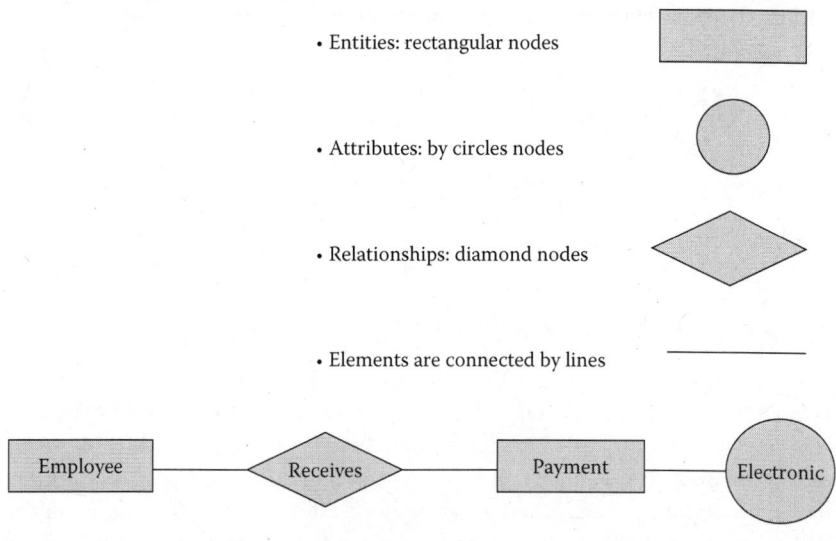

Figure 9.1 Symbols for E-R graphical elements.

entities (abstract classes) to the relationship symbol (diamond), and text identifying the relationship would be found inside the symbol, as in Figure 9.1.

Therefore, in E-R diagrams, a relationship is represented by a set of connections in the graph. Each instance of the relationship represents an association between zero or more occurrences of the first entity with zero or more occurrences of the second entity.

Similar to relationships, attributes are also part of UML. In E-R, the domain of an attribute is determined by the users and consists of all the possible legal values, categories, and operations that are permitted on the attribute. An attribute without a value is "null." A multivalued attribute has multiple values for a single entity instance. Multivalued attributes in E-R are often modeled as entities.

Simple or *atomic* attributes are not divisible into any simpler data unit. *Composite* or *group* attributes are formed by combining or aggregating related attributes. A postal address, for example, is a composite attribute. It is formed by combining atomic attributes such as house number, street name, city, and state.

Incorporating Semantic Networks into E-R

The E-R representation of the data model can be enhanced by incorporating concepts of relationships from semantic networks. The same notions are used in object-oriented analysis and design. However, many E-R diagrams do not include these notions explicitly.

The concept of a semantic network was developed by Quillian (1968) as a way to model human memory and to represent the meaning of words. The key notion is that the meaning of a word can be determined by its relationship to other words.

This concept that the meaning of a word is determined by its relationship to other words is similar to the way that mathematical logic gives meaning to words, as described in Chapter 2 ("Logical and Scientific Approach"). If the phrase "determined by its relationship to other words" includes "interpretation," then the concepts of semantic networks are closely aligned with those of mathematical logic. However, the concept of "the meaning of a word" remains stronger in mathematical logic because of the precision of predicate calculus and the strict definitions of terms such as sentence, model, and interpretation. One key difference between a language-based definition of "interpretation" and a model-based definition (per mathematical logic) is that individual words are not "interpreted" as individual entities in mathematical logic. Instead, they are interpreted holistically as part of a (logical) sentence. This means that in logic, the interpretation of an entity (e.g., a word) cannot be done in isolation. Instead, all the relationships that the word possesses from the sentence it belongs to must also be interpreted. This difference is at the heart of language- and document-based approaches to architecture and systems engineering.

Keeping in mind this distinction from mathematical logic, semantic networks have three structural primitives to describe relationships between entities. These three primitives can be used to better conceptualize and implement E-R modeling.

The first of the three abstractions used in semantic networks is *aggregation*. The aggregation abstraction is expressed by the relationship: IS-PART-OF. This is just a formal way of saying that an entity is constructed from components. Next, there are the *generalization* and *specialization* abstractions, which are expressed by the relationship IS-A, which is shortened to ISA. An entity subclass inherits the attributes of the entity superclass in addition to its own distinct attributes. The third, the *association* abstraction, is expressed by the relationship IS-ASSOCIATED-WITH. In this relationship, an entity can affect the attribute values of another entity. Each of these three abstractions is elaborated in the following text and illustrated in the graphical notation.

Entities can be combined into aggregation structures to form new entities using the IS-PART-OF relationship. Synonyms referring to the relationships of the aggregated entity to the aggregates are the following:

> has, holds, contains, consists of, comprised of, includes, incorporates, encompasses, embraces, embodies

Synonyms referring to the relationships of the aggregates to the aggregated entity are the following:

> element of, member of, segment of, member of

The symbol of the IS-PART-OF relationship is the same diamond used for all entity relationships. However, instead of writing IS-PART-OF inside the diamonds in the graphical structure, a synonym is usually written. Figure 9.2 illus-

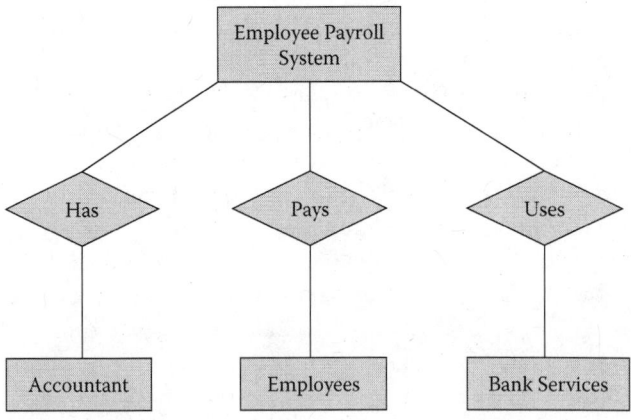

Figure 9.2   Aggregation structure.

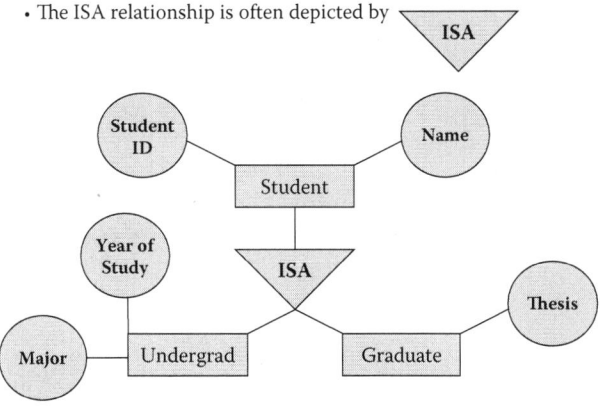

Figure 9.3 **Example of generalization–specialization structure.**

trates a simple aggregation structure for an Employee Payroll System that has three parts: an accountant, employees, and bank services.

Figure 9.3 uses the ISA relationship to illustrate the relationship of a university to its students and faculty. However, without use of a synonym such as "is comprised of," the meaning of completeness is not conveyed. So, in the E-R diagram in Figure 9.3, it is ambiguous as to whether the university has any parts other than students and faculty.

The generalization–specialization structure expresses the concept that entities can have subtypes and super types or, in the language of classes, subclasses and or super classes. The ISA structure is used to depict the generalization or specialization relationship between entities.

Specialized entities inherit the attributes of their generalizations. Each super type entity may have its own attributes (contrast with object-oriented constructs). The ISA relationship is often depicted by the triangular symbol used in Figure 9.3, even though it is an E-R relationship.

The association abstraction is used to depict relations between entities that have only a few attributes in common. It is used when entities cause attribute values to change. This relationship is expressed in terms of verbs such as:

request, obtain, rent, buy, order, use, register, distribute, assign, give, move, drive, fly, initialize, create, update, change, diagnose

The relationship can be expressed in terms of verbs and aggregate nouns, as illustrated in Figure 9.4.

As with UML, E-R diagrams express cardinality as a numerical mapping between two entities. Cardinality can be expressed as follows:

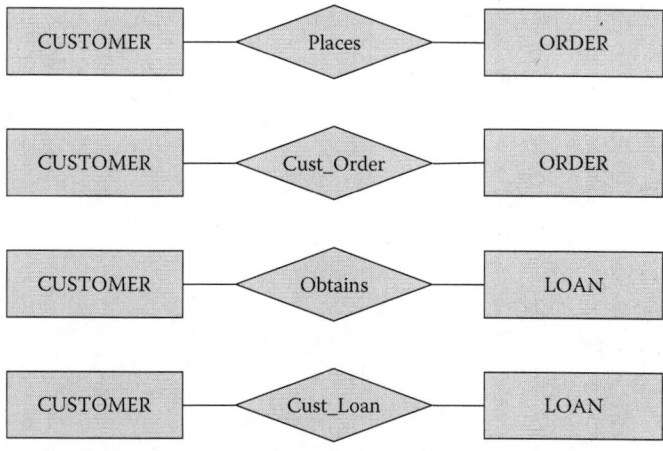

Figure 9.4   Example of association abstraction.

One-to-one (1:1)
One-to-many (1:N)
Many-to-many (M:N)

The notation for expressing cardinality is similar to but different from the notation used in UML. Figure 9.5 provides an example.

A recursive relationship exists when an entity participates in a relationship with itself. However, note that the relationship exists between instances of the same entity. In Figure 9.6, "employee" has a recursive relationship of "working with" to itself. But this does not mean the employee works with itself. Rather, it means that there are instances of employees who work with other employees.

- Cardinality is a numerical mapping between two entities
- Cardinality can be expressed as:
  - One-to-one (1:1)
  - One-to-many (1:N)
  - Many-to-many (M:N)

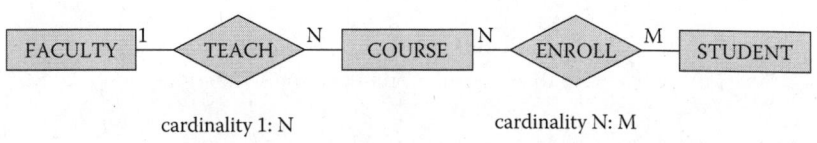

cardinality 1: N   cardinality N: M

Figure 9.5   Example of cardinality expressed in E-R.

- A recursive relationship exists when an entity participates in a relationship with itself

- Note that the relationship exists between instances of the same entity

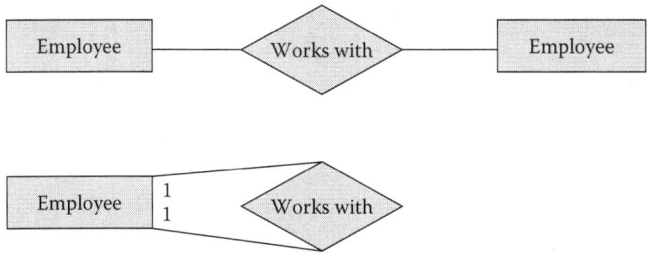

Figure 9.6   Example of E-R recursive relationships.

E-R Modeling Summary

E-R modeling is an approach that integrates the concepts of semantic modeling and object-oriented modeling. It contains key ideas that appear in object-oriented analysis and design. Elaboration of relationships gives further value to the model. E-R modeling is conceptually similar to the UML modeling of natural language but has distinct notational differences.

IDEF1x Standard

IDEF1x is part of the IDEF family of models maintained by the U.S. National Institute of Standards and Technology (NIST). IDEF1x is the Integration Definition for Information Modeling (IDEF1 Extended, 1993). It is a language (syntax and semantics) for developing a logical model of data and a graphical modeling technique for constructing semantic data models. An IDEF1x model comprises one or more views, definitions of the entities, and domains (attributes) used in the views. The model must contain a statement of purpose.

IDEF1x Concepts and Terminology

An IDEF1x *view* is a collection of entities and assigned domains assembled for some purpose. The views are graphical representations. An IDEF1x model consists of one or more such views along with the definitions of the entities and domains in the views. The views are not hierarchically structured pages of a multilevel representation as in the IDEF0 pages. Rather, think of an atlas that has maps of countries or states, regional maps, city maps, and maps that feature specific aspects: population, rainfall, elevation, etc.

The components of IDEF1x that appear in a view are the following:

Entities
Attributes or keys
Relationships (or associations)

This is very similar to E-R modeling, except that the concept of a "key" plays an important role in IDEF1x use of attributes. Entities are the representations of a set of real or abstract objects that share the same characteristics and can participate in the same relationships. An individual member of the set is referred to as an instance of the entity. *Note*: An individual is an instantiation of an entity. The distinction is between set elements and their representation. Entities in IDEF1x are similar to the E-R concept of an entity as a conceptual data unit and the UML concept of a class. Also, recall that in UML, objects are instances of classes.

Entities in IDEF1x

In the IDEF1x graphical notation, an entity is depicted by a box, as illustrated in Figure 9.7. Each entity has a unique name and identification (ID). The name is a noun phrase and the ID is a number. The format is NAME /NUMBER. An entity is referred to as *identifier independent* if each instance of the entity (i.e., object) is uniquely identifiable without reference to its relationship to other entities. The graphical notation for an independent entity is a rectangle with square corners. An entity is referred to as *identifier dependent* if each instance of the entity (i.e., object) is uniquely identifiable with reference to its relationship with other entities. The graphical notation for a dependent entity is a rectangle with rounded corners.

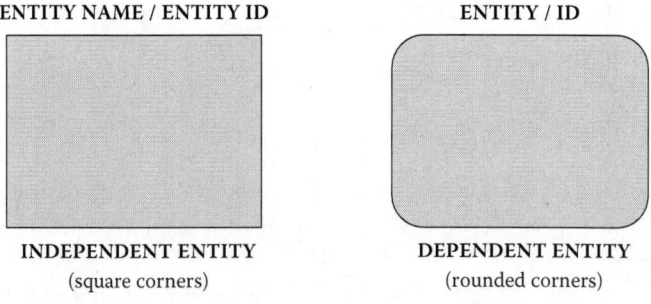

INDEPENDENT ENTITY
(square corners)

DEPENDENT ENTITY
(rounded corners)

Figure 9.7 Graphical representation of entities in IDEF1x.

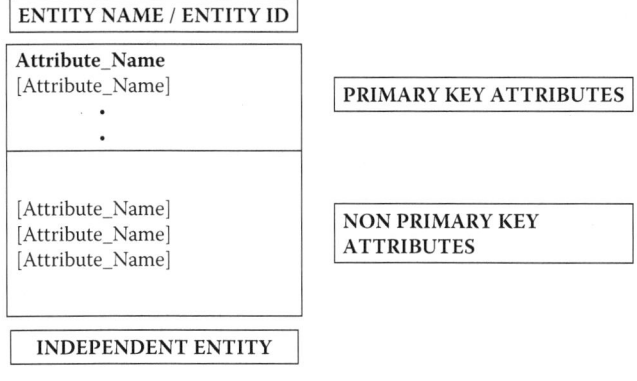

Figure 9.8   Attribute syntax.

Attributes in IDEF1x

An attribute in IDEF1x is a mapping from an entity in a view to a domain; that is, a domain when associated with an entity in a view becomes the range of values for an attribute. Domains are discussed in more detail in the following text. Every attribute must have a name (a noun phrase in singular form) that is unique among all attributes and entities in the model. An entity can possess any number of attributes. And each attribute is either possessed by an entity or has migrated to an entity through a specific connection relationship (association) or a categorization relationship.

As illustrated in Figure 9.8, attributes in IDEF1x are either primary-key attributes or non-primary-key attributes. The attributes of an entity are listed inside the rectangle that represents the entity. The syntax is "attribute_name" with primary key attributes being listed above a line in the box that divides primary from nonprimary keys. Foreign keys are attributes in entities that designate instances of related entities. For example, attributes in a parent entity that migrate to a child entity are foreign keys. A migrated key (attribute) may be used in the primary key, the alternate key, or the nonkey attributes of an entity.

Domains in IDEF1x

A domain is a set of values in which one or more attributes can have values. The range of values taken on by an attribute of an entity is central in rule modeling and necessary for developing models that are executable. The values of an attribute, for example, may reflect the characteristics of a physical system. An example of domain is the 50 states that comprise the United States. An entity may have an attribute that takes on values in this domain. The syntax of the attribute would be "state_name." The attribute in this case is the "state" and the name would take on a value from one of the fifty states {Alabama, Arkansas, Arizona … Wyoming}. The actual value of

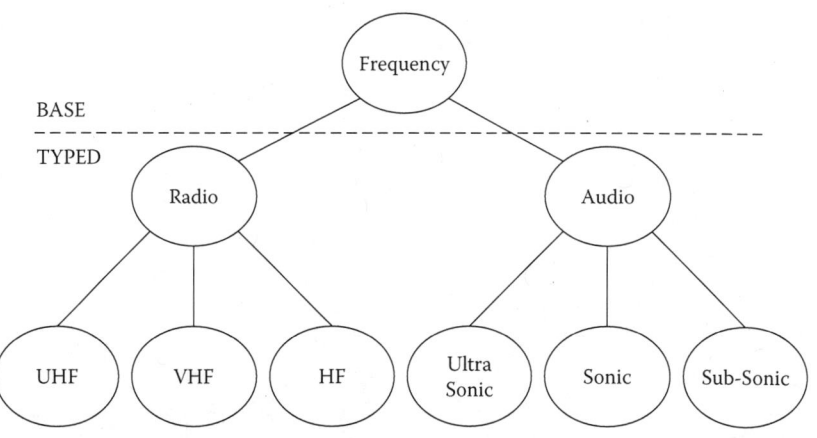

Figure 9.9 Example of base and typed domains.

the attribute could also be taken from an alternative representation of the names of the states as given by the postal abbreviations {AL, AK, AZ, ... WY}.

Domains in IDEF1x have data types and domain rules. A base domain is an IDEF1x domain whose data types are either alphanumeric or Boolean. The domain rule for a base domain is either a specified list of values of an attribute, such as the previous example for state_name, or a specified range of values. A typed domain is a base domain in which the values are further restricted. Typed domains can also have subtypes. For example, in the previous base domain comprising the fifty states, eight of them are commonly referred to as the Northeast. These states are represented by the values {CT, MA, RI, ME, NH, VT, NY, NJ}. This restriction defines a typed domain. A further restriction of the domain values to the list {CT, MA, RI, ME, NH, VT} defines a subtype domain that refers to the six states in the Northeast that are known as New England.

A domain hierarchy is a generalization hierarchy of its types and subtypes, as illustrated in Figure 9.9. Communications frequencies can be used to form a base domain. Among the many typed domains that could be specified from this base domain are radio frequencies and audio frequencies. These two typed domains could be further specified into the six subtypes at the bottom of the hierarchy in Figure 9.9. The types and subtypes used in this domain hierarchy are mutually exclusive, but, in general, domain types and subtypes need not be nested (as in the example of the states) or mutually exclusive (as in the example of the communications frequencies).

Relationships in IDEF1x

Lines indicate connection relationships (associations) between entities in an IDEF1x model. The connection relationship is directed; it establishes a "parent–child" association. A relationship has a name, a verb or verb phrase. As with UML, the name is placed beside the relationship line. The relationship is typically expressed in the parent-to-child direction; but alternatively, it can be expressed in the child-to-parent direction.

The connection relationship also has cardinality, which is the number of child entity instances associated with a parent entity. A variety of cardinalities are considered in IDEF1x:

Zero, one, or more (0 ...)
One or more (1 ...)
Exactly one (1)
Exactly n (n)
From n to m (n ... m)

As with UML and E-R diagrams, cardinality is called out on the relationship line and is placed near the appropriate entity box.

There are three types of connection relationships in IDEF1x:

Identifying and nonidentifying
Specific and nonspecific
Categorization

The null relationship is allowed when the type is nonidentifying. *Identifying relationships* transform the child entity in a parent–child relationship into a dependent entity. The primary key attributes of the parent entity become primary foreign key attributes of the child entity. The graphical notation is a solid line with a dot at the end. In the case of a *nonidentifying relationship*, primary key attributes in the parent entity migrate as foreign keys (i.e., non-primary-key attributes) to the child entity. The graphical notation for nonidentifying relationships is a dashed line with a dot at the end. Propagation of attributes between entities allows specification of the meaning of the data and content of the relationship.

A *specific connection relationship* is one in which a number of instances of one entity (child) can be related to zero or one instances of another entity (parent). A *nonspecific* connection relationship is one in which an instance of either entity can be related to a number of instances of the other entity.

A *categorization relationship* is a relationship between an entity called the *generic entity* and another called the *category entity*. An attribute of the generic entity is designated as the discriminator for a specific category cluster of that entity. The category entities are usually mutually exclusive. The category set may be complete (totally exhaustive), but is not required to be. A complete set of category entities is distinguished graphically by a short pair of parallel lines placed between the generic and category entities. Incomplete category sets are distinguished by a short single line (vice a pair of parallel lines). Categorization is a generalization–specialization relationship.

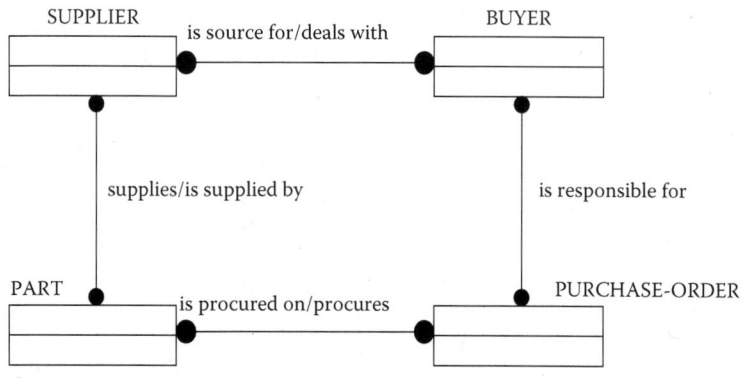

Non-specifying Relationships; N: M cardinality

Figure 9.10   Entity-Relationship Model diagram in IDEF1x.

IDEF1x modeling can be accomplished in five phases. First, define the purpose and viewpoint of the model. Next, identify the entities of the model by developing a pool of candidate entities that can be expanded as needed. Relationships should be defined between the entities after the pool has been created. The initial definitions of the relationships of the data model should be at the highest and broadest level. Detailed relationships should be defined initially through key attribute relationships and then other attributes.

This approach allows the architect or systems engineer to develop a diagram for the IDEF1x E-R Model, as illustrated in Figure 9.10. This model is useful for focusing on the entities and relationships rather than on large amounts of detail.

Developing the key-based model begins with the definition of keys. Define primary key attributes for each of the entity characteristics and properties of the data that uniquely distinguish one entity instance from another. Next, migrate primary keys to establish foreign keys. Validate relationships and keys. The key-based model depicts primary, alternate, and foreign keys. It distinguishes between independent and dependent entities and between identifying and nonidentifying relationships. All relationships are specified; *nonspecific* relationships (i.e., many-to-many) are not allowed.

Attribute definition is completed by introducing other nonkey attributes. Entity ownership of attributes must be established. The data structure must be validated and refined, and attribute domains defined. The result of the modeling process is a fully attributed model.

IDEF1x Summary

An IDEF1x view is a collection of entities and assigned domains assembled for some purpose. An IDEF1x model consists of one or more such views and definitions of the entities and domains in the views. IDEF1x is conceptually similar to the UML modeling of natural language but has distinct notational differences and uses the concept of keys as part of developing a fully attributed model.

References

Chen, P. 1976. The entity-relationship Model—toward a unified view of data. *ACM Transactions on Database Systems* 1:1:9–36.

National Institute of Standards and Technology (NIST). 1993. Federal Information Processing Standards (FIPS). http://www.nist.gov/itl/div897/pubs/by-num.htm. Publications 183 and 184.

> Publication 183: *Integration Definition for Function Modeling (IDEF0)*—93 Dec 21.
> Publication 184**: *Integration Definition for Information Modeling (IDEF1x)*—93 Dec 21.

Quillian, M.R. in Minsky, M. [editor] 1968. *Semantic Information Processing.* Cambridge, MA: MIT Press.

Sanders, G.L. 1995. *Data Modeling.* Danvers, MA: Boyd and Fraser.

---

* FIPS 184 contains an extensive bibliography on page 33.

Chapter 10

Introduction to DoDAF and MODAF Frameworks

Previous chapters have discussed techniques for developing and describing system architectures. There are many existing standards for architecture description, which take the form of architecture frameworks. In particular, defense communities around the world have developed frameworks with the intent of standardizing architecture description throughout their individual branches and promoting interoperability and communication between branches. This chapter will provide an overview of two such frameworks: the U.S. Department of Defense Architecture Framework (DoDAF) and the U.K. Ministry of Defence Architecture Framework (MODAF).

Key Concepts

Frameworks as structures and as common denominators
Types of frameworks
Architecture data bases

What Are Architecture Frameworks?

Examining the historical roots of the word *architecture* can help one understand its meaning when applied to an engineering context. Historically, architecture is "the art or practice of designing and constructing buildings" (Oxford 2008). A more generic definition is "Formation or construction resulting from or as if from a conscious act resulting in a unifying or coherent form or structure" (Merriam-Webster 2008). Many other definitions exist, but they share several common themes: structure, utility (or intention), and beauty. Similar to architecture, the word *framework* was not originally used in the context of systems engineering. The *Oxford English Dictionary* defines a framework as "A structure composed of parts framed together, esp. one designed for inclosing or supporting anything; a frame or skeleton." An alternate definition, provided by the *Merriam-Webster Dictionary,* is "a basic conceptional structure (as of ideas)." (*Conceptional* is defined by the *Merriam-Webster Dictionary* as "the capacity, function, or process of forming or understanding ideas or abstractions or their symbols" or "the originating of something in the mind." [Merriam-Webster 2008].) Again, there are many other definitions, but they share the common theme of a basic underlying skeleton, or a structure. Structure was defined in Chapter 3 ("Concepts, Standards, and Terminology") as "a stable instance of connecting or binding components of a system that explains how the junctures are organized to realize some property, behavior, or capability of the system."

Examining relational structures can give a more precise understanding of a framework. Pulling these definitions together, an architecture framework can then be described as a basic underlying relational structure for forming or constructing an architecture. Because of the basic natural relational tendency of the architecture framework, many of the artifacts can be viewed as models or model transforms, which will be discussed in more detail in Chapter 11 ("DoDAF and MODAF Artifacts").

There are many architecture frameworks in existence. These can loosely be classified into defense frameworks and nondefense frameworks. Some of the most well-known nondefense frameworks include The Open Group Architecture Framework (TOGAF) and the Zachman Framework. Two of the defense frameworks, the U.S. DoD's Architecture Framework (DoDAF) and the United Kingdom's Ministry of Defence Architecture Framework (MODAF), will be discussed in this chapter. In addition to these, NATO has the NATO Architecture Framework (NAF). Many other countries have defense architecture frameworks as well, although these are often based on MODAF, DoDAF, or NAF. The goal of this chapter is to introduce the reader to the basics of DoDAF and MODAF, not to provide a comprehensive understanding.

History of DoDAF

The U.S. DoD uses DoDAF as an architecture framework standard across the DoD. Examining the roots of the DoDAF can help clarify the underlying context and purpose behind the current specification. In the early 1990s, the Defense Sciences Board identified a need for a comprehensive framework across the DoD to aid in obtaining interoperable and cost-effective military systems. The Command, Control, Communications, Computers (C4) Intelligence, Surveillance, and Reconnaissance (C4ISR) Integration Task Force responded to this need and developed version 1.0 of the C4ISR Architecture Framework in June 1996 (Sowell 2000).

As the C4ISR architecture was being developed, government policy was being put in place to emphasize the importance of using an architecture-based approach. Several federal policies were established to increase architecture use within the government, including the Clinger-Cohen Act of 1996, the Office of Budget and Management's Circular A-130, and the E-Government Act of 2002. In December

1997, a year and a half after version 1.0 of the C4ISR Architecture Framework was released, version 2.0 was released. In February 1998, the Under Secretary of Defense (AT&L), the Acting Assistant Secretary of Defense (C3I), and the Joint Staff Director of C4 Systems (J6) mandated that the C4ISR Architecture Framework be used on all C4ISR architectures. In the same memorandum, they put forth direction to start examining the C4ISR framework as a common framework to use for all defense architectures (Sowell 2000).

The U.S. DoDAF originated from the C4ISR Architecture Framework. In August 2003, version 1.0 of DoDAF replaced the C4ISR framework, adding additional guidance, product descriptions, and supplementary information (Department of Defense 2007a). The application of DoDAF was broadened beyond C4ISR applications to include all DoD mission areas. The Core Architecture Data Model (CADM) was also added in the first release of DoDAF. In April 2007, version 1.5 of DoDAF was released. Version 1.5 is an evolution of version 1.0, which attempts to capture some of the experience obtained from use of the original specification. In particular, version 1.5 focuses on adding elements and guidance to aid in capturing net-centricity and net-centric concepts within the architecture descriptions. Version 1.5 is the present specification of DoDAF, but version 2.0 is currently being scoped and is expected to be released late in 2008.

DoD Architecture Framework (DoDAF)

The stated goal of DoDAF version 1.5 is to provide "a foundational framework for developing and representing architecture descriptions that ensure a common denominator for understanding, comparing, and integrating architectures across organizational, joint, and multinational boundaries." DoDAF is intended to support the development of systems, integrated, and federated architectures using a common set of products and views. An integrated architecture is defined by DoDAF to be an architecture "in which all elements are uniquely identified and consistently used across all products and views within the architecture." This implies a need for an underlying data structure to support the consistent use of architectural objects throughout the architecture description. A federated architecture is defined as "a distributed strategic information base, which defines the mission, the information and technologies necessary to perform the mission, and the traditional processes for implementing new technologies in response to the changing mission needs." Federated architectures are particularly useful in supporting decision making, gap/duplication analysis, interoperability analysis, and reusability analysis. Although DoDAF can theoretically be used to model any system, it was specifically designed to handle family of systems, system of systems, and net-centric concepts, and thus it is more suited to concepts in which many individual elements are geographically distributed and in which communication is a key factor in enabling the successful implementation of the concept.

Structure of DoDAF

DoDAF comprises two layers: the data layer and the presentation layer, as shown in Figure 10.1. The presentation layer includes architecture products and views that help to visually understand the system in various ways. The data layer consists of all data elements and their defining attributes and relationships. Each layer has a series of products, which provide a way of visualizing pieces of the architecture. Products can be graphical, tabular, or textual representations of the architecture and are categorized into four types of views: the All View (AV), Operational View (OV), System and Services View (SV), and Technical View (TV). A depiction of these views and how they interact is shown in Figure 10.2. Each type of view has a series

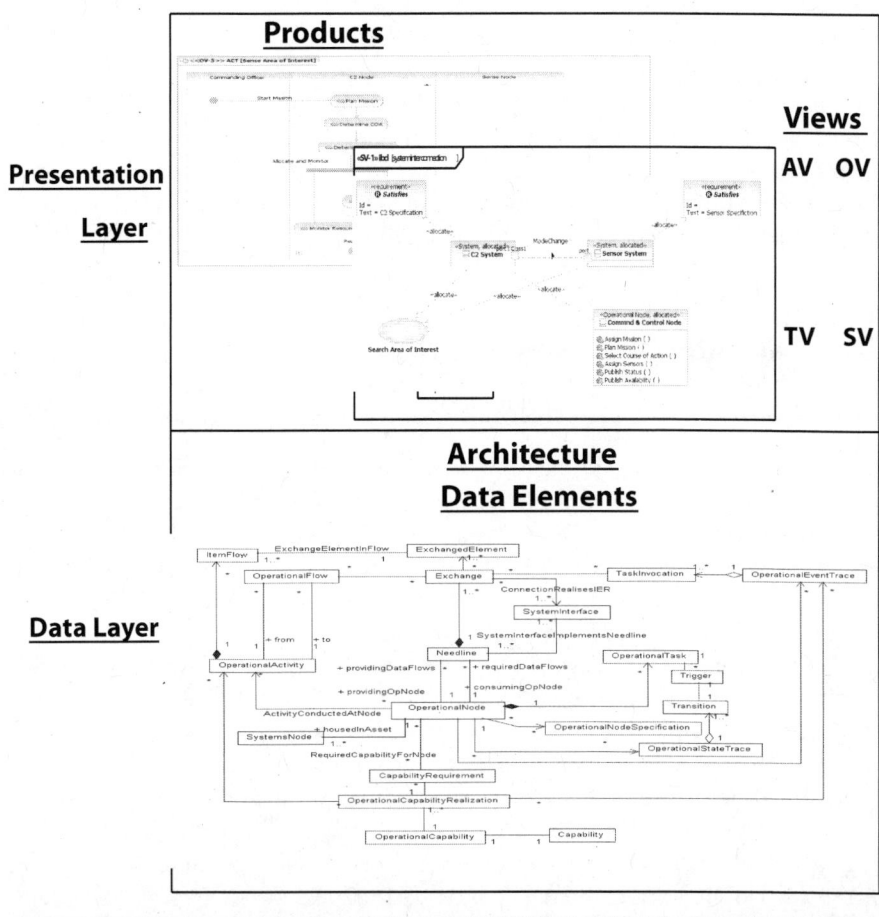

Figure 10.1 DoDAF layers. (Adapted from Department of Defense. 2007a. Version 1.5, *Department of Defense Architecture Framework (DODAF), Volume I: Definitions and Guidelines*.)

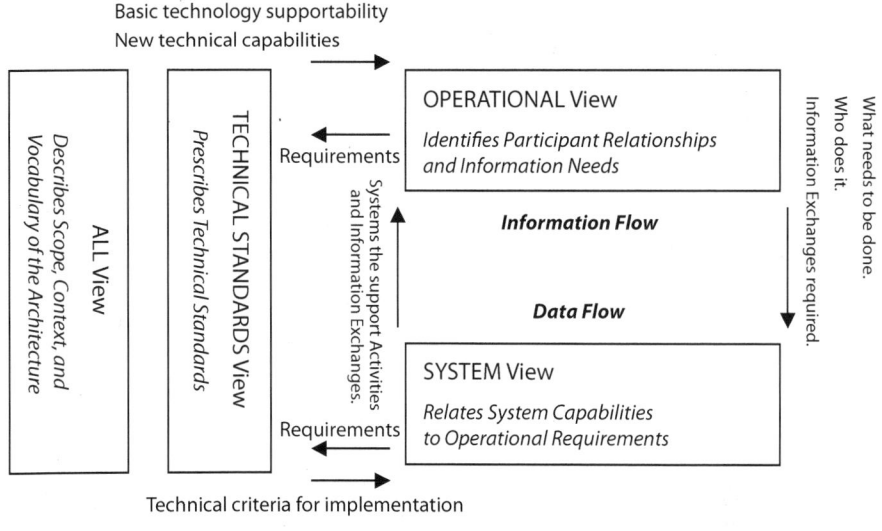

Figure 10.2   DoDAF views and their relationships. (Adapted from the Department of Defense. 2004. *Department of Defense Architecture Framework (DoDAF), Deskbook*.)

of specific products that can be included in the description of the architecture. The individual products associated with each view will be discussed in Chapter 11 ("DoDAF and MODAF Artifacts").

AV is intended to provide information that relates to all the views across the entire architecture description. This includes the scope, purpose, and intended uses of the architecture; the environment in which the architecture exists; and any analytical findings of the architecture development process. Additionally, AV contains an integrated dictionary having definitions of all terms used in all products of the architecture description. This helps maintain consistency and enables communication both within the development team and with outside parties. The AV products provide information concerning the entire architecture but do not represent a distinct view of the architecture (Department of Defense 2007b).

The OV contains information about the functional relationships of pieces of the architecture. The OV captures the operational nodes, the tasks or activities performed, and the information that must be exchanged to accomplish the mission. It is in the OV that the high-level operational concept is depicted. Operational nodes, their relationships and information exchanges, and the frequencies and natures of those exchanges are defined in the OV. Capabilities and operational activities are identified, and their relationships are depicted as well. Rules that constrain the operations are identified and described. Operational activities are arranged into event sequences that represent the way in which capabilities are accomplished. Business process rules and system data requirements are also documented in the

OV. The OV essentially represents a functional decomposition of the architecture. A pure OV is materiel (system) independent, although an OV may have materiel constraints and requirements that must be addressed. For this reason, it may be beneficial or even necessary to include some high-level SV architecture data as overlays or as augmenting information onto the OV products.

SV is a set of primarily graphical and tabular products that describe systems and services and interconnections providing for, or supporting, defense functions. The SV fully describes the physical architecture and its relation to the OVs. Many of the SV products focus on specific physical systems with specific physical (geographical) locations or areas of operation. It is important to realize that although each system has a specific geographical location, it is unlikely that all the systems will be geographically colocated. Often, in families-of-systems, systems-of-systems, and net-centric architectures, the systems are widely dispersed geographically and rely on an intricate communication network to maintain interfaces between the nodes. In the SV, the systems, services, interconnections, interfaces, and their respective functions and roles in the operational activities are identified. Additionally, the planned evolutionary steps for concept implementation or migration from the current concept are defined in the SV. Technology forecasting captures the emerging technologies that may possibly be integrated into future implementation of the architecture.

The TV describes current and emerging standards that apply or could potentially apply to the systems and services. The TV is the minimal set of rules governing the arrangement, interaction, and interdependence of system parts or elements. The TV provides the technical guidelines for systems implementation on which engineering specifications are based, common building blocks established, and product lines developed. These guidelines can be included for a variety of reasons, including the ease of interfacing, system security, compliance with regulations, or many others.

In addition to the four views, DoDAF also has a data model called the Core Architecture Data Model (CADM). CADM provides a way to represent data elements and relationships that support the creation of products. It captures the necessary data and data elements to enable integration across multiple architecture descriptions, thereby increasing communication across the various mission areas and associated agencies and partners. A detailed discussion of CADM is outside the scope of this book, but more information can be found in DoDAF version 1.5, Volume III.

DoDAF Architecture Development Principles, Expectations, and Process

One of the goals of DoDAF is to create architectures that can be integrated and federated across the DoD. To help achieve this goal, the DoDAF documentation provides a set of basic guiding principles for architecture development. The DoD recommends that these principles be adhered to throughout the development

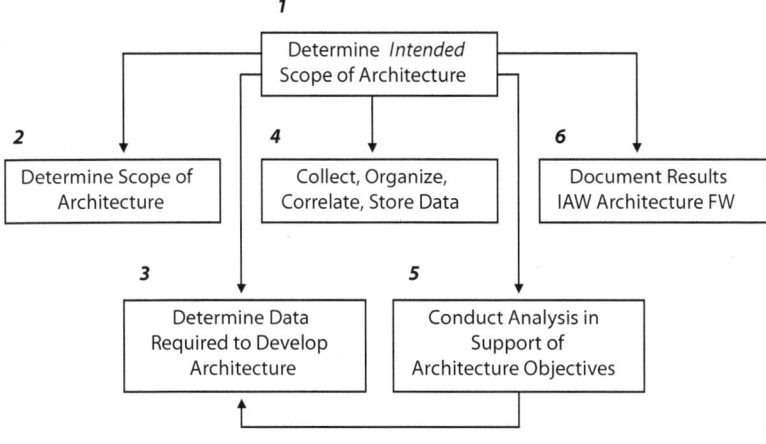

Figure 10.3 DoDAF architecture description process. (Adapted from Department of Defense. 2007a. Version 1.5, *Department of Defense Architecture Framework (DODAF), Volume I: Definitions and Guidelines.***)**

process to help support stakeholder requirements. These principles and guidelines are as follows (Department of Defense 2007a):

- The architecture should have a specific and commonly understood purpose.
- The architecting effort should be focused and scoped appropriately.
- Architectures should provide a clear representation of the information by using common terms and definitions as well as clear and easy-to-read views.
- Integration with other DoD architectures should be straightforward, and external interfaces to other DoD architectures should be clearly defined.
- Agility should be achieved by making the architecture modular, reusable, and decomposable.

In addition to the principles and guidelines, the DoDAF documentation also provides a recommended six-step architecture development process, which is depicted in Figure 10.3 (Department of Defense 2007b). Although this process is called the *architecture development process*, it is really a process for building up the architecture description and provides little insight into the actual design of the architecture itself. Use of this process can help organize and expedite the creation of DoDAF views for a given design or concept.

History of MODAF

The Ministry of Defence Architecture Framework (MODAF) was developed by the U.K. Ministry of Defence (MOD), but it adopts and adapts much of DoDAF.

In addition to the DoDAF views and products, MODAF has also added extensions that are currently not in DoDAF. Version 1.0 of MODAF was released in August 2005. MODAF version 1.1 was released in April 2007. The current version, MODAF 1.2, was released on June 20, 2008 (MOD 2008).

MOD Architecture Framework (MODAF)

Similar to DoDAF, MODAF uses a series of views to describe an architecture. However, MODAF uses a slightly altered terminology. DoDAF "views" become "viewpoints" in MODAF, and DoDAF "products" are "views" in MODAF. MODAF has a total of six viewpoints. The Operational Viewpoint (OVP), Systems Viewpoint (SVP), Technical Standards Viewpoint (TVP), and All Viewpoint (AVP) are the MODAF equivalents of the four DoDAF views. In addition to these, MODAF also has the Strategic Viewpoint (StVP) and the Acquisition Viewpoint (AcVP). The relationship between MODAF and DoDAF is shown in Figure 10.4.

The StVP has been introduced to support the capability management process. In particular, the StVP attempts to ensure that the capabilities of the architecture are in line with the strategic initiatives of the MOD. By decomposing the strategic policies into a capability taxonomy and describing capability-level measures of effectiveness, the StVP aids in capability auditing and gap analysis. The StVP can also allow capability dependencies to be shown, which can help with planning and enable trade-offs across multiple capabilities.

The AcVP describes programmatic details not handled in the more technical viewpoints. This includes aspects such as integration of capabilities across the MOD, roadmapping, and maturity assessment across the Defence Lines of Development

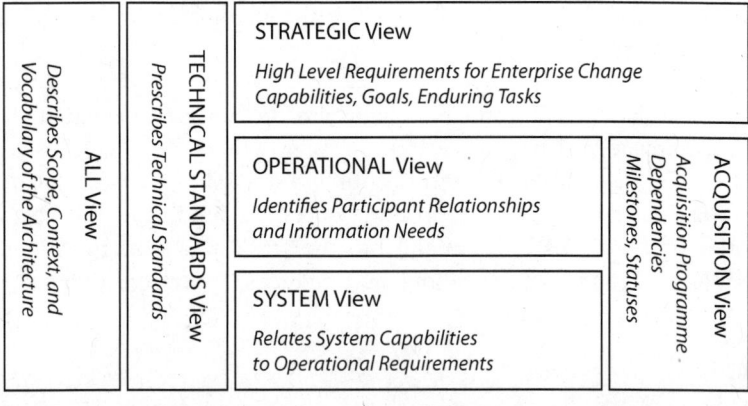

Figure 10.4   Relationship between DoDAF and MODAF. (Adapted from Ministry of Defence (MOD). 2008. *The Ministry of Defence Architecture Framework Version 1.2.* **http://www.modaf.org.uk/ [accessed June 26, 2008].)**

(DLoDs). The Acquisition Views (AcVs) provide important information to managers and acquisition personnel, and provide a link between the acquisition teams and customers.

Version 1.2 of MODAF added Service-Oriented Views (SOVs). These views describe services that are required to support operations not explicitly included in the other views. Services are defined by MODAF as "a unit of work through which a provider provides a useful result to a consumer." SOVs describe available services, how they interact and work together, and what information interfaces are provided.

Similar to DoDAF, MODAF also has a data layer, called the *MODAF Meta Model* (M3). M3 is done in UML 2.0 and is used to ensure consistency between views and improve the ease of integration, particularly between tools that help create MODAF products. M3 aids in defining what architectural objects are available for the creation of MODAF views, and can also help to identify the relationships between the objects. A detailed discussion of the M3 is outside the scope of this book, but more information can be found in the *MODAF Technical Handbook* (MOD 2008).

Summary

Architecture frameworks are a basic underlying relational structure for forming or constructing an architecture. DoDAF and MODAF are two defense architecture frameworks that provide guidance and rules for describing, representing, and understanding defense systems architectures. DoDAF is mandated by the United States for use on all U.S. DoD projects. It has four views: the OV, SV, TV, and AV. MODAF is used by the U.K. MOD and has two additional viewpoints: the StVP and the AcVP. The next iteration of MODAF will also include the SOV. Each view or viewpoint in DoDAF and MODAF has a set of associated products, which will be discussed in detail in Chapter 11 ("DoDAF and MODAF Artifacts"). Each framework has associated data models: DoDAF uses CADM and MODAF uses M3. Many other countries have their own specific defense architecture frameworks, although a large number of them are based on DoDAF and MODAF. In addition, NATO has the NATO Architecture Framework (NAF), which attempts to create a framework that is compatible across nations.

References

Department of Defense. 2007a. *Department of Defense Architecture Framework (DODAF), Volume I: Definitions and Guidelines.*

Department of Defense. 2007b. *Department of Defense Architecture Framework (DODAF), Volume II: Product Descriptions.*

Department of Defense. 2004. *Department of Defense Architecture Framework (DoDAF), Deskbook.*

Merriam-Webster. 2008. *Merriam-Webster Online.* http://www.merriam-webster.com/. Retrieved April 23, 2008.

Ministry of Defence (MOD). 2008. *The Ministry of Defence Architecture Framework Version 1.2.* http://www.modaf.org.uk/ (accessed June 26, 2008).

Oxford. 2008. *Oxford English Dictionary.* http://dictionary.oed.com/. Retrieved April 23, 2008.

Sowell, P.K. 2000. *The C4ISR Architecture Framework: History, Status, and Plans for Evolution.* McLean, VA: The Mitre Corporation.

Chapter 11

DoDAF and MODAF Artifacts

Kelly Griendling

As discussed in Chapter 10 ("Introduction to DoDAF and MODAF Frameworks"), U.S. Department of Defense Architecture Framework (DoDAF) Views and U.K. Ministry of Defence Architecture Framework (MODAF) Viewpoints are supported by a series of products (for DoDAF) and views (for MODAF). These products and views provide a placeholder for the depiction of information that aids in the understanding of each view or viewpoint. Products and views are developed in the course of the systems engineering life cycle and support capture, display, and sharing of data or information about a system or system of systems. This chapter provides a basic overview of the DoDAF and MODAF products and views, and their roles within their respective architecture frameworks.

Key Concepts

Architecture viewpoint
Representations of views
Views as relational structures
Consistency and concordance

Most of the architecture views in these frameworks introduce relationships between the elements, activities, functions, or attributes of a system or family of systems. As such, the views lend themselves to the logical concept of "model" that was discussed in detail in Chapter 5 ("Architecture Modeling Languages"). Some of the views also introduce transformations between the parameters used in the architectural descriptions, but do not address the preservation of relationships under the given parametric transform. As such, these views cannot be considered model transforms.

Introduction

As discussed in Chapter 10, DoDAF Views and MODAF Viewpoints are supported by a series of products (for DoDAF) and views (for MODAF). These products and views provide a placeholder for the depiction of information that aids in the understanding of each view or viewpoint. Products/views provide a way to visualize information about the architecture and can be graphical, tabular, or textual representations of the specific information contained within them. Each product can be viewed as a relational structure, and from this, logical models can be created. It is important to realize that not every project needs to create every product every time. Before choosing to take the time to create a product, care should be taken to understand how the product will be used, who the audience will be, and why the specific product is required. How to determine which products are required or useful for understanding a given architecture is outside of the scope of this book. Also, it should be understood that DoDAF and MODAF are not design processes, but rather, tools for representing or depicting a design. They can aid the designer in ensuring consistency and traceability as well as in communicating various aspects of the design, but they do not provide insight into how to design architectures.

DoDAF Products

The DoDAF products are summarized in Figure 11.1. The All View (AV) and Technical View (TV) have two products each, Operational View (OV) has seven products, and the System and Services View (SV) has eleven products, for a total of twenty-two products. The products are grouped according to which view they support, but information should flow logically between products, and several products can share sets or subsets of data. DoDAF does not specify the use of any one methodology for building products (e.g., structured analysis versus object orientation) or one notation over another (e.g., IDEF1X or Unified Modeling Language [UML] notation); however, it does make some suggestions as to how products might be depicted, and those will be noted as the products are discussed. The choice of which methodology to use is ultimately governed by the nature of the architecture being defined, the expertise and preferences of the architecture team, the needs of the customer, and the needs of the architecture end users. Owing to the difficulty associated with switching between notations (i.e., IDEF to UML, UML to IDEF), the architecture team must carefully select which notation to use, and stick with that notation as much as possible during product development (Department of Defense 2007a).

The AV and TV have two products each, the OV has seven products, and the SV has eleven products, for a total of twenty-two products. The products are grouped according to which view they support, but information should flow logically between products, and several products can share sets or subsets of data. DoDAF does not

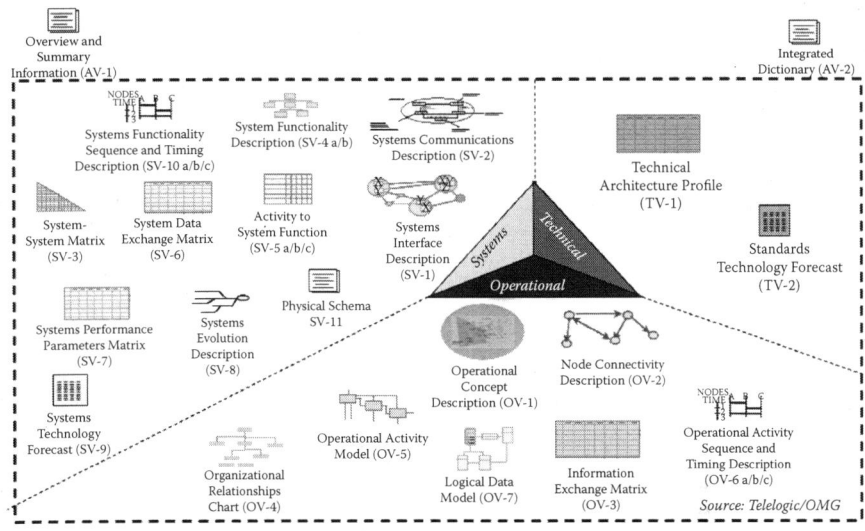

Figure 11.1 Overview of DoDAF products. (Reproduced from Hardy, D., OMG UML Profile for the DoD and MoD Architecture Frameworks. http://syseng.omg. org/UPDM%20for%20DODAF%20WGFADO-INCOSE%20AWG-070130.ppt.)

specify the use of any one methodology for building products (e.g. structured analysis versus object orientation) or one notation over another (e.g., IDEF1X or Unified Modeling Language (UML) notation); however, it does make some suggestions as to how products might be depicted, and those will be noted as the products are discussed. The choice of which methodology to use is ultimately governed by the nature of the architecture being defined, the expertise and preferences of the architecture team, the needs of the customer, and the needs of the architecture end users. Due to the difficulty associated with switching between notations (i.e., IDEF to UML, UML to IDEF), the architecture team must carefully select which notation to use, and use that notation as much as possible during the product development.

All View

AV products provide information pertinent to the entire architecture, but do not represent a distinct view of it. AV-1, Overview and Summary Information, describes the scope of the architecture, including the subject area and timeframe. AV-1 describes the context of the architecture, including doctrine; tactics, techniques, and procedures (TTP); relevant goals and vision statements; concepts of operations (CONOPS); scenarios; and environmental conditions. It also includes information about file formatting that will be used in creating other products. One of the most important aspects of an AV-1 is that it provides a place to document results, conclusions, observations, or lessons learned that are pertinent to the entire architecture.

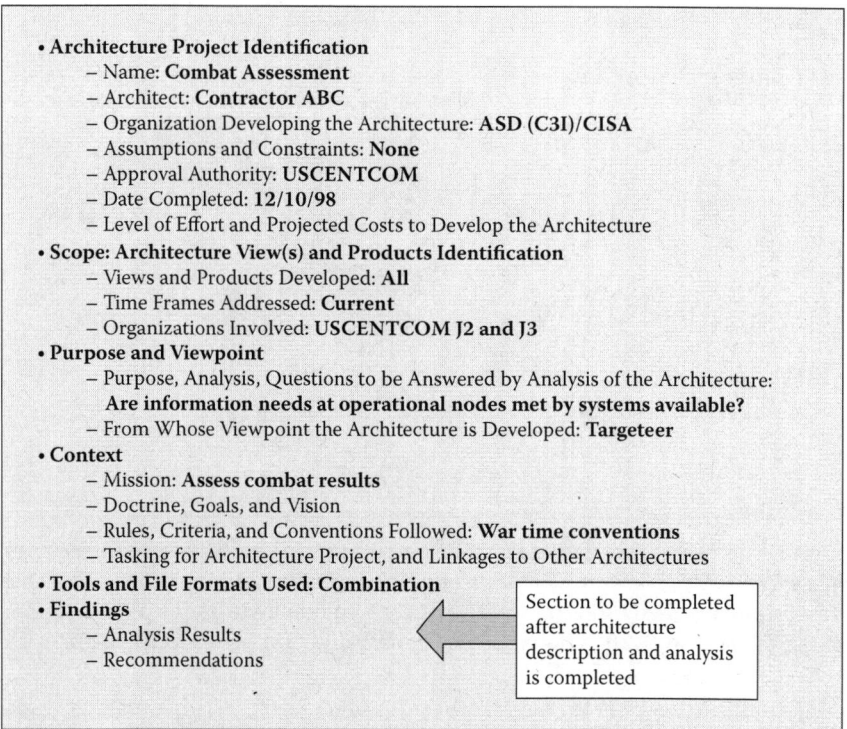

- **Architecture Project Identification**
  - Name: **Combat Assessment**
  - Architect: **Contractor ABC**
  - Organization Developing the Architecture: **ASD (C3I)/CISA**
  - Assumptions and Constraints: **None**
  - Approval Authority: **USCENTCOM**
  - Date Completed: **12/10/98**
  - Level of Effort and Projected Costs to Develop the Architecture
- **Scope: Architecture View(s) and Products Identification**
  - Views and Products Developed: **All**
  - Time Frames Addressed: **Current**
  - Organizations Involved: **USCENTCOM J2 and J3**
- **Purpose and Viewpoint**
  - Purpose, Analysis, Questions to be Answered by Analysis of the Architecture: **Are information needs at operational nodes met by systems available?**
  - From Whose Viewpoint the Architecture is Developed: **Targeteer**
- **Context**
  - Mission: **Assess combat results**
  - Doctrine, Goals, and Vision
  - Rules, Criteria, and Conventions Followed: **War time conventions**
  - Tasking for Architecture Project, and Linkages to Other Architectures
- **Tools and File Formats Used: Combination**
- **Findings**
  - Analysis Results
  - Recommendations

Section to be completed after architecture description and analysis is completed

Figure 11.2   Notional AV-1.

This means that it provides a location to pull together information from multiple products and views, as well as the Core Architecture Data Model (CADM) and any external modeling and simulation, prototyping, or other assessment or experience, and capture this information for mutual benefit and future use (Department of Defense 2007b). An example of an AV-1 is provided in Figure 11.2.

AV-2, the Integrated Dictionary, provides a dictionary of terms to facilitate consistent use of vocabulary and terminology. This includes a common set of definitions that can help alleviate confusion and ensure common word usage. It is particularly important for multidisciplinary development. Stakeholders, including decision makers, engineers, architects, and customers, can use an AV-2 to provide a common language that all players can understand. If implemented and utilized properly, an AV-2 can help prevent miscommunications due to differing taxonomies.

Operational View

As explained in Chapter 10 ("Introduction to DoDAF and MODAF Frameworks"), the OV captures the operational nodes, tasks or activities performed, information

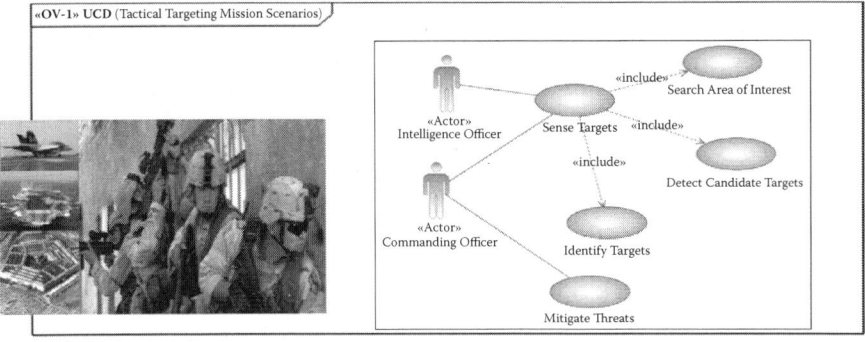

Figure 11.3   Example OV-1 in UML.

that must be exchanged, types of information exchanged, frequency of exchange, which tasks and activities are supported by the information exchanges, and the nature of information exchanges. OV products should be as materiel independent as possible, although an OV may have materiel constraints and requirements that must be addressed. In practice, it is common to include some high-level SV architecture data as overlays for adding information to the OV products.

There are seven OV products, but OV-6 has three distinct pieces, which means that there are nine depictions associated with the OV. OV-1, High-Level Operational Concept Graphic, depicts the overarching concept for operations. This includes the identification of major operational nodes, their role in the architecture, and what lines of communication must exist between these nodes. An operational node is defined as "an element of the operational architecture that produces, consumes, or processes information." See Department of Defense (2007b), Section 4.2.1. Key interfaces between nodes are identified in OV-1. Often, OV-1 also shows key external interfaces with which the nodes must exchange information. An example OV-1, drawn using UML, is shown in Figure 11.3.

Returning to the concept of model that was introduced in Chapter 5 ("Architecture Modeling Languages"), it is possible to look at OV-1 as a model. Nodes can be considered elements in the model, and communication lines and information exchanges are relationships between the elements. It is possible to capture more information in the OV-1 model than simply the nodes and communications. The way that the nodes are combined creates a model of the capability or capabilities being depicted in the OV-1 graphic. The OV-1 narrative provides the sentences that are interpreted into the nodal relationships and speak to higher-level concepts, such as capability. If implemented carefully, OV-1 should provide a capability model that captures the conceptual aspect of the solution by conveying the key components and the required relationships.

OV-2, Operational Node Connectivity Description, provides a representation of where a need for information exchange between operational nodes or organizations exists. This is usually depicted graphically, and typically includes nodes that are both internal and external to the architecture. Required information exchanges are represented using needlines, which represent what information obtained by one node is needed by another node, but not how that information is transferred. The needline also uses an arrow to show from which node to which other node the information is being transferred. OV-2 does not represent the actual communication network, but rather how the information needs of the system are fulfilled. For example, if one of the operational nodes was a sensing node and another node was a ground control station, there may be a needline to transfer sensor data from the sensing node to the ground control node. If, however, this data was transferred through a satellite relay, this would not be represented in OV-2 because the satellite relay would not actually need the data; it would simply be routing the data. The way in which the information transfer was actually implemented can be represented in SV-2, which will be discussed later in this chapter. An example of an OV-2, represented in UML Profile for DoDAF and MODAF (UPDM), is shown in Figure 11.4. It can be noted that some of the operational activities (detailed in OV-5) are represented with the node in which they are performed. This is a common practice that is recommended by the DoDAF documentation (Department of Defense 2007b).

Here again, OV-2 can be considered a model as presented in Chapter 4 ("Structure, Analysis, Design, and Models"). Similar to OV-1, the elements of this model consist of operational nodes. However, in this case, the relations are the needlines. It is important here to understand the difference between the communication and interface lines that make up the relations in OV-1 and the needlines that make up the relations in OV-2. The needline is independent of any communication or interface; it simply requires that there be some path for information exchange between these nodes. It is possible that when the systems are instantiated, these paths involve relays

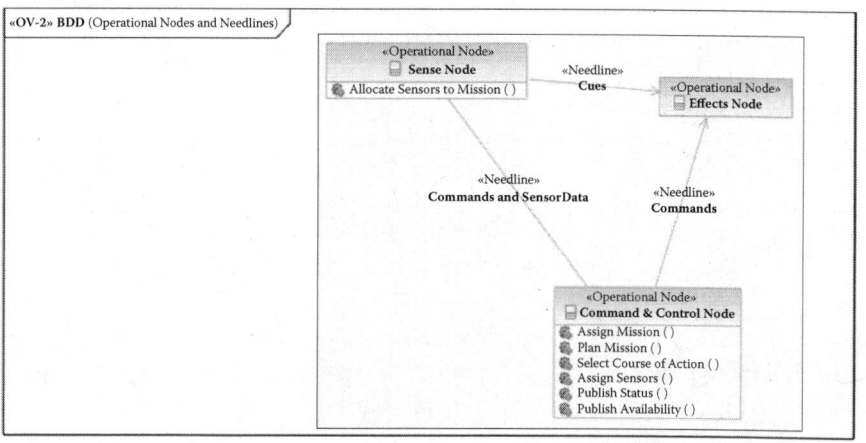

Figure 11.4   Example OV-2 in UML.

through other nodes. However, OV-2 is system independent, and thus provides a conceptual model of the information exchanges required for the architecture to function as desired.

OV-3, Operational Information Exchange Matrix, provides details regarding each of the information exchanges. These details include who exchanges the information, what information is exchanged, the reason why the exchange is necessary, and how the information will be exchanged. An information exchange is defined as "an act of exchanging information between two distinct operational nodes and the characteristics of that act, including the information element that needs to be exchanged and the attributes associated with the information element (e.g., Scope), as well as attributes associated with the exchange (e.g., Transaction Type)" (Department of Defense 2007b, Section 4.3.1). An information element is in turn defined as "a formalized representation of information subject to an operational process (e.g., information content that is required to be exchanged between nodes)." Information elements are not unique to a given information exchange and can be included in multiple information exchanges. It is important to realize that each needline in OV-2 may require one or many information exchanges in OV-3. Significant attributes of the information exchanges should also be noted in OV-3. OV-3 should include all pieces of the operational elements, including the operational activities, operational nodes, and the specific information flow between these elements. A partial example of an OV-3, represented in UML, is shown in Figure 11.5.

OV-4, Organizational Relationships Chart, shows the relationships or the command structure between human elements of the architecture, including key organizations or decision makers. This can include both internal and external human elements. There are many types of relationships that can be represented in an OV-4, which may include supervisory reporting, command and control relations, command–subordinate relations, and collaborative or cooperative relationships. OV-4 helps represent how various organizations or key players come together and interact to perform human elements required for the operational activities shown in OV-5. An example of an OV-4, shown in UML notation, can be seen in Figure 11.6.

OV-5, Operational Activity Model, is one of the most commonly created and used views within DoD (OSAD 2005). It depicts operational activities and in which operational node they occur. In addition, OV-5 often also depicts the inputs and

«OV-3» Matrix (Information Exchange Matrix)					
		Producer		Consumer	
Needline ID	Info Exchange ID	Sending Op Node	Sending Op Activity	Receiving Op Node	Receiving Op Activity
N1	IE1	Sense Node	Sense AOI	Effects Node	N/A
N2	IE2	Sense Node	Sense AOI	Command and Control Node	
N3	IE3	Command and Control Node	Allocate Sensors	Sense Node	Sense A
N4	IE4	Command and Control Node	N/A	Effects Node	N/A

Figure 11.5   Example OV-3 in UML.

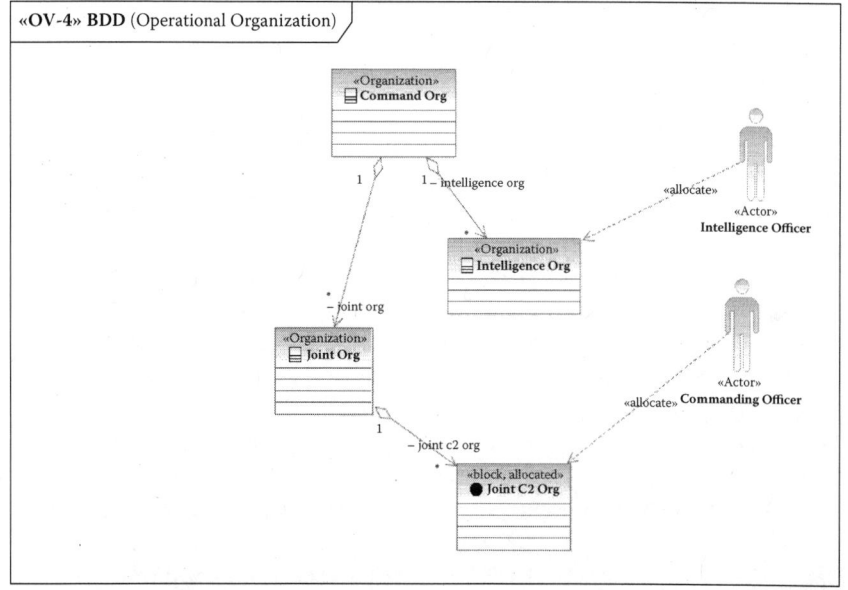

Figure 11.6   Example OV-4 in UML.

outputs to the activities. One of the most important results from OV-5 products is an understanding of the lines of responsibility for each of the operational activities. Presenting the activities in this way can also help avoid unnecessary redundancy, and to see where redundancy might be required. It can help identify the operational activities that are critical to operations (those that have critical inputs or outputs) and those that play a more supporting role. Inputs and outputs can also show where exchanges with the environment would occur and where the responsibility lies for managing that external interface. Although OV-5 has the word *model* in its title, it is often depicted as a hierarchical decomposition of activities. The hierarchical form is commonly used to improve traceability and clearly show which operational nodes are responsible for which operational activities and which interactions with the environment. However, hierarchy is just one type of relationship between the activities. Including the environment and representing the inputs and outputs between activities can increase the utility of OV-5. Elements of the model would include activities and the environment, and the inputs and outputs would establish the relationships between the elements. An example OV-5 is shown in Figure 11.7.

The link between OV-4 and OV-5 can be formalized using a model transform. As discussed previously in this chapter, OV-4 shows human interactions that are required to support OV-5. If OV-5 is considered a model of the operational activities and OV-4 is a model of the organizations that are players in the execution of OV-5, a model transform between OV-4 and OV-5 provides an understanding

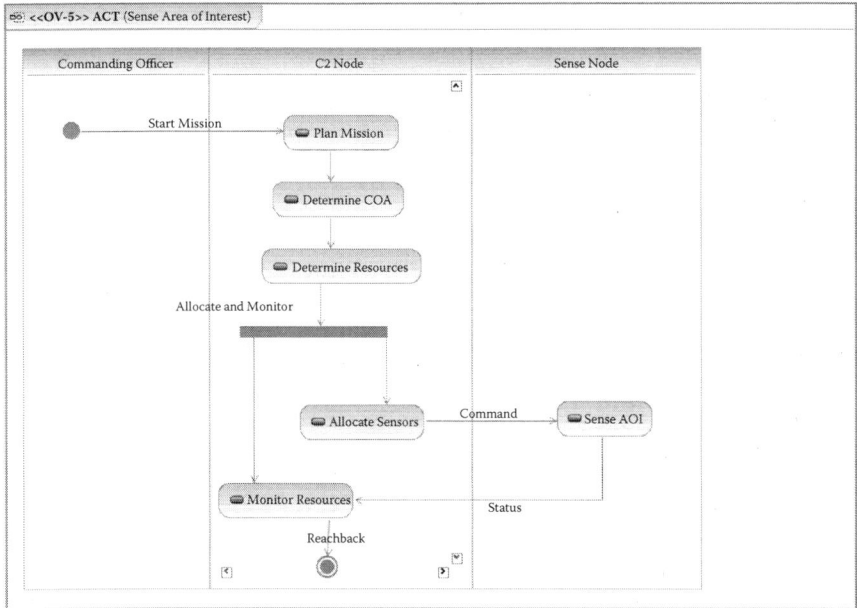

Figure 11.7   Example OV-5 in UML.

of the roles played by each organization. This can be very important to understanding, delineating, and clarifying responsibilities to encourage effective communication and interaction between key players.

OV-5 shares a close relationship with OV-2, as OV-2 shows the higher-level information flow between the nodes and OV-5 shows the activities that support that information flow. Additional annotations are often included as part of an OV-5, and they can be very beneficial in representing key information about the architecture. For example, estimated costs or risks associated with performing the activities, security standards that apply across inputs and outputs, business rules, or details on information to be exchanged can help strengthen OV-5. An example of an OV-5 with some annotations can be seen in Figure 11.8.

One of the most critical aspects of OV-5 is that it provides a key link to understanding how the architecture fulfills the overarching capability needs. According to the DoDAF Volume II, "A capability can be defined by one or more sequences of activities, referred to as operational threads or scenarios." OV-5 provides a place to depict the activities that are required to achieve desired capabilities, and also to understand the information flow that must occur to achieve the desired effects. Capability-related attributes can be added to an OV-5 to aid in understanding how each activity contributes to the overarching goals of the architecture. OV-5 is one of the most central products in the Operational View and, as such, is one of the most commonly used products. In fact, a 2005 survey by the DoD revealed that

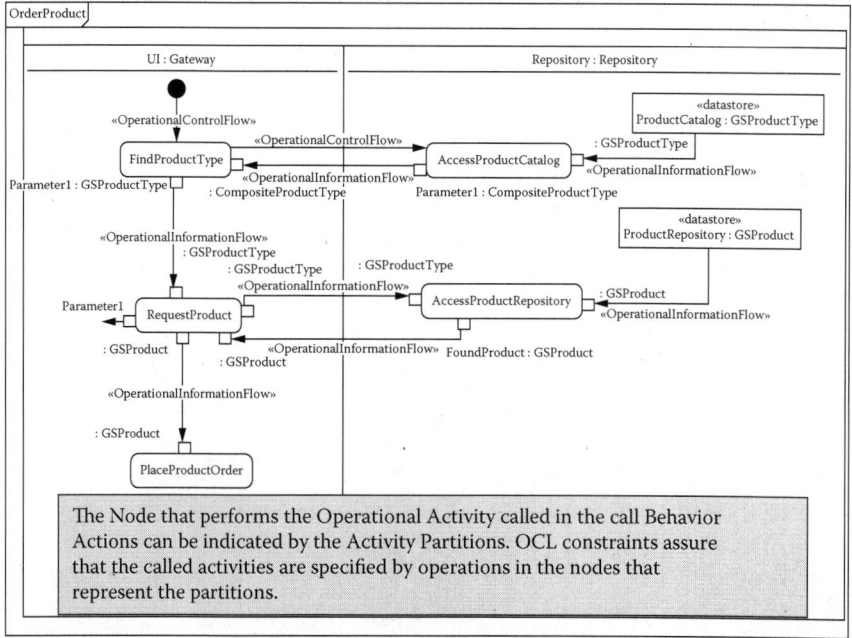

Figure 11.8 Example OV-5 in UML with annotations.

79 percent of DoDAF uses included an OV-5, making it one of the five most frequently used products. (OASD 2005)

OV-6 is designed to capture the dynamic, or time-dependent, behavior of architecture elements. OV-6 attempts to answer the question: In what order should the activities take place and when should they occur relative to each other? Modeling techniques beyond a UML diagram may be required to better understand the dynamic aspects of the architecture. The DoDAF documentation suggests logic data language, Harel statecharts, Petri nets, IDEF3 diagrams, and UML statechart and sequence diagrams as potential modeling alternatives. Understanding the timing characteristics of the architecture can be critical to design and the identification of potential failures. There are three products included in OV-6.

OV-6a, Operational Rules Model, is used to understand and identify operational and business rules that constrain and govern the architecture. In particular, this view describes the boundaries of what can be done and where specific effects are required in the presence of a given cause. Operational rules can be thought of as if-then statements that constrain or control the mission or the operational thread. Often, the rules in OV-6a are associated with specific operational activities. In these cases, the rules can govern the input–output relationship of the activity. These rules may have the form "if (these) inputs are given, provide (this) output." An example of an OV-6a is shown in Figure 11.9.

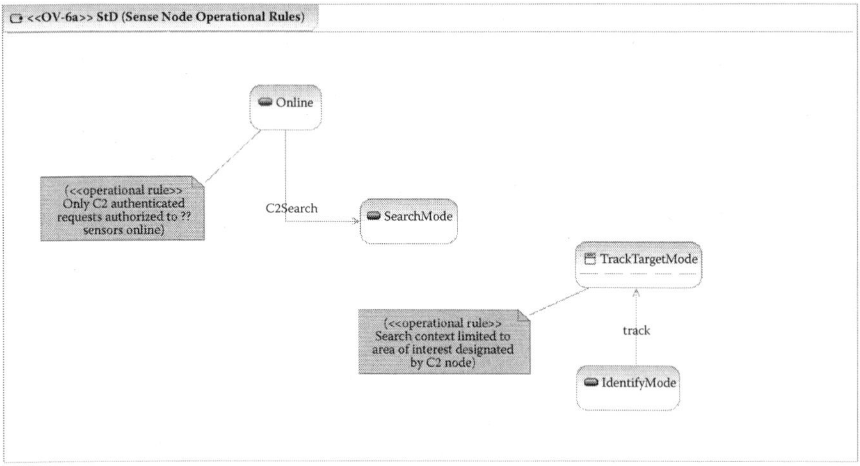

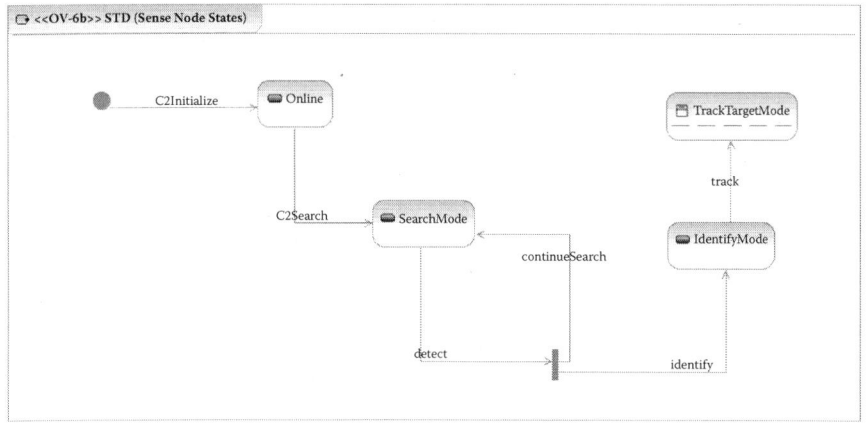

Figure 11.9   Example OV-6a in UML.

OV-6b, Operational State Transition Description, describes the possible set of state transitions that can occur in the operational nodes or activities. OV-6b should reflect all the possible behaviors of each node. In the example shown in Figure 11.10, the possible states of a sense node are shown, along with the event that will cause each state to be entered. For example, initializing the sense node will cause it to come online. Initiating a search will cause it to enter the search mode. Detection will cause the node to transition to either the identification state or to continue in the search state. In this way, OV-6b reflects the way in which the activities are sequenced and what the trigger is for transitioning between activities. In the same way, an OV-6b can be used to describe the transitions between states

Figure 11.10   Example OV-6b in UML.

for a given activity. The level of detail required for an OV-6b is dependent on the problem at hand and the use of the product. UML statechart diagrams can be used to depict OV-6b.

OV-6c, Operational Event-Trace Description, captures the time sequencing of operational activities for a given scenario. There are often several OV-6c artifacts that capture the behavior of the architecture in various scenarios. OV-6c describes the specific operational threads associated with each scenario, including the sequence of events and the information required for each of the operational activities. It is important to realize that the execution of a specific operation thread often represents a specific capability. OV-6c describes the required events, timing, and information flow required to achieve that capability for a given scenario. This product can be very useful in performing gap analyses or analyses of alternatives, because it can show both the critical path or event sequence for a given capability, as well as what additional activities may enhance the capability. An example of an OV-6c is shown in Figure 11.11. However, there is some debate over the most informative way to depict OV-6c. Alternatives recommended in the DoDAF guidance include the UML activity diagram and IDEF3.

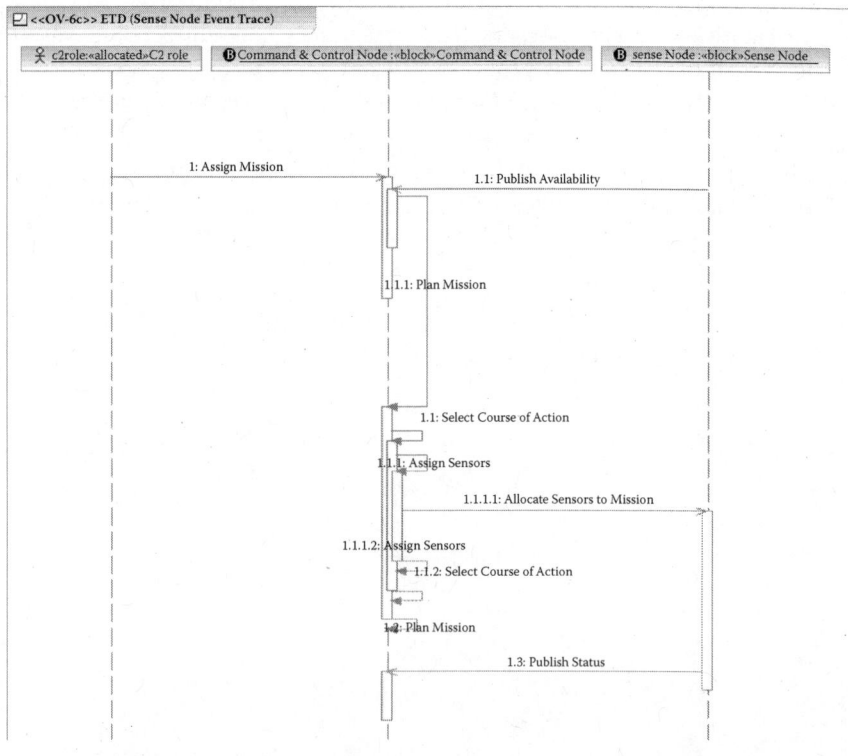

Figure 11.11 Example OV-6c in UML.

In the context of formal modeling, the elements of the OV-6a model would be the inputs and outputs of the OV-5 activities and the relations would be the specific rules that govern the inputs and outputs. These relations often have a cause-and-effect (and therefore dynamic) nature. The OV-6b model shows the relationships between events and state transitions (the elements), essentially creating a model of possible node behaviors. In the case of OV-6c, the events and activities are the elements in the model, and the sequencing and information flow are the relations. The OV-6c products represent models of each scenario, and thus are models of how capabilities are achieved to accomplish scenario goals.

OV-7, Logical Data Model, is the key product for understanding the data structure for the architecture. This includes the data types and the rules that govern the data in the system. One of the important roles of OV-7 is to provide a way of determining if and how an architecture will be able to interface with external systems, architectures, or organizations. Additionally, OV-7 can help ensure interoperability within the architecture between various organizations and nodes. This product is particularly important when involved organizations or systems use similar vocabulary to describe very different data types. For example, a sensor may be designed to output track data in a given format that uses a decimal notation to give track locations. A receiver receiving that data and translating it to a graphical track image may require the latitude and longitude to be given in minutes and seconds. This will lead to an incorrect depiction of the track, resulting in poor information for the decision makers in command and control. In the example shown in Figure 11.12, only data exchanges and their nature are shown. However, attributes can be associated with each information element to ensure compatible formats and clarify exactly what data is being transferred.

The need for OV-7 to be modeled formally is clear from the title of the product alone. As the "Logical Data Model" that models the data structure of the architecture, it is important to create a formal model that completely captures all the elements and relations of the data. Incorrect formulation of this model can lead to communication and information transfer deficiencies, which will likely significantly degrade the performance of the architecture as a whole, particularly in cases where there is geographical distribution among systems. Data elements and types are elements in the data model; data exchanges, rules about data exchanges, and data format transitions are all types of relations in the model.

System and Services View

The System and Services View (SV) is a set of graphical and textual products that describes systems and services and interconnections providing for, or supporting, defense functions. SV products focus on specific physical systems with specific physical (geographical) locations. The SV captures the functions of systems,

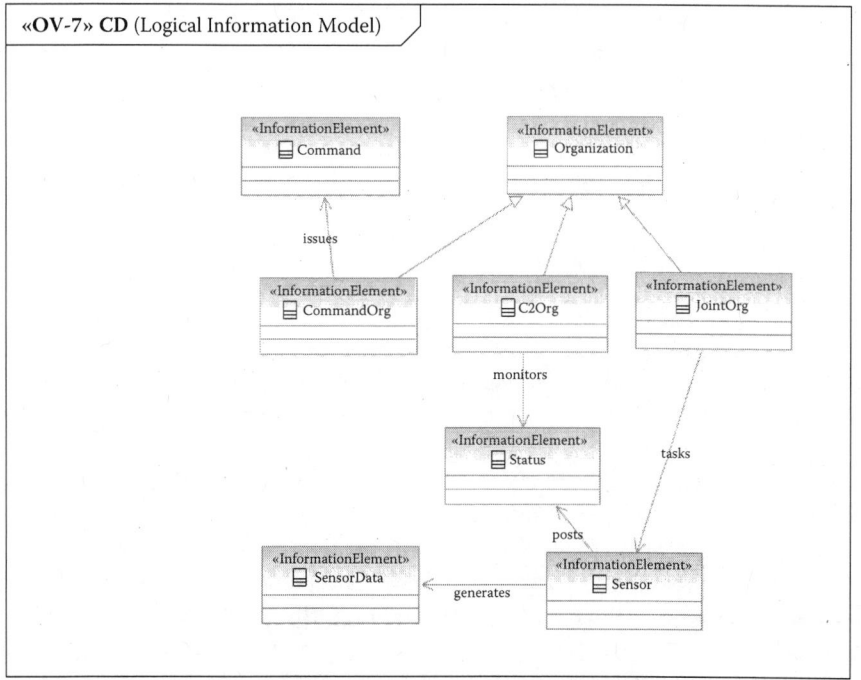

Figure 11.12 Example OV-7 in UML.

services, and interconnections to support OVs. There are 11 SV products; however, some of these have several parts, leading to a total of 16 depictions.

SV-1, Systems and Services Interface Description, is similar to OV-1 in that it depicts nodes and the interfaces between the nodes. However, whereas OV-1 is system independent, the SV-1 nodes are system nodes and include the specific systems and services that exist at each node. SV-1 depicts which systems support which operational nodes, what interfaces exist between these systems, and which interfaces are key interfaces. Key interfaces are interfaces where either "the interface spans organizational boundaries (may be across instances of the same system, but utilized by different organizations)," "the interface is mission critical," "the interface is difficult or complex to manage," or "there are capability, interoperability, or efficiency issues associated with the interface" (DoD 2007b).

Similar to OV-1, SV-1 can be viewed as a logical model. System nodes are elements in this model, and interfaces between nodes provide the relationships. Fundamentally, the SV-1 model is simply the system-level instantiation of the OV-1 model, and thus care should be taken to verify that these models agree. In fact, creating a relation-preserving transform between OV-1 and SV-1 could help ensure that the system nodes fully accommodate the conceptual needs of the architecture and that the overall operational concept and capabilities are realized in the system implementation.

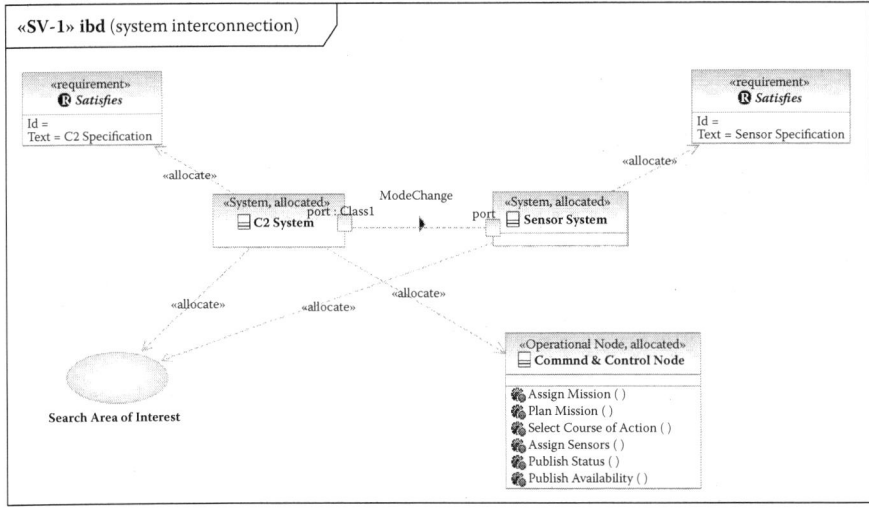

Figure 11.13   Example SV-1 in UML.

Systems are not necessarily a single piece of hardware or software; they can be groups of hardware and software that work together to serve a specific role or function. Interfaces in SV-1 are similar to needlines in OV-2 in that the interfaces are simplified versions of the actual communication path that is required to transfer the information. Additionally, each OV-2 needline will become one or more interface lines in SV-1. An example SV-1 is provided in Figure 11.13. As of 2005, SV-1 was the most-used DoDAF product in the DoD, with 88 percent of DoDAF implementations including an SV-1 (OASD 2005).

SV-2, Systems and Services Communication Description, provides a description of the actual communications systems, links, and networks that support the needlines and interface lines shown in OV-2 and SV-1. SV-2 provides a detailed description of how the communications needs of the architecture are actually met, which is extremely important for most architectures, and in particular, net-centric architectures. Additionally, the communications setup is going to play a key role in defining the overall infrastructure. For each interface, SV-2 should show a description of the way in which the interface is actually implemented, including the needed communications systems, their links, the networks that information is transmitted over, communications protocols used, and gateways (Department of Defense 2007b).

Communications systems include all systems that enable and control the transfer of data, such as routers and satellites. This does not include the systems that process that data for a specific use or translate the data into a specific format for a given piece of hardware or software. The communication link is the actual physical connection between two communication systems, for example, a cable or a radio

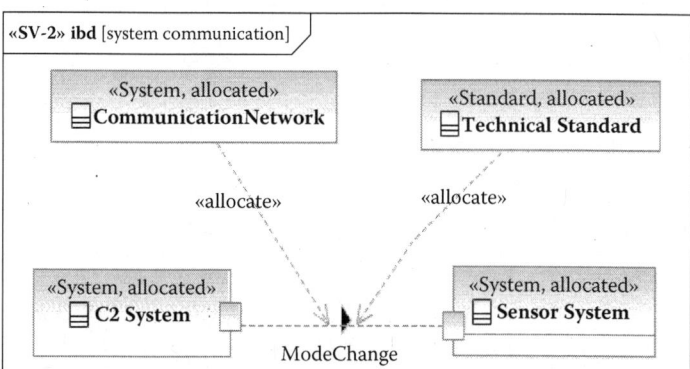

Figure 11.14 Example SV-2 in UML.

frequency. A path is made up of one or more links that are connected in such a way as to enable the transfer of information from one system or node to another. The interface lines in SV-1 represent one or more communication paths that are described in detail in SV-2. An example of an SV-2 is shown in Figure 11.14. Often, this type of graphical representation will be accompanied by supporting text or annotations to give more details about how the communications are implemented, and to provide traceability between the communication paths in SV-2 and the interface lines of SV-1.

Similar to OV-2, SV-2 is also a model. In the case of SV-2, elements are communication systems and relations are links and networks. In fact, SV-2 may look very similar to a network model in its depiction. Analysis of SV-2 as a network model may be able to provide information about the connectivity, vulnerability, robustness, and clustering of the architecture communications (Balestrini 2008). It is important to realize that SV-2 is the physical instantiation of OV-2. Using a model transform between OV-2 and SV-2 can ensure that all operational needlines are addressed by the physical implementation of the communication systems.

The SV-3 includes the Systems-Systems, Systems-Services, and Services-Services Matrices. As suggested by the title, this product is depicted in matrix form and provides information on the interfaces between systems and services. The primary purpose of this product is to provide an overview of the interfaces and to enable rapid assessments of commonalities and redundancies. Typically, the systems and services are depicted in the rows and columns of the matrix, and the intersecting cells represent a potential interface. Where an interface exists, some notation is made in the cell. The SV-3 can be used to depict many different aspects of the interfaces, and it is not specified which information should be depicted in any given situation. Some suggested interface characteristics include status, purpose, classification, means, standards, and whether or not the interface is a key interface. Several characteristics can be depicted in a single matrix by using a set of colors or symbols. Additionally, annotations and notes can be used to show how elements of the SV-3 correspond to elements of the SV-1 and SV-2. UML does not have a diagram that can be used to depict an SV-3. The SV-3 is often created using spreadsheets (Department of Defense 2007b).

The three elements of the SV-3 can be described as logical models. The systems-to-systems matrix provides relations on the systems (as elements) by describing the interfaces. The services-to-services matrix provides a similar model with the services as elements. The systems-to-services matrix provides the interfaces as relations between the system elements and service elements. Using a clear annotation scheme to show how these interfaces correspond to SV-1 and SV-2 elements and relations can help check for consistency between views and prevent relations from being overlooked.

SV-4 has two pieces: SV-4a (Systems Functionality Description) and SV-4b (Services Functionality Description). Because these two products are very similar, they will be discussed together. It should be noted that SV-4b was added in version 1.5 of DoDAF specifically to help support net-centric architecting. SV-4 products describe the functions performed by systems and services, as well as the functional hierarchies. Essentially, this product represents a functional decomposition at the system level. SV-4 performs a similar role for systems as OV-5 does for operational nodes. Operational activities and system functions serve a similar role, namely, describing tasks with specific inputs and outputs. The main distinction between system functions and operational activities lies in the way in which the tasks are carried out. System functions are typically automated tasks executed by some automated system, whereas operational activities describe business tasks that can be completed by humans, automated systems, or some combination of the two. SV-4 shows the required data flow for the input and output of each system and from where and to where these data flows occur. The functions represented in SV-4 can include system functions, human–computer interface (HCI) functions, and graphical user interface (GUI) functions. Data flow diagrams and tree diagrams are often used to represent the SV-4 products. In UML, class diagrams can be used to represent SV-4.

The functional decomposition that is central to SV-4 is usually represented as a hierarchical functional description. Similar to OV-5 for activities, it is important for providing functional traceability through the products. In addition to the hierarchical relationships established between the system functions, SV-4 can be used as a logical model to capture the data flow, that is, inputs and outputs for each system, specifically from where the data flows originate and where they terminate. In this case, the systems will be the elements in the model and the data flows will provide the relations.

SV-5 has three pieces: SV-5a (Operational Activity to Systems Function Traceability Matrix), SV-5b (Operational Activity to Systems Traceability Matrix), and SV-5c (Operational Activity to Services Traceability Matrix). SV-5 provides important traceability between the operational activities depicted in OV-5 and the systems and functions depicted in the system view products. SV-5a and SV-5b are used to describe the relationship between operational activities and systems and system functions. OV-5 and SV-5 combine to present crucial information to decision makers. Using OV-5 and SV-5, capability needs can be traced to operational activities, operational activities to system functions, and system functions to systems, thus providing the mapping from capabilities to specific systems. This can help decision makers identify stovepiped systems, capability gaps, and the capability impact of choosing to implement or not implement a specific system. Decision makers can also see where systems are redundant or duplicated, or where systems can perform multiple needed functions. This type of information is extremely useful in aiding acquisition and investment decisions. Additionally, understanding the mapping between system functions and capabilities can help with requirements definition for new systems that are needed to help achieve capabilities.

There is no UML diagram that specifically supports SV-5. The information required to create an SV-5 can typically be developed from the information in OV-5 and SV-4, and therefore can be extracted (perhaps via a script) from the UML representations of these products. However, the SV-5 products are typically represented by a series of matrices and can be created in a spreadsheet. In SV-5a, the rows of the matrix correspond to system functions and the columns correspond to operational activities. The cells of the matrix correspond to the relationship between system functions and operational activities. In the simplest SV-5a, an "x" would be placed in any cell where the system function supports the operational activity. It is important to realize that the mappings are not one-to-one. Multiple system functions can support an operational activity, and a system function can support multiple operational activities. It is also important to realize that more information can be embedded in the cells than simple identification of the presence of a relationship. A color or shape code can add information, for example, about the nature or state of the relationship. Another possibility is that the "x's" can be replaced with a number to represent how well or to what degree a system function supports an operational activity (Department of Defense 2007b).

In SV-5b, matrix rows correspond to systems and columns correspond to operational activities. However, this is often done by simply extending or rearranging SV-5a such that system functions are grouped by the system that performs them. Similarly, operational activities can be grouped by the capability they support. By using this kind of extension, one can see the mapping from system to capability using this product (Department of Defense 2007b). It is this feature, perhaps, that makes SV-5 one of the more commonly used SV products (OASD 2005). SV-5c is similar to SV-5b, except that instead of systems, the operational activities are mapped to services. SV-5c was specifically added in version 1.5 of DoDAF to support development of net-centric architectures (Department of Defense 2007b).

SV-5 represents a series of three transforms, and thus merits further discussion. Transforms are particularly useful in providing traceability because they provide relationships between elements of different models, thus giving consistency and showing a clear linkage between the elements and relations of one model and the elements and relations of another. SV-5a is a transform between the operational activities (described in OV-5 and OV-6) and the system functions (described in SV-4). This transform shows how specific systems and functions physically carry out the operational activities that combine to provide capabilities. SV-5b is similar to SV-5a, but groups the system functions in terms of the systems that perform them, so the transform is between the operational activities and the actual systems that contribute to their execution. SV-5c performs a similar transform to SV-5b, but instead of using systems, it uses services, and transforms between the services and the operational activities they support. However, as currently defined in the DoDAF, it is important to note that the three SV-5 transforms are not model transforms, because they are not required to preserve the relationships contained in SV-4 and OV-5 (and OV-6).

The purpose of SV-6, Systems and Services Data Exchange Matrix, is to detail the characteristics of the data exchanges, particularly the automated ones, between systems. This product is particularly important for understanding the potential costs associated with the implementation and upkeep of the data elements of the architecture, as well as constraints on data exchange that are imposed by the limitations of the physical systems. SV-6 describes how each data exchange is implemented, including which data elements are used, data format and media type, how often and what size data is exchanged, how much data can be exchanged (e.g., bandwidth), what security measures protect the data, and information assurance and timeliness of the exchange. Estimated accuracy and units can also be included in the SV-6 matrix. The SV-6 data exchanges can be mapped to the corresponding information exchanges of OV-3, essentially giving additional details about the automated information exchanges. SV-6 provides the details about the actual exchange, whereas OV-3 documents the operational context of the exchange. The specific data exchange characteristics to be included in the SV-6 will vary depending on the program, so there is no one correct set of information that needs to be included. Typically, each row in the SV-6 matrix will represent a specific system interface and data exchange, and the columns will correspond to the characteristics of that exchange being documented. Simply put, SV-6 is a table of key information about the data exchanges. In UML, a logical service realization diagram can be used to capture at least some of the data for an SV-6. However, a table is the simplest, and arguably the clearest, way to depict an SV-6.

The SV-6 model uses systems and services as the elements and data exchanges as the relations. Using a model transform between OV-3 and SV-6 can help ensure that the information exchanges called for in OV-3 are carried out in the actual implementation described in SV-6. Thus, a model transform between OV-3 and SV-6 can be used to preserve the information exchanges in the physical construct.

SV-7, Systems and Services Performance Parameters Matrix, details the current, future expected, and required performance estimates of systems (including both hardware and software), interfaces, and functions. This product will often not be developed early in the design process, although early estimates of expected performance requirements (if they can be obtained) can be extremely useful in developing hardware and software requirements and also in selecting systems for inclusion in the architecture. This product will also be extremely important to any modeling and simulation activities or other quantitative analyses, as this will help identify what needs to be modeled and measured in the simulation environment. This product is critical to understanding the actual abilities of the architecture and for assessing whether the architecture will meet performance needs.

Because the content of this product will depend heavily on the systems under consideration, no set of parameters is recommended for inclusion. Instead, an example will be given. For the hardware element "kitchen chair," performance parameters would include weight limits, attractiveness, comfort level, sturdiness, and reliability. SV-7 would include the value measured or observed for each of these parameters, the value desired, and any changes to the current value that are expected with changes in technology in the future. In the chair example, if the current chair is a wooden chair, comfort level may be increased in the future if seat cushions are added. Typically, SV-7 is represented in a tabular format, for which there is no UML equivalent. SV-7 itself is one of the few DoDAF products that is not a model. The information contained in SV-7, however, is important for describing the performance of other products. SV-7 provides the metrics required to quantify the parameters of the models, and may be used to compare one model to another as a candidate solution in the architecture.

SV-8, Systems and Services Evolution Description, describes the plan for how the systems or overall architecture will evolve over time. This evolution could be the result of technology improvements, a phased budget, an ongoing transition that requires an evolutionary development approach, or a host of other possibilities. Because the DoD acquisition policy calls for phased implementation of programs, most architectures are going to be created using an evolutionary approach. SV-8 describes what changes or evolutions are expected and when they are expected (or planned) to occur. This is often represented by simply using a timeline, although there are many possibilities for the representation of this information. This author has previously depicted an SV-8 by overlaying a color scheme on SV-5 products to show in what phase the systems, services, or functions are expected to come online. In this way, not only was the way in which a capability was to be implemented described, but also when the capability would be implemented. It may be surprising to realize that SV-8 is actually a model. The elements are the systems and technologies, and the relations are the timing and sequencing of their implementations.

SV-9, Systems and Services Technology Forecasts, supports the SV-8 by describing expected future technological improvements. SV-9 cannot cover every possible technological development, but should instead focus on technologies that are both likely to develop in a reasonable timeframe and will have a large impact on the capabilities of the architecture. This product will largely contribute to the evolution plan depicted in SV-8 by showing both the potential improvements and the expected timeline for the technology to be developed. One way to depict an SV-9 would be to create a table of functions or services and expected technology improvements (if any) in predefined timeframes (e.g., the next year, the next five years, and the next ten years). Another possibility might be to list future technologies, their current level of maturity, the expected timeline for maturation, the expected performance, and the uncertainties associated with the expected performance and timeline. It may also be useful to include the expected cost to aid acquisition decision making.

SV-10 has three components: SV-10a (Systems and Services Rules Model), SV-10b (Systems and Services State Transition Description), and SV-10c (Systems and Services Event-Trace Description). Similar to OV-6, SV-10 describes dynamic characteristics, including timing and event sequencing. Whereas OV-6a describes operational rules, SV-10a describes the system rules that are often derived from the operational rules. This generally results in performance constraints or requirements on the systems. There are several broad categories of system rules. Structural assertion includes rules that are generally related to business needs, such as specified file structures or use of a particular schema in a database. Action assertions govern dynamic aspects of system behavior, and are often manifested in if-then type rules. Derivations include rules that are the derived result of other rules or facts. It is recommended that SV-10a be represented using a formal and well-documented language for rule modeling.

SV-10b is a graphical product that is used to describe the state transitions that occur as the result of particular events. SV-10b is similar in nature to OV-6b, and can similarly be represented using UML statechart diagrams. SV-10b is particularly useful for identifying missing connections that, if not caught, could lead to errors in the behavior of systems. If modeling and simulation of dynamic characteristics are required, using a Petri net or similar method for representing SV-10b can be useful for both modeling the behavior and visually depicting the product.

SV-10c is similar to OV-6c in that both products depict event traces. SV-10c shows the sequence of events and the data exchanges that occur during the event sequence. The primary purpose of SV-10c is to show, in detail, the sequence of events, functions, and data exchanges that must occur for the overall functionality to be achieved. This acts as a check to ensure that each element will get the required information at the needed time and will be able to serve its purpose in a timely fashion. It should be possible to map system-level event sequences back to the operational activities they accomplish, and therefore back to the sequence of activities that provides a given capability in a given scenario. Similar to OV-6c, UML sequence diagrams are recommended for representing SV-10c.

Just as OV-6a, b, and c were considered models, so are SV-10a, b, and c. The elements of the SV-10a model would be systems and services, and relations would be the specific rules that govern the inputs and outputs. SV-10b shows the relationships between events and state transitions (the elements), essentially creating a model of possible node behaviors. In the case of SV-10b, the state transitions occur between the states of a system or service, rather than the operational nodes of OV-6b. In the case of SV-10c, the events and activities are the elements in the model, and the sequencing and information flow are the relations. SV-10c products represent models of each scenario, and are thus models of how systems and services contribute dynamically to perform capabilities that accomplish scenario goals. A model transform between the OV-6 products and the SV-10 products can be used to ensure that the activities, events, and sequences required by OV-6 are actually carried out through the implementation of the physical systems and services.

SV-11, Physical Schema, is the systems-view version of OV-7. This product gives specific details regarding the system data structure. This effectively takes OV-7 a step further, showing the actual implementation details of the information requirements in OV-7. Among other things, SV-11 can include such information as the formatting for messages, file types used, file structures used, transfer protocols, and standards. SV-11 plays an important role in helping to verify and ensure interoperability throughout the systems. A UML class diagram is suggested for representing SV-11, and in particular, the Object Management Groups (OMG) Database Profile is recommended (Department of Defense 2007b).

Similar to OV-7, SV-11 is a formal model by its nature, and thus care should be taken to represent it as such. SV-11 is essentially the system-specific version of OV-7 and, as such, models the actual implementation of the OV-7 data structure. Elements in the SV-11 model include data elements, file types and structures, and data storage and transfer systems, among other things. Relations include file transfer protocols and standards across data interfaces.

Technical View

The Technical View (TV) is the minimal set of rules governing the arrangement, interaction, and interdependence of system parts or elements. The TV provides the technical systems implementation guidelines on which engineering specifications are based, common building blocks established, and product lines developed. There are only two products in the TV. TV-1 (Technical Standards Profile) describes the current collection of technical standards, implementation conventions, standards options, rules, and criteria. In addition to describing the standards, TV-1 also helps identify relevant portions of the standards and to what aspects of the architecture they are applicable. TV-2 (Technical Standards Forecast) describes emerging technical standards, implementation conventions, standards options, rules, and criteria. One of the important uses of TV-2 is to understand how the emerging standards will change the needs of the systems and services, and how the architecture will have to evolve to accommodate changing standards. Both products provide information required for creating the rules models in SV-10, and TV-2 provides forecasting information important for developing SV-8. Typically, both products are depicted in a tabular format (Department of Defense 2007b).

MODAF

As discussed in the Chapter 10 ("Introduction to DoDAF and MODAF Frameworks"), MODAF is derived from DoDAF, and thus shares the DoDAF products. However, MODAF has added some additional viewpoints and views and made several modifications to the DoDAF views. This section focuses only on the views that are unique to MODAF and the tweaks that have been made to DoDAF products. A discussion of the MODAF recommendations for representing DoDAF products will not be included. Information on this topic can be found on the MODAF Web site www.modaf.org.uk. (MOD 2008).

In MODAF, OV-1 has three parts. OV-1a is the graphic that depicts the overall concept of operations. OV-1b is a text description providing more details about the nodes and communication links shown in OV-1a. OV-1c is a tabular representation of key parametric data associated with the scenario (often showing evolution of capability over time). These should reflect the capability requirements that the architecture is being designed to fulfill.

Strategic Viewpoint

The MODAF Strategic Viewpoint (StV) has six supporting views. These have been introduced to support the capability management process. StV-1, Enterprise Vision, provides a strategic context for the architecture beyond the concept of operations. StV-1 shows the needed capabilities and the enterprise goals for the entire enterprise, and these are constant for the entire enterprise over its life span, and can apply to many specific architectural descriptions. It provides a place to put the capabilities delivered by a specific architecture into the greater operational context at the enterprise level. StV-1 can be useful in understanding how an enterprisewide transformation might occur and helps to prevent stovepiping. MODAF recommends representing StV-1 using structured text, a UML class diagram, a UML composite structure diagram, or a SysML structural diagram.

StV-2, Capability Taxonomy, gives the hierarchy of capabilities and what enterprise phase each supports. This is intended to be a complete list of capabilities for an enterprise, and therefore is likely to span more than one architecture. It can help depict how capabilities achieved using different architectures come together to accomplish the enterprise-level goals. StV-2 can also be extremely useful in capability gap analysis and in developing Key User Requirements (KUR). In general, the capabilities are very abstract, although at the lowest level some qualifiers or metrics may be added. MODAF suggests depicting StV-2 using a tabular description, and hierarchical graphic, or a UML class diagram.

StV-3, Capability Phasing, provides the time element to the capabilities in StV-2. This provides the overall plan for phasing in capability over time across the enterprise to meet enterprise goals. StV-3 supports the capability audit process that is used for gap and duplication identification. The phasing-in of capabilities is typically associated with a set of milestones that represent when a new capability is anticipated to come online. In addition, StV-3 can represent when capabilities might be phased out. StV-3 is usually represented with a tabular depiction, although an alternative might be to use a Gantt chart.

StV-4, Capability Dependencies, describes the dependencies between capabilities and "clusters" of capabilities. Capability clusters consist of capabilities that are logically grouped, supporting similar enterprise goals. Capability clusters also often represent groups of capabilities that are tightly related, having a large number of dependencies between them. The dependencies represent interactions between capabilities, clusters, or acquisition programs that aid in achieving the enterprise goals. Candidate depiction formats for StV-4 include nested box diagrams, UML class diagrams, UML composite structure diagrams, SysML structural diagrams, and N^2 diagrams. Different types of dependencies can be noted by color-coding the lines or by using different dash patterns.

StV-5, Capability to Organisation Deployment Mapping, maps capability configurations to organizations. Capability configurations are a specific solution to a given capability. For example, a deployable Web blogger is a capability configuration for the capability "Front Line News Reporting," and is used by the organization "Air Assault Brigade." StV-5 is used for planning of capability fielding, capability integration planning, options analysis, gap or redundancy analysis, and identification of deployment shortfalls. StV-5 can be represented using a tabular description or a structured timeline view.

StV-6, Operational Activity to Capability Mapping, maps the operational activities in OV-5 to the capabilities in the StVs. This is useful in showing traceability between capabilities and operational activities, and can also show how external capabilities (from elsewhere in the enterprise) can be used to support operational activities, and how operational activities in a given architecture can help support other capabilities within the enterprise. In the same way that SV-5 shows the traceability between the OV and the SV, StV-6 shows the traceability between the StV and the OV. This is typically shown using a mapping matrix with capabilities represented in rows and operational activities represented in columns.

Acquisition Viewpoint

Acquisition Views (AcVs) have been introduced to describe programmatic details, such as the dependencies between projects and how capability integration occurs across all the Defence Lines of Development (DLoDs). The Acquisition Viewpoint is important to integrating acquisition and development and aids in fielding of capabilities. There are two Acquisition Views. AcV-1, Acquisition Clusters, shows the dependencies between programs. Programs are grouped by organization, and the dependencies between organizations are also depicted. AcV-1 describes how organizations and programs need to come together to achieve capabilities and enterprise goals. It can be depicted using topological graphics, nested box diagrams, UML composite structure diagrams, or UML class diagrams.

AcV-2, Programme Timelines, adds the time element to AcV-1. It shows the schedule for programs, key milestones, and when dependencies occur. This is an important view for program management, risk analysis and identification, managing dependent programs, portfolio management, technology roadmapping, and through life management planning (TLMP). AcV-2 can be depicted using timelines or Gantt charts. Annotations or colors can be overlaid onto the Gantt chart to reflect risk level, status, or outstanding issues. A textual addendum can provide additional details where necessary.

Service-Oriented Viewpoint

The Service-Oriented Viewpoint is a new addition in MODAF version 1.2. This viewpoint describes services that are required to support operations as depicted in the OV. A service has a very broad definition, which is "a unit of work through which a provider provides a useful result to a consumer." The MODAF service views are based on the NATO NAF version 3.0 service views, although the set of views is not the same. Although NAF is outside the scope of this book, it should be noted that there are equivalent NAF views. More information on this relationship can be found on the MODAF Web site. There are five Service-Oriented Views (SOVs); but because SOV-4 has several parts, there are seven total depictions. Because the SOV is evolving with the new version of MODAF, information is less available about this viewpoint than the others. Service elements were also added to some of the OVs, including OV-2, OV-5, and OV-7c, but these changes are outside the scope of this overview.

SOV-1, Service Taxonomy, depicts a hierarchy of services. Each level in the hierarchy shows specializations of the services in the level above it. It contributes to a standard library of services that is compatible at the enterprise level. SOV-1 can be used for identifying service gaps, governing service-oriented architectures (SOAs), and planning. In addition to showing the services, SOV-1 also shows services attributes and service policies or constraints. A UML class diagram is the recommended approach for representing SOV-1 because it naturally includes the generalization–specification concept, but tabulation and hierarchical graphics are also options.

SOV-2, Service Interface Specification, defines service interfaces. Specifically, these are the interfaces with the users of the service, and those interfaces used by the service. This gives an understanding of which services are compatible with one another and what service interfaces are available for external use. Additionally, it provides an interface standard for new service providers. At a minimum, SOV-2 should include the service names, the interface names, which interfaces are provided and which are used, the operation name, and the parameters. Operations are the methods of access, and parameters are the data that must be passed to or produced by the service. SOV-2 is typically depicted in a tabular form, although it can be depicted in UML.

SOV-3, Capability to Service Mapping, shows which services contribute to which capabilities. This is particularly important in SOAs that use or create network-enabled capabilities. SOV-3 is depicted using either a mapping matrix or a UML object diagram.

SOV-4 has three components: SOV-4a (Service Constraints), SOV-4b (Service State Model), and SOV-4c (Service Interaction Specification). SOV-4a specifies constraints on the implementation of services. These constraints include, for example, required availability, and are important to the performance of the overall architecture or enterprise. SOV-4b acts as a state model for the services. It includes what the potential states of the service are and the transitions between the states. UML statechart diagrams or other state transition models can be used to depict SOV-4b. SOV-4c details how and when services interact with external agents. It does not give specific sequences of services, but rather potential sequence paths between services with compatible interfaces. This is typically depicted using a UML sequence diagram.

SOV-5, Service Functionality View, describes the functions the services perform. This helps understand the role of the services and can also help with requirements definition. This basically describes what the services are supposed to do, but does not provide information on how the services should perform those functions. A functional diagram is suggested for depicting SOV-5.

Summary

DoDAF and MODAF provide guidance and rules for describing, representing, and understanding architectures using a series of views and products. Each of these products was discussed at an overview level in this chapter, such that the reader should have a clear understanding of what each product is, when and why it is used, and how it might be represented. Additionally, DoDAF products that can be considered models or transforms according to the definitions presented in Chapters 4 ("Structure, Analysis, Design, and Models") and 5 ("Architecture Modeling Languages") were discussed, including what the elements and relations of each model are. Where available, example products were provided in UML.

References

Balestrini, S. July 2008. Graph Theory: A Tool for Understanding Complexity. Georgia Institute of Technology, Atlanta, GA.

Department of Defense. 2007a. *Department of Defense Architecture Framework (DoDAF), Volume I: Definitions and Guidelines.*

Department of Defense. 2007b. *Department of Defense Architecture Framework (DoDAF), Volume II: Product Descriptions.*

Hardy, D. 2007. OMG UML Profile for the DoD and MoD Architecture Frameworks http://syseng.omg.org/UPDM%20for%20DODAF%20WGFADO-INCOSE%20 AWG-070130.ppt.

Ministry of Defence (MOD). 2008. The Ministry of Defence Architecture Framework (MODAF) version 1.2. Information Superiority Architectures (CISA) Worldwide Conference, Omaha, NE http://www.modaf.org.uk/ (accessed June 26, 2008).

Office of the Assistant Secretary of Defense (OASD) for Networks and Information Integration. *The State of DoD Architecting.* Winter 2005 Command. December 1, 2005. www.modaf.com/file_download/17

Chapter 12

Other Architecture Frameworks

Architecture frameworks have been developed for many domains, not just for defense systems. The use of frameworks crosses international boundaries as well. The rise of information technology over the past decades has resulted in the development of frameworks for the architecture of enterprises. Large organizations, whether commercial or government based, are information-intensive systems. Thus, the focus of enterprise architecture frameworks tends to be driven by information and information technology. This chapter will review two well-known enterprise architecture frameworks and indentify a number of other frameworks. The focus of an architecture framework does not need to be limited to the organization of information about a system or enterprise. For this reason, the chapter will conclude with a review of High-Level Architecture (HLA), which is focused on how a collection of system simulations can be federated into an architecture that can be used to simulate a system of systems.

Key Concepts

Logical structures
Data models
Enterprise architecture
Information technology
Federated architecture

Enterprise Architecture Frameworks

Enterprise architecture frameworks have been developed by several open source organizations; also, they have been extensively developed by commercial organizations. The two frameworks reviewed in the following text and the short reviews of other frameworks in this section are intended to give a brief introduction to the subject and a sense of the breadth of the subject. However, they are by no means the only significant frameworks that have been developed.

185

The Open Group Architecture Framework (TOGAF)

The Open Group Architecture Framework (TOGAF) was developed by the members of The Open Group, working within the Architecture Forum. The Open Group is a nonprofit open source organization. TOGAF is freely available for online viewing without a license and can be found at www.opengroup.org/architecture/togaf. Terms for licensing and conditions of use can also be found at the TOGAF Web site.

Under development for over a decade, TOGAF is an example of a government-developed standard transitioning into an open source group. The original development of TOGAF version 1 in 1995 was based on the Technical Architecture Framework for Information Management (TAFIM), which was the result of millions of dollars of investment by the U.S. Department of Defense. TOGAF is currently updated on an annual basis. The review presented here is based on TOGAF v8.1.1.

TOGAF includes a method and a set of supporting tools for developing information technology (IT) enterprise architectures. The goal is to create an industry-standard method for developing architecture descriptions that is independent of any tools or technologies.

The term *enterprise* has many interpretations. When used in TOGAF, it refers to an organization or group of organizations linked together by common ownership with a shared bottom line. For commercial enterprises, the bottom line is strongly related to profit. The technical advantages of a good enterprise architecture should bring bottom line business benefits such as a more efficient IT operation, faster and more flexible procurement processes, better return on existing investment, and reduced risk for future investment.

TOGAF embraces but does not strictly adhere to ANSI/IEEE Std 1471-2000 (IEEE 2000a) terminology, which defines *architecture* as follows:

> The fundamental organization of a system, embodied in its components, their relationships to each other and the environment, and the principles governing its design and evolution.

In TOGAF, the term *architecture* has two meanings, depending on its contextual usage:

1. A formal description of a system, or a detailed plan of the system at component level to guide its implementation.
2. The structure of components, their interrelationships, and the principles and guidelines governing their design and evolution over time.

TOGAF endeavors to use the terminology of architecture in a way that is consistent with the ANSI/IEEE Std 1471-2000 standard. However, TOGAF also seeks a balance between the concepts and terminology of ANSI/IEEE Std 1471-2000 and other commonly accepted terminology that is familiar to the majority of the TOGAF readership.

An architecture description in TOGAF is a formal description of an information system, organized in a way that supports reasoning about the structural properties of the system. An architecture framework is a tool that can be used for developing a broad range of different architecture descriptions. It should be noted that TOGAF speaks of different architectures, whereas a standards-based use of the term would consider the enterprise as having one architecture and many views (i.e., viewpoint models: see Chapter 16 ["Model Driven Architecture"]).

TOGAF divides enterprise architecture into four types of architecture:

Business
Application
Data
Technology

Business Architecture is concerned with the processes that an enterprise uses to meet its business goals. Application Architecture is concerned with the specific software applications used by the enterprise, how they interact, and how they are designed. Data Architecture is concerned with how enterprise data stores are organized and accessed. Technology Architecture is concerned with the hardware and software infrastructure that supports applications and their interactions.

TOGAF comprises three main parts: the Architecture Development Method (ADM), the Enterprise Continuum, and the TOGAF Resource Base. The TOGAF Resource Base is a set of guidelines, templates, and background information to help the architect in using ADM.

ADM is a process for creating an architecture. Requirements Management is at the center of the process; and throughout the development process, frequent validation of results is required. In TOGAF, an enterprise architecture is developed using the eight phases of ADM. The process begins with defining what is called an "Architecture Vision." Next, the Business, Information Systems, and Technology Architectures are developed. The Information Systems Architecture includes the Application and Data Architectures. The first four phases lead to defining enterprise Opportunities and Solutions. The last three phases of ADM are Migration Planning, Implementation Governance, and Architecture Change Management. The ADM cycle is a closed loop that ends with the Architecture Change Management phase feeding into the Architecture Vision phase (to start the cycle again).

Each phase of ADM is performed using a prescribed series of steps. For example, the steps for the Technology Architecture phase are as follows: create a baseline, consider views, create an architecture model, select the services to be implemented, confirm business objectives, determine criteria, define the Technology Architecture, and conduct a gap analysis.

Zachman Framework

The Zachman Enterprise Architecture Framework is the result of work by John Zachman in the practice of enterprise architecture and his personal enquiry (if not quest) as to how an enterprise architecture can be best represented. He offers the following definition of an Enterprise Architecture Framework:

> ... a logical structure for classifying and organizing the descriptive representations of the enterprise that are significant to the management of the enterprise, as well as to the development of the enterprise's systems.

See Inmon, Zachman, and Geiger 1997, and Sowa and Zachman 1992. The Web site www.zifa.com of the Zachman Institute for Framework Advancement provides further information about the Zachman Framework not presented in this brief review.

The Zachman Framework is portrayed in a rectangular grid consisting of 36 cells that is derived from the intersection of six fundamental interrogatives with five fundamental stakeholders in an enterprise and the elements of a functioning enterprise. The six fundamental interrogatives in their simplest form are What?, How?, Where?, Who?, When?, and Why? Each question becomes a column in the framework, and each column is a generic model. The five fundamental stakeholders are the planner, owner, designer, builder, and subcontractor. Each cell in each of the generic models (columns, i.e., a question) specializes the model from the viewpoint of the given stakeholder. The cells in each column are intended to be related to one another. The enterprise architecture is considered complete only when the model for each cell is complete.

The Zachman Framework is intended as a general construct that enterprise architects can adapt to their particular problem. The logic is generic and recursive. There are some high-level rules such as cells in columns being related to each other. Other rules include the following: do not add new rows or columns to the framework, do not change the names of the rows or columns, do not classify any meta concept into more than one cell, and do not create diagonal relationships between the cells. Detailed prescriptive rules are not part of the framework (as, e.g., was the case with the TOGAF). However, the following discussion is useful in giving insight as to how the cells in the framework might be developed.

The Zachman Framework for IT enterprise architecture would give additional meaning to the following six fundamental interrogatives. "What?" becomes the data used by the enterprise. Models of the data are then created from the perspective of the five fundamental stakeholders. The model for the planner would be a list of data and information important to the business. This model is the contextual model and establishes scope and objectives. The next row is from the perspective of the enterprise owner, and the cell in the "What?" column specializes the generic data model, turning it into the conceptual data model. This is where the relationships between the entities listed in the contextual data are initially defined. It is through these relationships that the data becomes a model. The third fundamental stakeholder is the designer, who uses the conceptual data model to create the logical data model. The physical model is the domain of the fourth fundamental stakeholder, the builder. In general, the builder develops a technology model; and in the data column this becomes a model of the physical data, for example, the technological details of implementing a database or a software application that uses the data. These details might include computer language, computing platforms, and operating systems. The last stakeholder, the subcontractor, would seek a detailed model of the enterprise data and would be seeking models that include data definitions, etc.

The other five questions would also have interpretations from the IT enterprise viewpoint. For example, "How?" becomes the system functions (to be implemented by software applications); the "Where?" becomes the enterprise network; "Who?" becomes the people or possibly the organizational units of the enterprise; "When?" becomes time (e.g., key business cycles); and "Why?" becomes the motivation (e.g., the business goals that the data support). Each of these other five columns would lead to further specialization (within the cells of the column) as was illustrated with the data column (i.e., "What?").

Other Architecture Frameworks

The 1996 Clinger–Cohen Act in the United States precipitated the use and generation of architecture frameworks by many agencies of the government. The DoD Architecture Framework (DoDAF) traces back to the DoD response to the Clinger–Cohen Act. Specifically, the act requires the use of performance-based management principles for acquiring IT and places responsibility on an Agency's chief information officer for "developing, maintaining, and facilitating the implementation of sound and integrated information technology architecture." The framework developed for the U.S. federal agencies outside of the DoD is called the *Federal Enterprise Architecture Framework* (FEAF). It can be found at www.whitehouse.gov/omb/egov/a-1-fea.html.

Within the broader international community, FEAF is not the only government enterprise architecture. Numerous other countries have architecture frameworks. For example, a common framework has been legislated for use by the departments of Queensland, the government of Australia, which can be found at www.qgcio. qld.gov.au/02_infostand/gea.htm. And France has created a framework similar to DoDAF and MODAF, which is required for all major weapon and IT system procurements. The Atelier de Gestion de l'ArchiTEcture des systèmes d'information et de communication (AGATE) can be found at the Web site www.achats.defense. gouv.fr/article33349. An example of a purely commercial framework is the one developed by Capgemini for business, information, and technology. This framework can be found at the Web site www.capgemini.com/services/soa/ent_architecture/iaf.

High-Level Architecture (HLA)

HLA is a technology for distributed simulation that was first developed in the early 1990s to facilitate interoperability among simulations and promote reuse of simulations and their components. Based on the premise that no one simulation can satisfy all users or uses, it is used for federating simulations and has been used to create federations that can simulate the behavior of a system of systems. The component simulations in an HLA-compliant simulation are referred to as *federates*. HLA federations with hundreds of federates have been demonstrated in computer runtime. A key limiting factor is the number of object interactions that can occur within the federation. The granularity of computing timescales can quickly be driven to the level of picoseconds in a computing environment in which federates operate at greatly differing granularities.

HLA standards are defined under IEEE Standard 1516 (IEEE 2000b). HLA is defined by rules, interface specifications, and the Object Model Template, which specifies a common format and structure for documenting HLA object models. The HLA rules describe the responsibilities of federations and the federates that join. The interface specification describes the functional interface between the federates and the HLA Runtime Infrastructure (RTI). The elements of an HLA federation are the Simulation Object Models (SOMs), which are the basic components; the Federated Object Model (FOM); and a structured six-step process referred to as the *Federation Development and Execution Process* (FEDEP). SOMs are combined to form a FOM through the structured FEDEP process.

Enabling communication among simulation components that have been developed by different organizations at different times for different purposes poses a significant challenge. Communication among the federates is enabled by the Data Interchange Format (DIF), which provides a stable common format that has been designed to preserve the previous investment in the simulation components.

An excellent introduction to HLA can be found in Dahmann et al. (1997). Future research in HLA can be expected to include extensions of XML for the support of FOM and SOMs, exploration of using the Web Services Description Language (WSDL) for application programming interface (API) support, and exploration of modular FOMs.

Summary

Large organizations, whether commercial or governmental, are information-intensive systems. Enterprise architecture frameworks support an organization's management of information and IT. The 1996 Clinger–Cohen Act passed by the U.S. federal government requires the use of performance-based management principles for acquiring IT and managing information. In the United States, this led to the development of FEAF for nondefense agencies and DoDAF for defense agencies. Governments and commercial interests in numerous other countries have also developed architecture frameworks.

Architecture frameworks need not be limited to the organization of information in an enterprise. The HLA is an example of how a collection of system simulations can be federated into an architecture that can be used to simulate a system of systems.

References

Dahmann, J.S., Fujimoto, R.M., and Weatherly, R.M. 1997. The Department of Defense High Level Architecture. *Proceedings of the 1997 Winter Simulation Conference. 7–11 December 1997, Atlanta, GA*. Washington, DC: IEEE Computer Society.

Inmon, W. H., Zachman, J.A., and Geiger, J.G. 1997. *Data Stores, Data Warehousing, and the Zachman Framework: Managing Enterprise Knowledge*. New York: McGraw-Hill Publishers.

Institute of Electrical and Electronics Engineers (IEEE). 2000a. IEEE 1471. IEEE recommended practice for architecture description of software-intensive systems.

Institute of Electrical and Electronics Engineers (IEEE). 2000b. IEEE Std 1516-2000. IEEE standard for modeling and simulation high level architecture (HLA).

Sowa, J.F. and Zachman, J.A. 1992. Extending and formalizing the framework for information systems architecture. *IBM Systems Journal*, vol. 31, no. 3, pp. 590–616.

USING ARCHITECTURE MODELS IN SYSTEMS ANALYSIS AND DESIGN

3

Chapter 13

Modeling FBE-I with DoDAF

Fleet Battle Experiment-India (FBE-I) was an early success of the use of architecture to analyze operational tests of Network-Centric Operation (NCO). The Department of Defense Architecture Framework (DoDAF) was used to develop architecture descriptions of the time-sensitive targeting (TST) part of the experiment to assess the interoperability of the system of systems (SoS) used for the TST support of the NCO tactics.

Key Concepts

Network-Centric Operation (NCO)
Time-sensitive targeting (TST)
Activity models
Interoperability
Operational capabilities

This chapter will introduce some of the key architecture products developed for FBE-I at a level of abstraction that simplifies the actual architecture so as to make the concepts more understandable to a general audience without a background in defense systems. As the chapter unfolds, it will become clear how and why the particular DoDAF products were developed and how they were used in interoperability assessment. Further background on the early use of architecture in defense systems and its role in capabilities-based acquisition can be found in Dickerson et al. (2003).

The Fleet Battle Experiments (FBEs)

Just over ten years ago, the U.S. Chief of Naval Operations (CNO) initiated a series of experiments called *fleet battle experiments* (FBEs) designed to investigate NCO as a framework for development of new doctrine, organizations, technologies, processes, and systems for future war fighting. Each experiment in the series was designated by a letter, starting with A. The FBE was combined with naval phonetics for the letter, for example, alpha (for A), so that the first experiment was referred to as FBE-Alpha (when spoken) or simply FBE-A (when written). Thus, FBE-I (spoken as FBE-India) was the ninth in the series. The previous experiment, FBE-H, had marked a turning point in the CNO's objective to develop and demonstrate NCO capabilities. A more detailed discussion of Network-Centric Warfare (NCW) and its relation to NCO and Command, Control, Communications, Computing, and Intelligence, Surveillance, and Reconnaissance (C4ISR) is presented in Chapter 15 ("Toward Systems of Systems and Network-Enabled Capabilities").

The vision of FBE-I was to "operationalize" NCW by building and maintaining a C4ISR architecture that provided Joint Forces with wide-area connectivity, enhanced bandwidth, and reach-back capability. The success of the NCO concepts verified in FBE-I was demonstrated in actual combat by coalition land forces in Afghanistan after the attacks of September 11, 2001 and again by the Marine Corps in their swift march to Baghdad during the invasion of Iraq from March 18 to May 1, 2003. This was a marked departure from a centuries-long tradition of how the marines had fought.

FBE-I was also a pilot experiment for the then recently established chief engineer of the U.S. Navy to assess if architecture-based methods of mission capabilities engineering were actually useful. The detailed architecture model for TST capability in FBE-I affirmed that architecture could be used in this way, which is not to say that all defense systems architects practice their art in a way that is as successful as this early experiment.

Modeling NCO

Despite the emphasis in this book on logical modeling, architecture frameworks, and tools, some of the best work in architecture and systems engineering starts with informal discussions, hand drawings on paper or whiteboard, and loosely defined diagrams. And this was the case when the operational concept for the NCO of TST was developed for FBE-I.

The informal discussion began by taking a step back from the TST problem to understand the bigger picture, the doctrinal context for the mission capability sought. This was at a time when military thinking was shifting to what is now called "effects-based warfare." The existing doctrine already had three key operational effects that were sometimes labeled as D3: deny, delay, or destroy the threat. This became a touch point for rethinking what was needed to change from the existing doctrine to realize NCO in effects-based warfare.

The key elements of the operational concept for warfare are depicted in Figure 13.1. A military force operating under a command authority with an area of responsibility is assigned to engage a threat. When the concept of *engage* is shifted

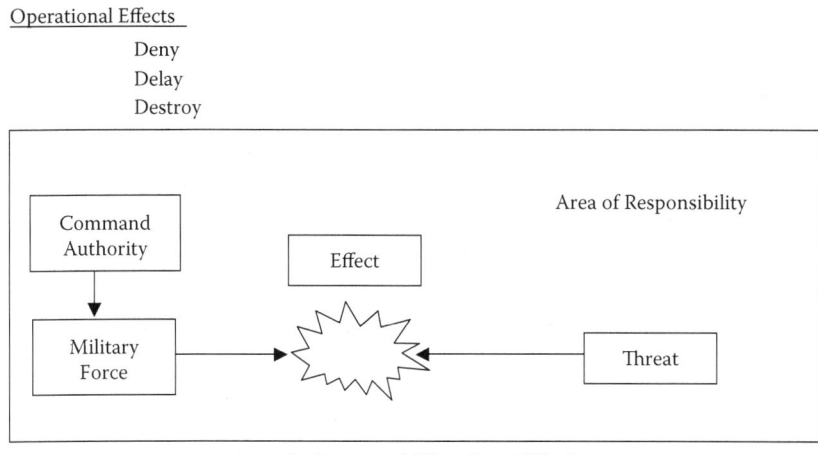

An early diagram of *Effects Based Warfare*

Figure 13.1   Key elements of the operational concept for warfare.

to D3 effects, then the diagram in Figure 13.1 becomes effects based. Despite the use of the "explosion" in the diagram, the shift in thinking had been made to effects-based warfare.

NCO is more about the capabilities achieved through the interoperation of systems than it is about networks. The networks simply enable the interoperation. Figure 13.2 illustrates the conceptual shift from a platform-centric system architecture to a network-centric system architecture. Platform-centric operations usually involve a sectored battlespace as a means of controlling weapon systems and engagements. Platforms carry sensors, processors, and weapons (or combinations), the effectiveness of which can be increased dramatically by SoS integration enabled

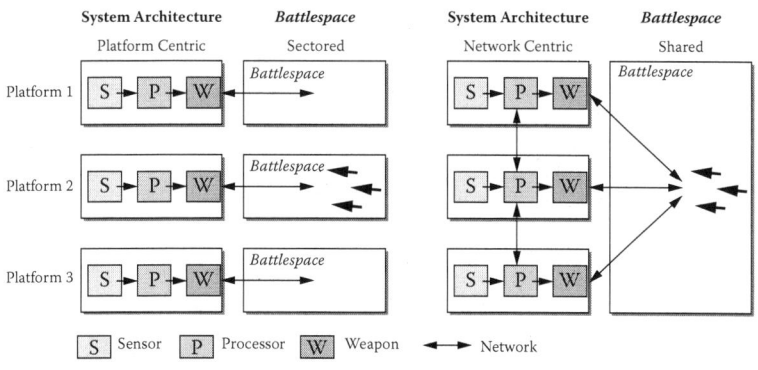

Figure 13.2   System of systems (SoS) integration through networks.

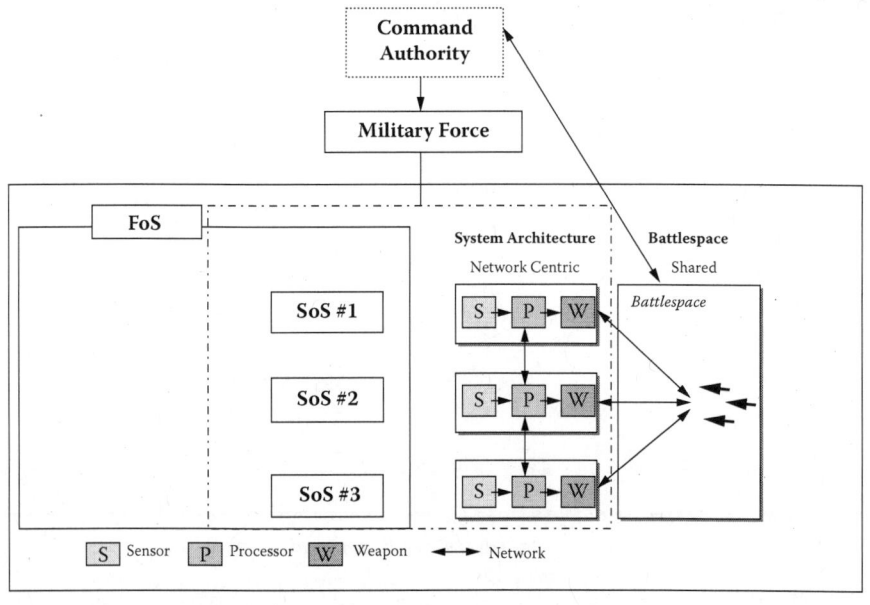

Figure 13.3 Network-centric aspect of the military force.

by a network-centric architecture. Figure 13.2 is an example of a key concept of NCO cited by Alberts, Gartska, and Stein (1999), namely, the effective linking of dispersed and distributed entities in the battlespace. The importance of real-time fusing of multiple sensor outputs as a driver for the target engagement architecture cannot be overemphasized. This change in architecture brought about by networking the sensors is also illustrated in the figure. A key result of this shift from platform-centric operations to NCO is the breaking down of sectors in the battlespace. There is now a shared awareness of the battlespace among the systems in the SoS as well as by the command authority, as illustrated in Figure 13.3. NCO are based on sharing a common tactical picture.

The implications for change in the nature of combat illustrated in Figure 13.3 are profound. On a single platform, it is relatively easy to close the loop to observe, orient, decide, and act (OODA). The challenge in NCO is to enable multiple OODA loops that span space and time as effectively and as rapidly for dispersed force elements as for a single platform, particularly when some sensors may be involved in multiple loops. Any sensor or processor with useful data or information should be able to provide it to any element of the military force that can use it.

The TST operational concept used in this chapter derives the Joint Targeting Doctrine illustrated in Figure 13.4. The doctrine is referred to as "illustrative" because operational concepts for targeting currently used by the Joint Staff have, of course, evolved. Because this chapter is intended to illustrate the architecture

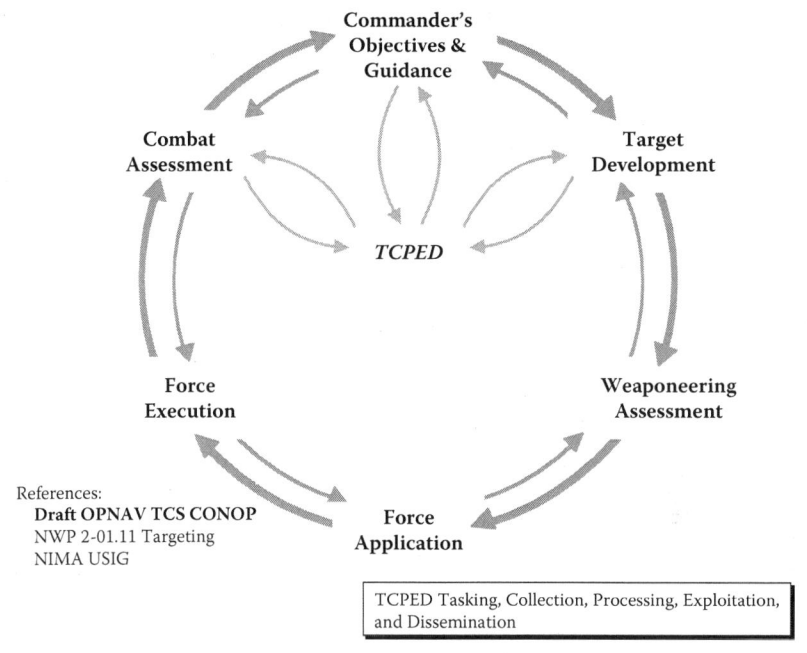

Figure 13.4   Illustrative Joint Targeting Doctrine.

methodology that was practiced rather than current doctrine, the illustrative doctrine will be sufficient for this purpose.

However, the diagram in Figure 13.4, as well as the next diagram, were actually used in the development of the FBE-I TST operational concept. The outer ring of the diagram in Figure 13.4 represents the six major activities of joint force mission planning and execution: commander's guidance and objectives, target development, weaponeering assessment, force application, force execution, and combat assessment. Each activity proceeds in a direct flow around the ring, but there are also arrows flowing backward. These are the feedback loops through the chain of command. "Reach-back" was one of the CNO's objectives for NCO. It was realized through a tightly coupled group of activities referred to as TCPED, which was the acronym for tasking, collection, processing, exploitation, and dissemination of sensor data not indigenous to the military force in theater. The pairs of arrows from TCPED to combat assessment, commander's guidance and objectives, and target development indicate where in the targeting cycle reach-back had been realized by the time of FBE-H.

A key enabler of TST through NCO in FBE-I was the capability of reach-back at the force planning level. This had to be integrated with indigenous tactical sensing capability for time-sensitive targets that were detected during force execution but had not been detected before the target development activity in the

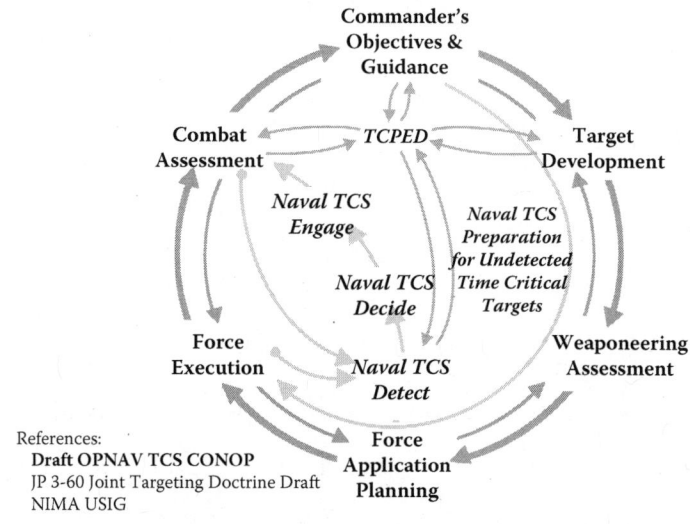

Figure 13.5 Joint Targeting Doctrine augmented for TST.

mission planning cycle. Thus, both the reach-back and indigenous observations could spawn responses to time-sensitive targets. However, these were not ad hoc responses. The commander's guidance actually planned for these unplanned events. The naval concept of Time Critical Strike (TCS), at the time of FBE-I, for the tactical responses to time-sensitive targets was based on a tactical sequence of "detect, decide, engage." All of these relationships are shown in Figure 13.5.

Operational Concept for FBE-I

The previous two sections have summarized a body of knowledge that had been assembled in the form of doctrine and loosely defined diagrams. This was the basis for the operational concept of the NCO TST experiment in FBE-I. It was at this point that it was appropriate, indeed necessary, to introduce a more formal and precise representation of the concepts developed. The development of the DoDAF-based architecture began with three products: OV-1, OV-4, and OV-5. Chapters 10 ("Introduction to DoDAF and MODAF Frameworks") and 11 ("DoDAF and MODAF Artifacts") provide the details of the frameworks and nomenclature used in the remainder of this chapter.

The High-Level Operational Concept Graphic (OV-1) for a mission should integrate OV-4 and OV-5 in context and highlight details of special interest. Scenario events can be related to operational activities. This allows OV-1 to be regarded as part of the data repository for the integrated architecture so that it can be used to preserve the conceptual integrity of the mission solution. Figure 13.6 is the actual

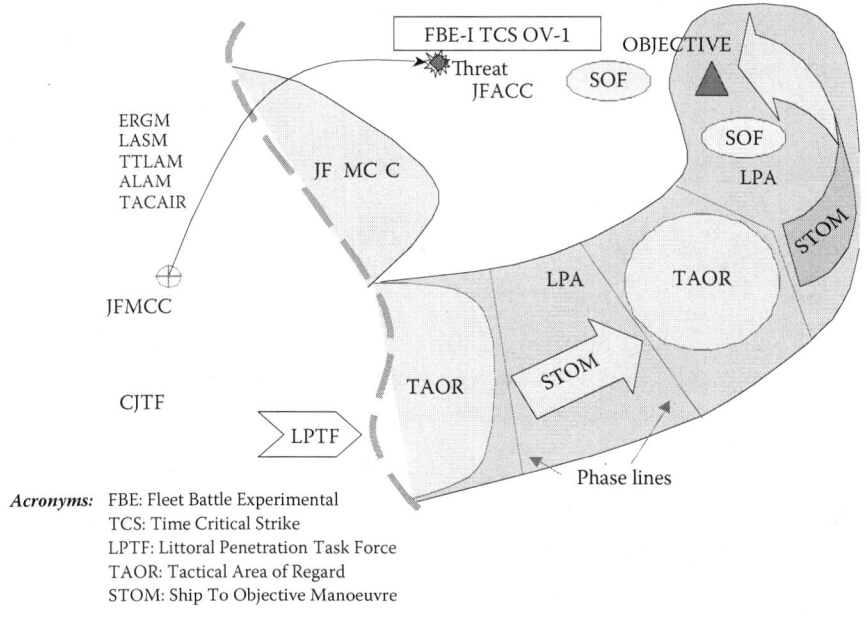

Acronyms: FBE: Fleet Battle Experimental
TCS: Time Critical Strike
LPTF: Littoral Penetration Task Force
TAOR: Tactical Area of Regard
STOM: Ship To Objective Manoeuvre

Figure 13.6   FBE-I operational concept.

FBE-I OV-1, including what was then referred to as TCS in the experiment. The term TCS was associated with naval operations and was later replaced with the term TST as the terms of joint doctrine evolved.

In the FBE-I OV-1, the theater is divided into two parts by a littoral interface depicted by a dashed line: (1) land (to the right of the line) and (2) sea (to the left). The Littoral Penetration Task Force (LPTF) penetrates from the maritime region of the theater into the land region using classical fire and maneuver tactics to reach the mission objective. An area of fire (i.e., LPTF having direct engagement with the threat) is referred to as a Tactical Area of Regard (TAOR). An area of maneuver (i.e., the LPTF either moves toward its mission objective or avoids direct engagement with the threat) is referred to as a Littoral Penetration Area (LPA). The fires and maneuvers to reach the mission objective in this way are referred to as a Ship to Objective Maneuver (STOM). The graphic in OV-1 depicts the alternating fires and maneuvers that comprise the STOM.

When the LPTF reaches the mission objective, it is supported by Special Operations Forces (SOF), which are a key element of the TCS that is enabled by NCO. The STOM and TCS are the two major capabilities depicted in OV-1. Each will have an architectural description that is initially an OV-5, that is, a functional decomposition of the operational activities that comprise the STOM and the TCS capabilities used in the operational concept for the mission. This is what is meant by saying that OV-5 should be integrated into the context of OV-1. The details of OV-5 for TCS capability are developed in the next section of this chapter. The integration of OV-4 into the context of OV-1 and the relationship between OV-4 and OV-5 are discussed next. It is important to note that only the names of the capabilities and not the details of the underlying activity models (the OV-5s) are included in OV-1.

In Figure 13.6, a command structure is called out, the Commander of the Joint Task Force (CJTF) who has authority over the Joint Force Maritime Component Commander (JFMCC) and the Joint Force Air Component Commander (JFACC). The details of this command structure are given in a separate OV-4, and are mentioned in the discussion of Figure 13.7. What is important in OV-1 is that the JFMCC has two areas of responsibility. One is to protect the flank of the LPTF as it penetrates from the maritime region of the theater into the land region. The other is to support the TCS capability in conjunction with the JFACC. The association of the JFMCC and JFACC with areas of responsibility in the theater is called out in the graphic of OV-1. In the graphic, the placement of JFMCC in the maritime side of the theater and the identification of special weapons of interest (Extended Range Guided Munition [ERGM], etc.) indicates that the JFMCC is providing fires from the sea.

Figure 13.7 illustrates OV-4 for the TCS capability in FBE-I, calling out the hierarchy of command authorities and the tactical-level activities under their control, that is, the detect, decide, engage activities that were introduced in the extension of Joint Targeting Doctrine to include naval TCS. OV-2 is provided to indicate that the organizational relationships will be instantiated in the operational concept by operational nodes (of the same name in this case). The different graphical symbols and arrangements indicate that OV-4 is the hierarchical structure of control, whereas the structure of OV-2 only indicates connectivity between operational nodes (i.e., no implication of the structure of control).

Further details of the operational concept for FBE-I can be found in Chapter 15 ("Toward Systems of Systems and Network-Enabled Capabilities"). The remainder

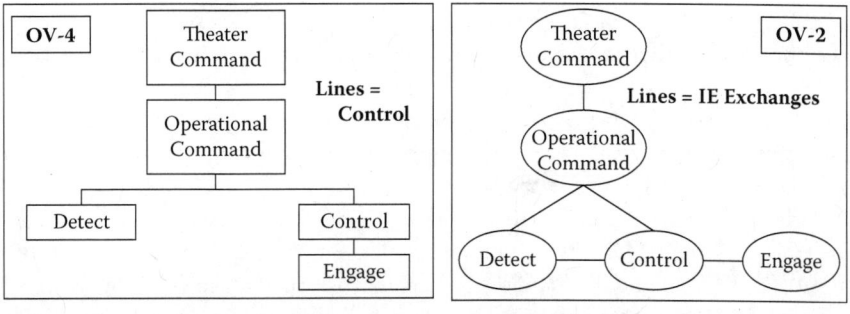

Figure 13.7   OV-4 organizational relationships chart and OV-2 operational node connectivity.

of this chapter will focus on the architecture descriptions necessary to assess the interoperability of the systems used for targeting the FBE-I TCS capability.

Using DoDAF Architecture to Model Targeting in FBE-I

Recall that the architectural artifacts presented in this model are simplified representations of actual work. So, the reader who is knowledgeable regarding the details of military doctrine and systems related to targeting or FBE-I itself is warned against getting lost in the details of the remainder of this chapter. The simple architecture presented in this chapter is built around six nodes and eleven connections. The focus will be on the interoperability assessment of a single SoS of the type used in the TCS experiment of FBE-I. The actual architecture had hundreds of nodes and thousands of connections and was maintained in a large database by one of the U.S. Navy System Commands. This and the next section of the chapter are based on Chapters 3 and 4 of Dickerson et al. (2003).

Figure 13.8 provides a high-level description of the hierarchy of activities that comprise OV-5 for naval TCS and OV-4 for FBE-I that were called out in OV-1. The three high-level TCS activities of detect, decide, and engage were later replaced by the joint targeting sequence of six high-level activities: find, fix, track, target, engage, and assess (F2T2EA). The joint targeting sequence has been mapped onto the TCS hierarchy for the sake of comparison. For example, "detect" in the naval TCS model corresponds to "find" in the joint targeting model. Because the F2T2EA model is at a greater level of detail than the three activities of TCS, some of the activities in the joint sequence are associated with the first level of TCS

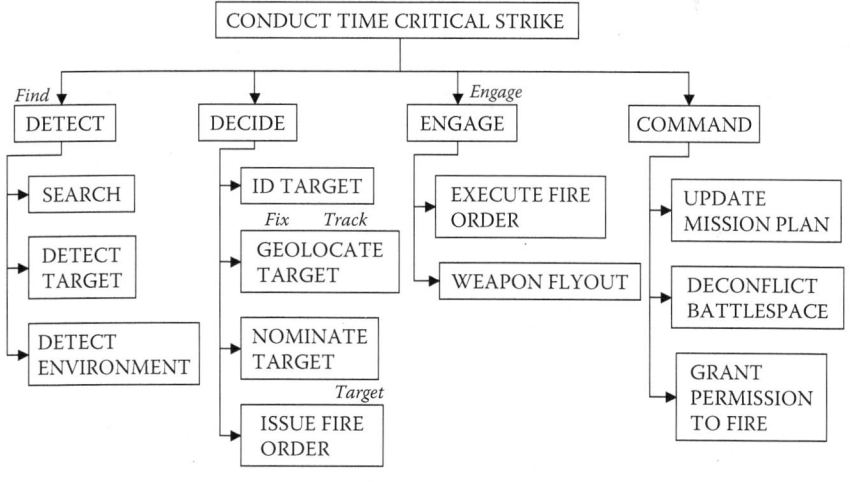

Figure 13.8 Functional decomposition (high-level OV-5 hierarchy).

decomposition . For example, the "fix" and "track" activities of F2T2EA are associated with the "geolocate target" activity in the first-level functional decomposition of "decide". Note that the TCS model ends with "engage" so that the assessment activity in the joint model is not mapped onto the TCS OV-5.

The systems architect who developed the actual TCS OV-5 for TCS in FBE-I refined the high-level operational activities using functional decomposition based on standardized operational activities and guidance from domain experts. For joint operations, the standard operational activities are provided by the Universal Joint Task List (UJTL). This provided authoritative configuration control over the operational activities used in the DoDAF OV products that were developed for the experiment. The UJTL is also used for training. A defense systems architect using the UJTL for the description of operational activities therefore provides the military customer a second benefit with no additional effort in architecture development.

The command and control activity (which is the domain of OV-4) has been labeled simply as "command." The vertical nature of functional decomposition in any activity or process model, such as an OV-5, can be misleading. Capabilities such as TCS are actually realized through the execution of threads of activities that weave their way through the vertical hierarchy created by functional decomposition. Sometimes, capabilities are spoken of as a horizontal integration of activities (of OV-5). And although "command" (as per OV-4) actually sits above the detect, decide, engage sequence of TCS in the sense of authority, when the three high-level TCS activities are executed in sequence under the authority of command, the twelve activities displayed in the first-level functional decomposition in Figure 13.8 will have horizontal threads that lace through the command activities.

The DoDAF architecture developed for the FBE-I TCS experiment was a C4ISR architecture that was tied to the TCS mission capability. Therefore, the transport of information elements was the focus of the NCO system flow. The first level of coherence in the activity model is the precedence order of the execution of the twelve activities. This precedence order is determined by examining the information being exchanged between the activities. At the highest level, the four major activities are producing and consuming information elements. The outputs of the "command" activity, for example, are a collection plan (for the sensors), a weapon target (pairing) plan, and the permission to fire. The "detect" activity produces sensor reports, which include target detections and features. The targets' environment and features are also detected. The targets must be sorted from the environment. The outputs of the "decide" activity are target nomination and the associated fire order, which includes the fire control solution against the target. The "engage" activity must produce the effect against the target (i.e., deny, delay, or destroy) and deliver a weapon launch report to the command authority.

When these information elements are compared with the information needs of the twelve activities in the first-level decomposition, a precedence order is automatically set. At this level of description, there is no choice. One activity that requires the output of another activity must necessarily follow that activity in the ordering of the execution of the activities. This is called the *precedence order* of the activities. In fact, this natural ordering of the activities is a test of the completeness of both the activity hierarchy and the list of information elements in the system flow. It is also a test of the necessity and sufficiency of the list of information elements. If an element is not an input to one of the activities in the decomposition, then either it is not necessary or a necessary activity may have been overlooked. Similarly, if there are any activities in the decomposition with information needs that are not met by the list of information elements, then the list is not sufficient to support the activity model.

Figure 13.9 is the result of applying this method of precedence to the twelve activities in the first-order decomposition in Figure 13.8. The result is a form of OV-6c from DoDAF and is called a *key thread*, that is, the precedence relationships between the leaf-level operational activities in OV-5. The activity flow that was used to model the naval TCS capability in FBE-I is triggered by reception of a task order and guidance, and ends in the delivery of an effect (into the battlespace). Note that the "update mission plan" occurs twice in OV-6c. The second occurrence is in response to the detection of the (previously undetected) time-sensitive target. The first occurrence was in response to the initial task order and guidance.

From the viewpoint of DoDAF, operational capability becomes the ability to execute the appropriate operational activity sequence end to end in the context of a mission environment. This is significant because it provides an architectural basis for modeling and analyzing operational capabilities. Because operational capability

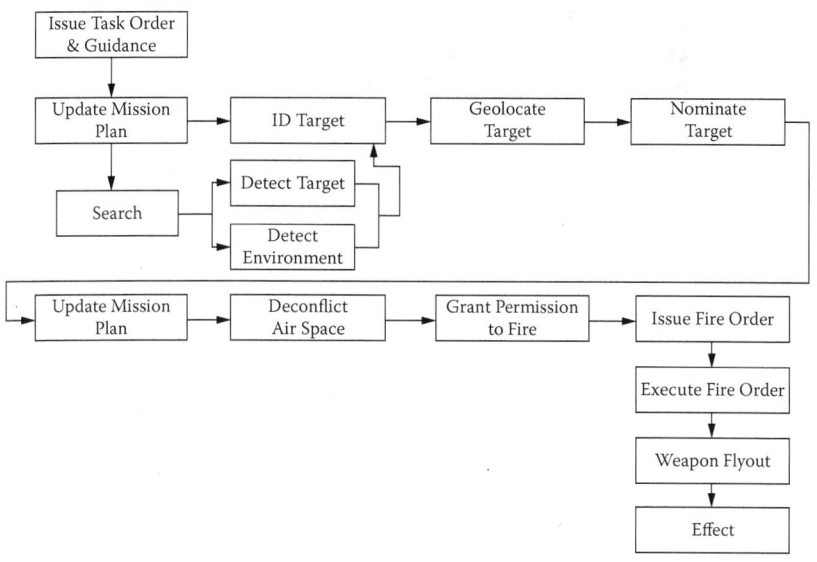

Figure 13.9   OV-6c event trace diagram (an architectural model of capabilities).

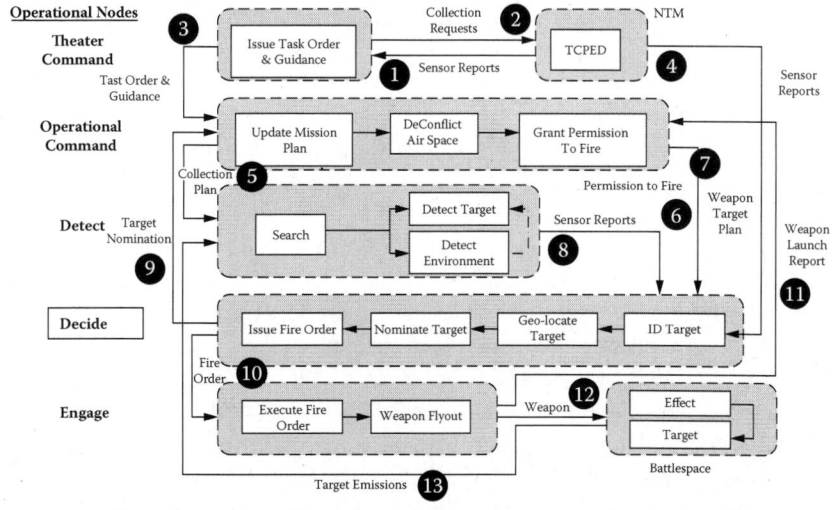

Figure 13.10 Relationships of architecture, interoperability, and capability.

also depends on the interoperability of the SoS enabling the execution of the end-to-end activity sequence, this viewpoint also gives insight into the relationship between interoperability and operational capability. Figure 13.10 integrates these concepts into a single graphic.

Using the architectural model of capability provided by the DoDAF OV-6c, integration of OV-6c with OV-2 highlights the relationship of interoperability to operational capability. The operational nodes in OV-2 are groupings of activities (as illustrated in the figure) that will be instantiated and given communication capabilities with other nodes. The dependencies of the operational capability on interoperability become traceable through the operational nodes in OV-2, which have aggregated the leaf-level operational activities of OV-5. By comparison, the system nodes in the SV-1 aggregate system functionality.

Interoperability exists at the nodal level and enables operational capabilities. In Figure 13.10, each numbered "dot" in the diagram is a point of such enablement. Recall that earlier in this chapter, it was asserted that the simple architecture for naval TCS in FBE-I would be represented using six nodes and eleven connections. The six nodes of the TCS capability are each one of the nodes in the figure, including a "reach-back" node. Note that the battlespace is treated as a seventh node with which the TCS capability interacts. The eleven connections are each one of the dots in the diagram, the twelfth and thirteenth dots being included to model the interaction of the TCS capability with the battlespace. This representation of OV-6c supports concordance between simulation of capabilities and the assessment of interoperability.

This representation of OV-6c also supports automated generation of the Operational Information Exchange Matrix (OV-3) in the DoDAF. Every line in the OV-3 for the C4ISR architecture represented in Figure 13.10 can be read directly from the diagram using the numbered dots as reference. Information elements being exchanged are called out in the diagram along with their association as inputs and outputs of operational nodes and the operational activities involved in capability. Because OV-6c in the nodal form of Figure 13.10 directly induces OV-3, information exchanges are directly linked to capabilities.

Because most definitions of capability in some way refer to the ways and means of delivering an effect, it is appropriate to end this section of the chapter with the DoDAF SV-1, the System Interface Description; because it links the operational nodes with the system views of the architecture. The systems belong to the means by which capability is enabled. When SV-1 is properly mapped onto the nodal form of the diagram in Figure 13.10, the result is a mission thread.

Figure 13.11 is an actual SV-1 that was used in the FBE-I TCS architecture. The four columns represent the three major activities of naval TCS used in the experiment as well as the assessment activity of the F2T2EA activity model. Each column is a swim lane for the systems that were identified by icons. The first two swim lanes are for systems that participated in the detect/assess activities. The first swim lane is associated with the planned engagement, and the second is for the time-sensitive

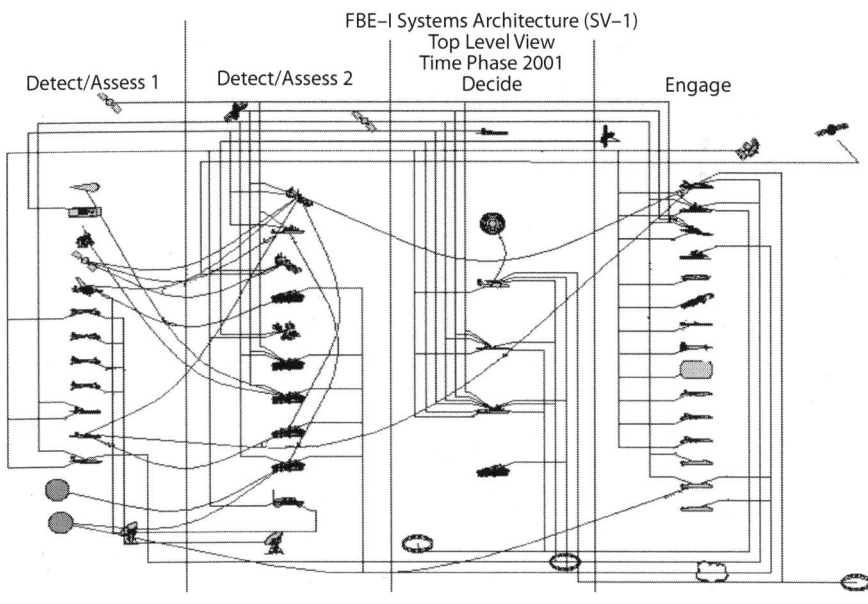

Figure 13.11   Top-level FBE-I TCS SV-1.

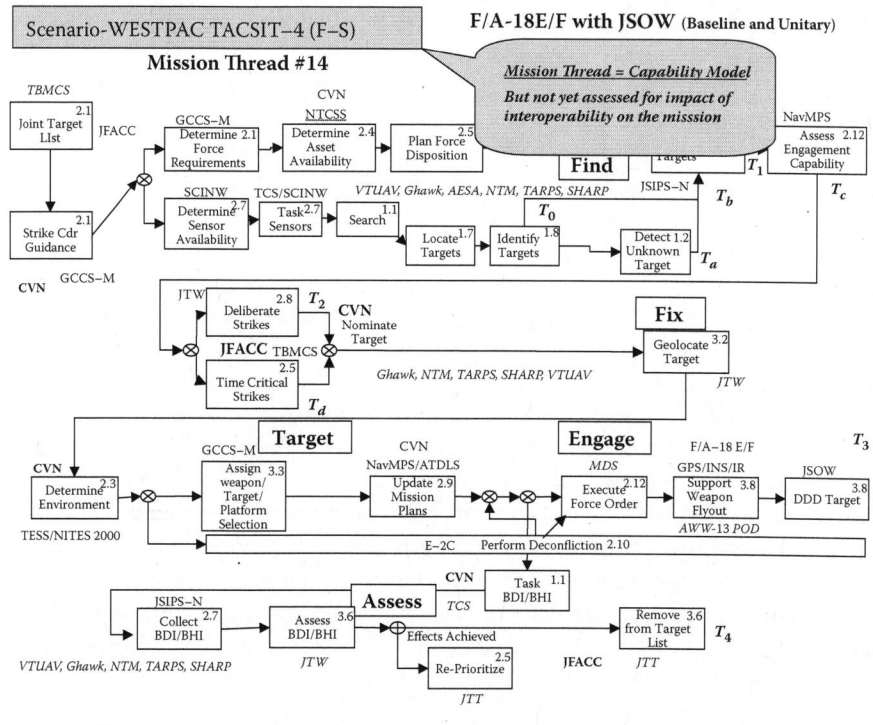

Figure 13.12 Scenario-WESTPAC TACSIT-4 (F-S).

target. The third and fourth columns are the swim lanes for systems that participated in the decide and engage activities. The lines are the connectivity between the systems.

With the guidance of an operational subject matter expert, the systems can be mapped onto the operational activities of an OV-6c by considering the inputs and outputs of each activity and each system. A system is a candidate to be mapped onto an operational activity if it can provide an output required of the activity. As with the precedence relation of the activity model, the mapping of systems onto activities will cause precedence relations between the systems. It is not sufficient to just map systems appropriately to operational activities. The implied interfaces between the systems must also be considered. SV-1 should provide the system interfaces that are supported.

Figure 13.12 displays an example of a mapping of systems and other assets onto a key thread and the resulting mission thread that is similar to the mission threads of FBE-I TCS. In this mission thread, the systems were sorted according to the following types: networks, sensors, processors, weapons, and platforms. The mission thread reflects this sorting and is a first step toward understanding how to realize new operational capabilities, such as those that were sought in FBE-I.

Using DoDAF Architecture for Interoperability Assessments

Several other DoDAF products in addition to SV-1 and those involved with developing the mission threads are needed before an interoperability assessment can be performed on the SoS. These include SV-3 (Systems^2 Matrix), which can be used in the alignment of systems to system functions; operational activities; other systems; SV-4 (Systems Functionality Description), which provides a list of systems functions that will be used to enable or execute operational activities; and SV-5 (Operational Activity to System Function Traceability Matrix), which aligns individual system functions with the individual operational activities that they enable. These DoDAF products will not be presented because that would involve getting into the engineering details of the systems, which is beyond the scope of this chapter.

SV-4 and SV-5 also present another challenge. Unlike the operational views in DoDAF, which can be built from the UJTL activities and the more detailed Services Mission Essential Task Lists, there is no accepted standard for things such as system functions. For this reason, these DoDAF products can be disruptive to the design of operational capabilities. They are needed to perform the engineering analysis needed for interoperability assessment, but they cannot be applied across the broad scope of joint or coalition forces. However, within a proper bounding of the engineering domain, the SV products can, and should, be applied to the requisite engineering problem.

However, there are large bodies of technical standards that can be applied to the analysis of engineering problems, and these can be captured in the DoDAF Technical Standards Profile (TV-1). For C4ISR architecture, technical standards typically fall into one of two categories: (1) logical standards associated with the information elements being exchanged (data), or (2) logical standards associated with the system function (application) processing the data.

Standards associated with the data itself can include those from military, government, or commercial bodies. Examples are as follows:

MIL-STD-6016 for J-series bit-oriented messages
MIL-STD-6011for M-series bit-oriented messages
Joint Publication 604 for U.S. Message Text Format (USMTF)
Imagery formats such as GIF or JPEG
Text formats such as Hypertext Markup Language (HTML)

Examples of applications standards for data processing and transport include the following:

Defense Information Infrastructure Common Operating Environment (DII COE) correlation services
Simple Mail Transport Protocol (SMTP, i.e., e-mail)
Simple Network Management Protocol

When the information from the DoDAF products discussed in this chapter are part of an integrated architecture, the key information needed by the systems architect and engineers to perform interoperability can be summarized in the DoDAF System Data Exchange Matrix (SV-6), which can be used to create end-to-end views of system information and service exchanges. An example of an SV-6 that can be applied to the capability model in Figure 13.10 that was used to represent the FBE-I architecture is displayed in Figure 13.13.

The detailed SV-6 is the basis for the static interoperability assessment. When the developing C4ISR architecture, the architect must work with engineering to assess the certification of each operational information element (IE): source node activity system, system function, and protocol and the destination node transport standard, activity, system, system function, and protocol.

In the U.S. Navy strike architecture used for planning the 2004 fiscal year, SV-6 had 2857 lines (connections). Because the data was stored electronically in a database and organized using an integrated DoDAF architecture, it was possible to use SV-6 to assess mission threads such as the one displayed in Figure 13.12. Although the SoS assembled in these mission threads were fully operational, the architecture and engineering analysis was able to pinpoint problem areas so that solutions could be developed. Figure 13.14 is a marking of Figure 13.12 that pinpoints where the Link 16 interoperability problems were. Each numbered starburst calls out where a problem exists. The results of this assessment will be revisited in Chapter 14 ("Capabilities Assessment") as part of an analysis of alternatives study at the SoS level.

SV-6 for the legacy architecture represented in Figure 13.13 had thousands of connections because it was heavily based on point-to-point communications. Recall that N elements in a point-to-point architecture can have a number of connections on the order of N^2. This is one reason why the defense communities have taken a strong interest in networks. If the same system of N elements had the interoperability to communicate seamlessly through a single network, for example, then only N connections would be needed. Figure 13.15 is provided to show that DoDAF architecture products are being used as part of the move toward network enablement of capabilities.

Summary

The vision of FBE-I was to operationalize NCW by building and maintaining a C4ISR architecture that provided joint forces with wide-area connectivity, enhanced bandwidth, and reach-back capability. The success of FBE-I was a significant step forward in the development of NCO capabilities.

FBE-I was also a pilot experiment for the then recently established chief engineer of the U.S. Navy to assess if architecture-based methods of mission capabilities engineering were actually useful. The detailed architecture model for the TST

OV-6c Int. ID	Source Node					
	Source Node	IE	Activity	Source System	Source System Function	Source System Protocol
1	NTM	Sensor Reports	TCPED	NTM	TCPED	NITF, USMTF
4	NTM	Sensor Reports	TCPED	NTM	TCPED	NITF, USMTF
2	Theater Command	Collection Reqs.	Issue Task Ord./Guid.	JDISS	Task Sensors	USMTF
3	Theater Command	Task Order/Guid.	Issue Task Ord./Guid.	JTF TBMCS, GCCS	Update Mission Plan	USMTF
7	Op. Command	Wpn. Tgt. Plan	Update Mission Plan	JFACC GCCS	Weapon Target Assoc.	USMTF
5	Op. Command	Mission Plan	Deconflict Airspace	JFACC TBMCS	Air Pic. Int./Dyn. Deconflict.	USMTF
5	Op. Command	Collection Plan	Update Mission Plan	JFACC GCCS	Collection Planning	USMTF
6	Op. Command	Permission to Fire	Grant Perm. to Fire	JFLCC GCCS	Decision, Tgt. Priority	USMTF
8	Detect (Guardrail)	Sensor Reports	Detect Tgt./Environ.	Guardrail CTT	Single Sensor Sense	USMTF
8	Detect (UAV)	Sensor Reports	Detect Tgt./Environ.	UAV SYERS/ASAS/IRIS	Single Sensor Sense	Binary
5	Control	Collection Plan	Control	GSM UHF Radio	Collection Execution	Voice
5	Control	Collection Plan	Control	CARS/DCGS	Collection Execution	Binary
9	Control	Target Nomination	Nominate Target	TOC GCCS	Decision, Tgt. Priority	USMTF
10	Control	Fire Order	Issue Fire Order	TOC AFATDS	Issue Fire Order	USMTF
11	Engage	Wpn. Launch Rpt.	Weapon Fly-Out	MLRS FDC AFATDS	Wpn. Init. and Launch	USMTF

OV-6c Int. ID	Destination Node					
	Inter-nodal Transport	Destination Node	Activity	Destination System	Dest. System Function	Dest. System Protocol
1	IBS	Theater Com.	Issue Task Order/Guid.	JDISS	Decision Planning	NITF, USMTF
4	IBS	Control	ID Target	TOC CTT, ASAS, TROJAN	ID Target	NITF, USMTF
2	JWICS	NTM	TCPED	NTM	Multi-Sensor Sense	USMTF
3	SIPRNet, JWICS	Op. Command	Update Mission Plan	JFACC TBMCS, GCCS	Update Mission Plan	USMTF
7	SIPRNet	Control	Issue Fire Order	TOC AFATDS	Issue Fire Order	USMTF
5	SIPRNet	Control	Control	TOC GCCS	Mission Execution	USMTF
5	SIPRNet, JWICS	Control	Control	TOC GCCS	Mission Execution	USMTF
6	SIPRNet, JWICS	Control	Issue Fire Order	TOC AFATDS	Issue Fire Order	USMTF
8	CDL	Control	ID Tgt./Geolocate Tgt.	GSM ASAS	Multi-Sens. Sense, Feat. Ext.	USMTF
8	CDL	Control	ID Tgt./Geolocate Tgt.	GSM ASAS, TROJAN	Multi-Sens. Sense, Feat. Ext.	Binary
5	UHF LOS Radio	Detect (Guardrail)	Detect Tgt./Environ.	Guardrail, UHF Radio	Single Sensor Sense	Voice
5	CDL	Detect (UAV)	Detect Tgt./Environ.	UAV SYERS/ASARS/IRIS	Single Sensor Sense	Binary
9	SIPRNet	Op. Command	Grant Perm. to Fire	JFACC GCCS	Dec., Tgt. Prioritization	USMTF
10	MSE	Engage	Weapon Fly-Out	MLRS-FDC, AFATDS, ATACMS	Wpn. Init. and Launch	USMTF
11	MSE	Op. Command	Update Mission Plan	JFACC AGCCS	Collection Planning	USMTF

Figure 13.13   Example of a detailed SV-6.

capability in FBE-I affirmed that architecture could be used in this way. With proper organization, DoDAF can be used to support modeling of operational capabilities and the requisite interoperability assessments.

Capabilities are derived from the individual operation and collective interoperation of systems. Interoperability and SoS performance assessments must

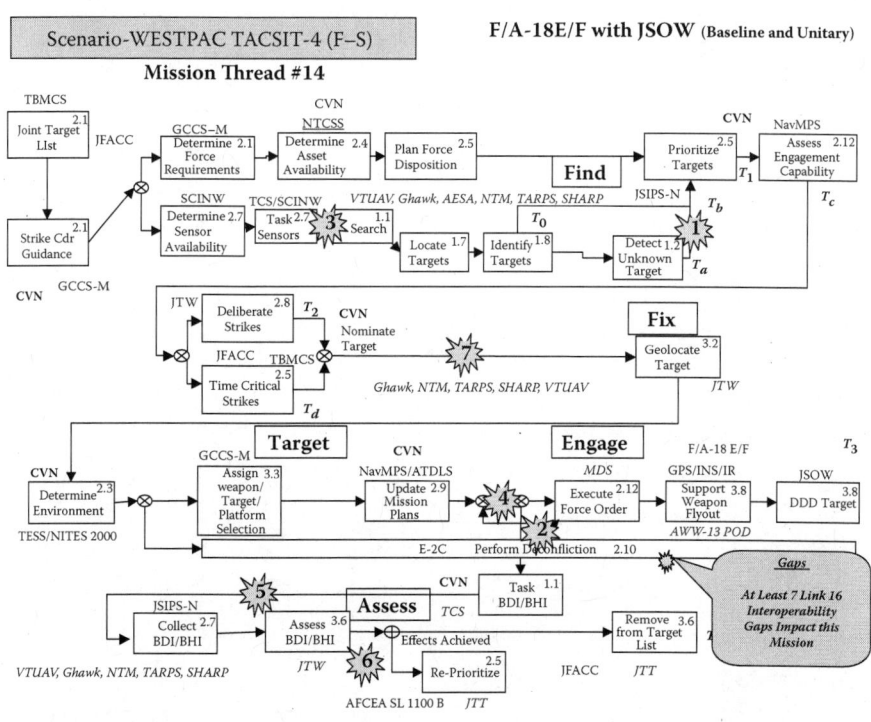

Figure 13.14   Scenario-WESTPAC TACSIT-4 (F-S) interoperability assessment.

DoD Integrated Architecture Check List

DoDAF Product	DoDAF Product Name
AV-1	Overview and Summary Information
OV-2	Operational Node Connectivity Description
OV-4	Organisational Relationships Chart
OV-5	Operational Activity Model
OV-6c	Operational Event-Trace Description
SV-4	Systems Functionality Description
SV-5	Operational Activity to Systems Function Traceability Matrix
SV-6	Systems Data Exchange Matrix
TV-1	Technical Standards Profile

OSD Net-Centric Check List

Core Enterprise Service	Related Attributes in the Net-Centric Check List
Discovery	visible, available, accessible, network connectivity
Collaboration	available, heterogeneous, trustable, interoperable
Mediation	understandable, trustable, cross-domain exchange
IA/Security	authentication, privileges, encryption, wireless
ESM	data management, open architecture, SOA
Applications	open architecture, SOA, scalable, available
User Assistance	responsive to user needs, available, accessible
Storage	data management
Messaging	12 transport tenets, decentralized ops & mgmt

Figure 13.15   Net-centric readiness checklists.

be in concordance with one another to properly assess operational capabilities. Architecture can be used to model and assess SoS capabilities and facilitate the concordance between SoS performance and interoperability. These types of assessments can also support capabilities-based acquisition.

References

Alberts, D.S., Gartska, J.J., and Stein, F.P. 1999. *Network Centric Warfare: Developing and Leveraging Information Superiority.* Washington, D.C. C4ISR Cooperative Research Program.

Dickerson, C.E. et al. 2003. *Using Architectures for Research, Development, and Acquisition.* Office of the Chief Engineer of the Navy, Assistant Secretary of the Navy. Defense Technical Information Center: (www.dtic.mil) AD Number ADA427961.

Chapter 14

Capabilities Assessment

Concept definition and analysis is a key part of systems design. Architecture-based methods can give new insights into concept refinement from a capabilities viewpoint. Interoperability is key enabler of operational capability. Commonly practiced methods of system trade-offs used in systems analysis and design can be elevated to the system of systems (SoS) level.

Key Concepts

Concept refinement
Architecture-based
  Analysis and design
  Interoperability assessments
  Trade-off studies
System of systems

Motivation

In recent years, the shift in the U.S. DoD and U.K. MOD from requirements-based system acquisition to capabilities-based acquisition has brought new challenges to the defense acquisition community. An excellent example of the challenges can be seen in the vision of the Chief of the U.S. Air Force in recent years to develop a new acquisition strategy for the Air and Space Operations Center (AOC). This new strategy was intended to lead to the development of the AOC not just as a command and control system, but also as a weapon system. Command and control systems, such as the AOC, had never been regarded as weapon systems prior to this vision. Historically, the AOC has been a family of systems (FoS) assembled ad hoc to meet specific mission objectives. The AOC equipment was comprised of various workstations that employed a variety of applications that processed inputs from diverse communications and sensor feeds. The vision to acquire the AOC and give it behaviors more resembling an actual weapon system, for example, an F-15 fight aircraft, was a significant undertaking by the U.S. Air Force. Some of the components of the AOC were under the direct control of the Air Force, whereas many others were not. The physical facility of the AOC, which was deployed in the theater of operations, was just the tip of the iceberg. Most of the systems that enabled the command and control capabilities of the AOC were not even located at the physical facility. They were like the mass of the iceberg that lies below the waterline and remains unseen.

Acquisition policy based on capabilities rather than systems must meet the challenge of effectively managing all of those systems that "lie below the waterline." How can systems architecture be used to address FoS problems such as this? This chapter builds on the results of early successes on the FoS problem by the U.S. Navy that were presented in Chapter 13 ("Modeling FBE-I with DoDAF"). It will use extensions of the early results to show how architecture can be used to address the types of problems encountered in the design and development of an FoS like the AOC.

Key concepts and terminology related to FoS and systems architecture were reviewed in Chapter 3 ("Concepts, Standards, and Terminology"), and as in Chapter 13, an (operational) capability will be modeled using sequence diagrams that describe the ability of an FoS to execute a specified course of action (sequence of operational activities).

An important distinction between the focus of traditional systems engineering and family of systems engineering (FoSE) can be understood by considering extensions of a conventional definition of systems engineering to FoSE. For example, if the well-known definition of systems engineering put forth by Blanchard and Fabrycky (1990) is extended from the system level to the FoS level, it would be considered that FoSE relates to the design and development activities in the life cycles of an FoS assembled to achieve specified operational capabilities through the individual operation and collective interoperation of the systems in the FoS. This has striking similarities to the definition of system put forth by Hitchins and used in the logical model of system in Chapter 3 ("Concepts, Standards, and Terminology").

Concept Refinement and Capabilities Analysis

The defense community of any nation faces a significant challenge when trying to acquire families of systems that can be readily assembled to deliver military capabilities. The acquisition structures of every defense community have been highly tuned to deliver individual systems. However, the concept refinement and capabilities analysis case study presented in this section can help to bridge the gap between the acquisition of capabilities and the acquisition of systems. In Chapter 15 ("Toward System of Systems and Network-Enabled Capabilities"), we will see that there is an intimate relationship between SoS or FoS operational capabilities and network enablement.

Process for Concept Definition and Refinement

The concept definition and refinement process illustrated in Figure 14.1 takes advantage of this relationship. There are four components to the realization of network-enabled operational capabilities: the mission scenario, proposed operational capabilities, operational concept, and the system and network architecture used to realize the operational concept. The mission scenario is an environmental model (in the sense of structured analysis) that provides a short description of the FoS purpose (to include mission objectives), a context diagram, and an event list. The proposed capabilities express *what* the FoS *must do* in the mission scenario. As discussed in Chapter 13 ("Modeling FBE-I with DoDAF"), the operational concept integrates the capabilities (by name only) with the command and control structure (OV-4 in the DoDAF) in the context of the mission scenario. The operational concept of the mission can also highlight details of special interest, and scenario events can be related to operational activities.

The capabilities bundled together in the operational concept are physically instantiated through the system and network architecture of the proposed FoS. It is important to understand that the FoS without an architecture is just a collection of systems. The command and control structure (OV-4 in the DoDAF) and the information exchanges between and the roles of the operational nodes (OV-2 in the DoDAF) each are part of the definition of the architecture that gives the FoS a structure that enables operational capabilities.

In Figure 14.1, there is a loop between three of the components of the diagram: the operational capabilities, the operational concept, and the system and network architecture of the FoS. This is the *concept refinement loop*. A given operational concept, which bundles specific capabilities and is physically instantiated with a specific system and network architecture, may, on analysis or further consideration, be determined *not* to be the final solution. In this case, any one or all three of the components in the concept refinement loop are subject to change (i.e., refinement). This loop can be iterated, provided there is time, until an agreeable solution is reached. Architecture provides a basis for concept refinement and capabilities analysis.

Sensor Fusion Node Case Study

This approach to concept definition and refinement will now be illustrated for a fictitious mission scenario. The scenario is set in the west-central Sudan, which is west of the Red Sea. The mission is to hold a fleeting and deceptive target at risk. In this scenario, a SCUD missile becomes the fleeting and deceptive target. It has been determined to be a weapon of mass destruction (WMD). The SCUD is one of 13 threat vehicles assembled in the operating area.

An Expeditionary Strike Group (ESG) and Joint Special Operations Task Force (JSOTF) are embarked to the Red Sea. These theater assets are tasked to locate the

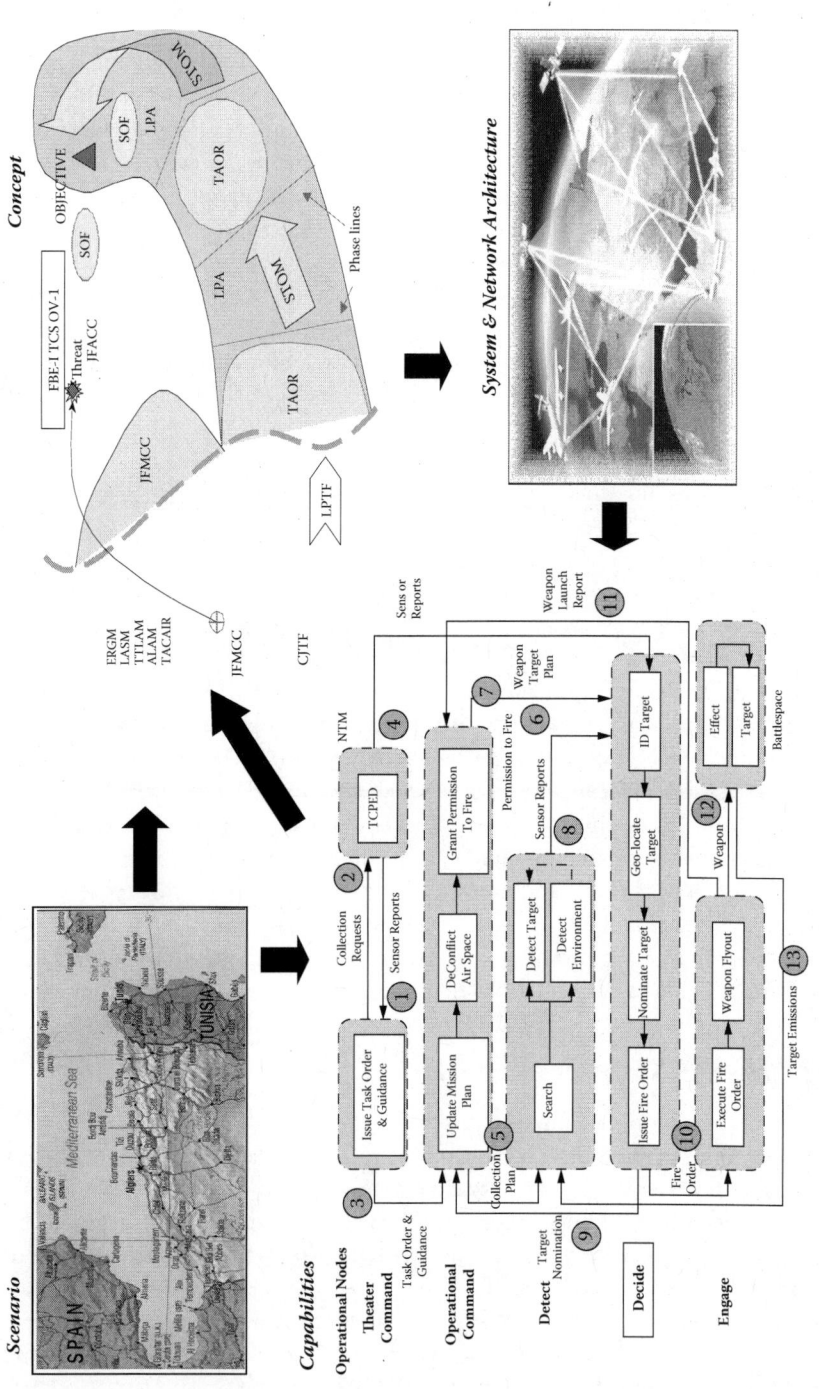

Figure 14.1   Realizing network-enabled operational capabilities.

target and hold it at risk. In theater command and control over the ESG, strike assets are provided by the Supporting Arms Component Commander (SACC), which is colocated with the Joint Intelligence Center (JIC). Surveillance assets are tasked to support the objective with wide-area observation. The sensor capability for the mission includes Global Hawk and Predator, which are unmanned airborne vehicles (UAVs) that carry airborne sensors, and the Modular Integrated Unattended Ground Sensor (MIUGS). The latter is a program developed under the U.S. Defense Advanced Research Projects Agency (DARPA) that transitioned in the year 2003. MIUGS is an emplacement of a field of acoustic and seismic sensors, each with detection range less than one kilometer. The Global Hawk was used to provide moving target indication (MTI) from its on-board radar, and the Predator was used to provide electro-optic (EO) video imagery. There is a cueing hierarchy between these sensors: MIUGS detections are used to cue the Global Hawk, and the Global Hawk MTI is used to cue the Predator. None of these sensors is capable of exclusively supporting the mission. In the baseline sensor architecture, each of these three sensors has individual "feeds" into the JIC where cueing and tasking decisions are made.

Capability Model

The operational capability model used for capabilities analysis was based on the commonly used activity model for targeting: Find, Fix, Track, Target, Engage, and Assess (F2T2EA). Because the operational objective is only to hold the WMD target at risk (i.e., engage it), only the first half of the model was needed in the capability analysis. Track time was selected as a key performance parameter (KPP): if the WMD target can be identified and tracked continuously, then it will be considered to be at risk.

The KPP metric for continuous track was the maximum length of time the WMD target was tracked (identified and located) without interruption. The criterion for continuous track time was set at a minimum of 30 minutes. Classification confidence was set at >0.9. A secondary KPP was also considered. This was track time with interruptions as measured with the metric total track time. No criteria were selected for this KPP because the amount of interruption in track time that can be tolerated depends on weapons timelines, scenario events, and other factors, which were beyond the scope of this case study. However, this secondary KPP would be the subject of concept refinement. The mission events for the fleeting and deceptive target were mapped against the F2T capability model (Figure 14.2) as follows:

Vehicles detected by MIUGS [Cue]
Vehicles acquired by MTI sensor [Find]
EO sensor retasked to establish ID [Fix]
Target location updated/ID maintained [Track]

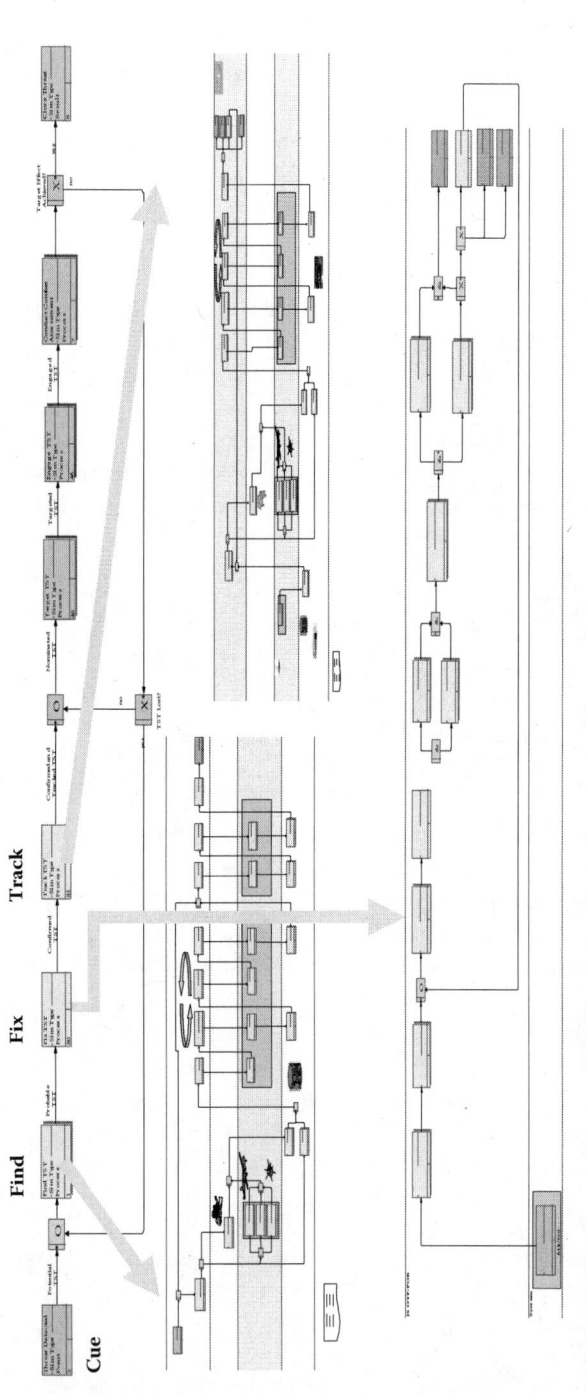

Figure 14.2 Capability model F2T.

The capability model assumes a cueing activity that normally triggers the F2T2EA targeting sequence.

So, at this point in the example, three of the four components of the concept definition and refinement process illustrated in Figure 14.1 have been assembled (scenario, capability model, and concept). The fourth building block (system and network architecture) will be the focus of the architecture-based analysis and design for the sensor FoS.

Design Change to the Sensor Architecture

A change to the baseline sensor architecture will be made by the creation of a new node for sensor fusion. The purpose of this new node is to improve track continuity (as measured by the primary KPP). Figure 14.3 illustrates an external view of the sensor fusion node and its relation to the communications architecture. The fusion node is physically distributed over two ground stations and the SACC/JIC. At each of the ground stations the new node receives high bandwidth data over the common data link (CDL), processes the data, and generates track data, which is transmitted over low-bandwidth theater Satellite Communication (SATCOM) to the component of the fusion node at the SACC/JIC where the final track fusion is performed.

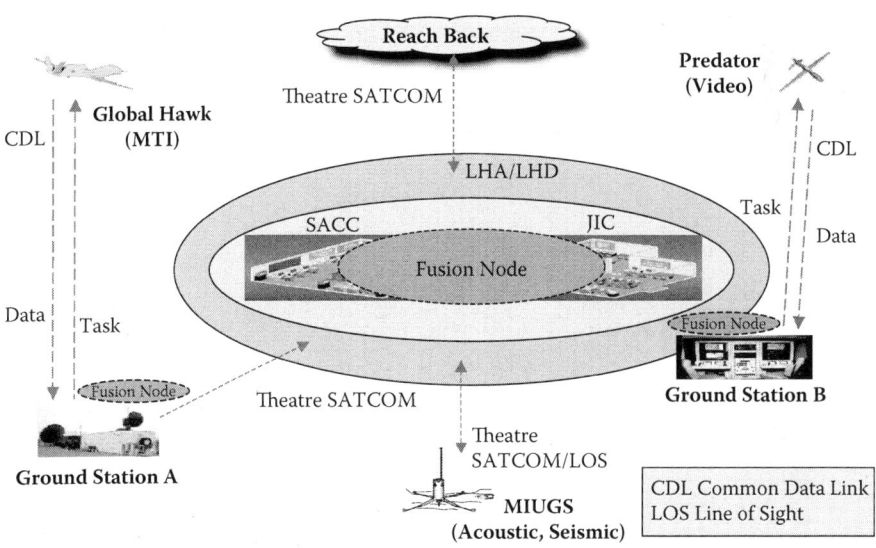

Figure 14.3　Sensor fusion node description (external view of the node).

Architecture Analysis

The architecture analysis must evaluate the loading on the communications architecture that is attributable to the fusion node, and it must assess the interoperability of the sensor architecture elements. The connections most challenged will be the SATCOM links between the ground stations and the JIC. Without reduction in data flow provided by the components of the fusion node at the ground stations, the SATCOM links would never be able to support the communications loading from the UAVs.

The loading on the communications architecture and interoperability assessment was simulated using a high-fidelity simulation called OPNET. Figure 14.4 displays a graph of an OPNET output for the SATCOM link between the ground stations and the JIC. [UHF SATCOM was necessary for the communications architecture but had the least throughput capability in the architecture (providing 16–20 channels of only 2.4 kbps throughput each).] The figure shows the increased loading on the SATCOM attributable to the track data transmitted to the JIC from the fusion node at the ground stations. The increased loading from the MIUGS– Global Hawk–Predator sensor suite had peaks at 736 bps. There were no criteria for acceptable increases in loading, but the peak increases are <20 to 30 percent of one SATCOM channel. The OPNET simulations also provide a preliminary validation of interoperability.

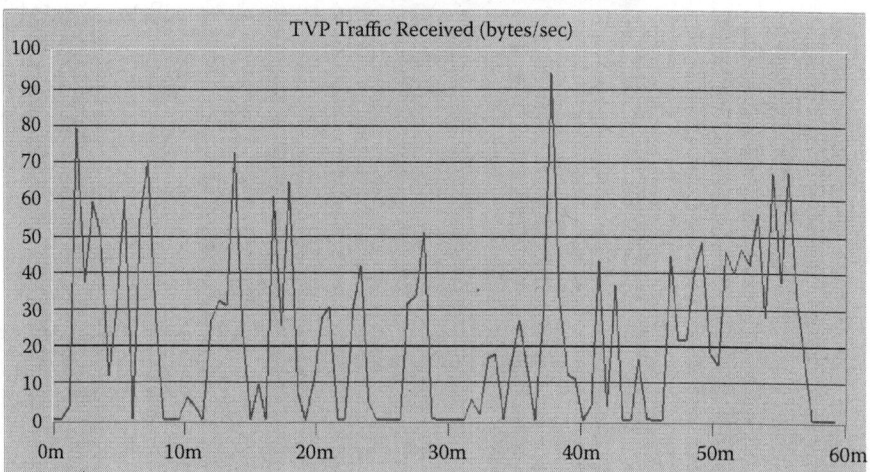

OPNET modeling provided communications loading estimates and preliminary validation of interoperability.

Figure 14.4   Communications architecture supports fusion node.

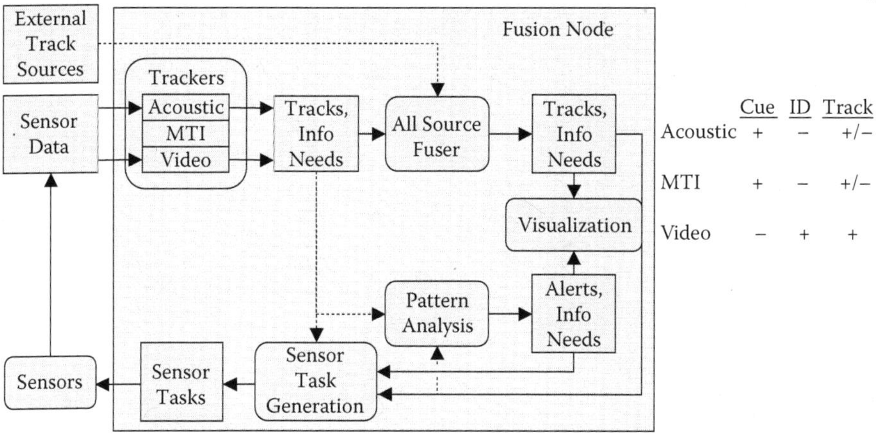

	Cue	ID	Track
Acoustic	+	−	+/−
MTI	+	−	+/−
Video	−	+	+

BAE-AIT Fusion Performance Model Was Used
To Predict Fusion Node Performance.

Figure 14.5   Sensor fusion node (internal view of the node).

The sensor fusion node collects multiple sensor feeds to create a more accurate picture of theater activity and provide greater track continuity. It also works in concert with the sensor manager to task and retask sensor assets. The architecture for the fusion node in Figure 14.5 allows for both a fast processing loop, as indicated by the dashed lines, and a slower loop, as indicated by the solid lines. The selection of processing loop was part of the trade-offs because the fast loop, although providing a higher quality of tracking service, placed greater loading on the communications architecture. The slower loop was used in the analysis that follows. The Advanced Information Technology unit of BAE Systems (BAE AIT) provided models and analyses that were used to assess the sensor fusion node. The BAE–AIT Fusion Performance Model was used to predict tracking performance provided by the sensor fusion node using different combinations of the three available sensors. Multihypothesis trackers utilized the Mori, Chang, Chung algorithm for calculating the probability of track report misassociation. Kalman filters were used for updating track states. This model has been validated by high-fidelity simulations under the DARPA Dynamic Tactical Targeting program.

The table to the right of the fusion node schematic in Figure 14.5 shows the functionality enabled by each of the sensor data feeds. No single feed can provide full tracking functionality. The fusion node aggregates tracking functionality through both data fusion and sensor task management.

Capability Analysis

The performance of the sensor fusion node for five different combinations of the sensors was evaluated using 25 Monte Carlo runs of the scenario based on a script of

	MIUGS Only	Global Hawk Only	MIUGS & Global Hawk	Global Hawk & Predator	MIUGS, Global Hawk & Predator
Time to Find (min)	0.7	5.3	0.5	5.3	0.7
Time to Fix (min)	1.0	5.6	0.8	5.7	0.9
Total Time in Track (min)	15.9	35.1	42.2	44.3	50.2
Maximum Track length	9.2	9.2	9.4	15.5	16.5
Track Length following ID	-	-	-	15.3	16.5
Time of first ID Declaration	-	-	-	22.7	20.4
ID Confidence	-	-	-	0.96	0.99
Tracking Accuracy (m)	56.0	8.3	8.3	6.4	6.4

Key Performance Parameters (KPPs):   *Primary* – Maximum Track Length –Classification ID Confidence   *Secondary* – Total Time in Track

Figure 14.6 Measures of performance.

the scenario that contained a variety of targets and confusers. Figure 14.6 provides representative results that were generated from one of 25 Monte Carlo runs. The Monte Carlo results for sensor fusion node performance using the full suite of sensors are shown in the far right-hand column of the table in Figure 14.6. The three highlighted rows in the table are the primary and secondary KPPs. The last two columns (on the right) are the only combinations of sensors for which ID confidence was recorded. This is because the EO imagery from the Predator was required to ID the target. It is clear from the table that the fusion node did not meet the 30-minute minimum tracking criteria, the primary KPP (continuous track, as measured by the metric "maximum track length"). That is, the capability to hold the target at risk was not met. However, the performance of the other sensor combinations makes clear the significant contribution of the sensor fusion node to mission capabilities.

Concept Refinement

Another favorable result from the table is that it does show that the criteria were met and exceeded if applied to the secondary KPP (interrupted track, as measured by the metric total time in track). The secondary KPP and some of the other details of analysis summarized in the table could be used as the basis for concept refinement, which would entail a second loop through the lower right-hand corner of the process displayed in Figure 14.1.

Summary of Concept Refinement and Capability Analysis

The straightforward process illustrated in Figure 14.1 for the realization of network-enabled capabilities can be used as part of an architecture-based approach to the analysis and design of a FoS. FoSE relates to the design and development activities of an FoS assembled to achieve specified operational capabilities through the individual operation and collective interoperation of the systems in the FoS. It is the system and network architecture of the FoS that transforms the collection of individual systems into a structure that enables operational capabilities. In the sensor fusion node case study, the change in the sensor architecture clearly enabled measurable operational capabilities as represented by the F2T activity model. Also, it was possible to assess the supportability and interoperability of the new node within the existing communications architecture.

Finally, the sensor fusion node case study also demonstrated how it may be possible to attribute changes in operational capabilities to changes in the system and network architecture. This introduces a new viewpoint on operations analysis, one that could be referred to as architecture-based operations analysis.

Analysis of Alternatives (AoA) at the FoS Level

Analysis of alternatives (AoA) at the FoS level can help to bridge the gap between the acquisition of capabilities and the acquisition of systems, and is complementary to concept refinement and capabilities analysis. This section will review an architecture-based trade study conducted in the year 2003 by the Office of the Chief Engineer for the U.S. Navy that demonstrated a method of analysis at the FoS level for trading off cost and operational capability. Although this exploratory work was straightforward and successful, broadly accepted standard methods for capability-based AoA at the FoS level still need to be established.

Background and Attribution

The trade study presented in this section involved three major organizations within the U.S. Navy. One was the Deputy Assistant Secretary of the Navy (DASN) for Integrated Weapon Systems (IWS). Another was the Space Warfare Systems Command (SPAWAR). The third was the U.S. Navy Center for Tactical Systems Interoperability (NCTSI), which was conducting a trade study on Interface Change Proposals (ICPs) for Link 16 (a tactical command and control communications link). Based on some of the early successes of the Office of the Chief Engineer in architecture for mission capabilities, DASN (IWS) proposed a collaboration of the Chief Engineer and SPAWAR with NCTSI to explore an architecture-based capability analysis of the Link 16 ICP trades to see if new insights could be gained and to develop a simple example of the architecture method that could be shared with the systems engineering community. The material in this section is a summary of the results of the architecture-based trade study proposed by DASN (IWS). This material has been presented at a number of public forums and instructional classes on systems architecture. It is offered as the second half of this chapter because of its complementary nature to the concept refinement and capability analysis approach in the previous section.

Problem Statement

The U.S. Navy has a process for reporting issues with deployed systems. In the case of Link 16, NCTSI maintains a summary of the reported issues and seeks to find cost-effective solutions. From the viewpoint of structured analysis and design of information-intensive systems, the Link 16 problem can be considered as a problem of information flow. In the case of Link 16, the information elements in the flow were called "J-series messages." NCTSI had developed a large spreadsheet that viewed the major platforms, their Link 16 equipment, and the platform subsystems that relied on Link 16 for mission-essential information as nodes of an architecture. From this viewpoint, it was straightforward to assign a score to the Link 16 connections between all the types of nodes in the architecture. Figure 14.7 provides a small sample of the actual spreadsheet that NCTSI had constructed as part of the trade study. Each row in the spreadsheet represents a J-series message type and each column is a (type of) node in the architecture. Each cell in the spreadsheet was color coded using a red/yellow/green scheme. A green cell indicated that the node had no reported problems with the associated J-series message. Yellow and red indicated increasing levels of problems reported that could impact mission capabilities. The color white (i.e., no color fill in the cell) indicated that there was no relationship between the node and the J-series message.

NCTSI also had a set of 137 ICPs that had been proposed to address the reported problems. Each ICP description included an estimated cost to implement the proposed change. Of the 137 ICPs, 76 mapped to applicable J-series messages using the Operating Standard 516.2 and the ICP descriptions. These 76 ICPs became the subject of the Link 16 trade study.

Some of these ICPs could resolve one or more reported problems, but there were reported problems that required more than one ICP to achieve a satisfactory resolution. The NCTSI trade study was considering the impact of various combinations of the ICPs on the distribution of the color codes in the spreadsheet. In the collaboration with the Office of the Chief Engineer, it was hoped that a mission capability viewpoint would give further insights as to how to choose the best combination of ICPs. At the time, the Office of the Chief Engineer had just finished working with the Office of the Chief of Naval Operations to develop over a dozen mission

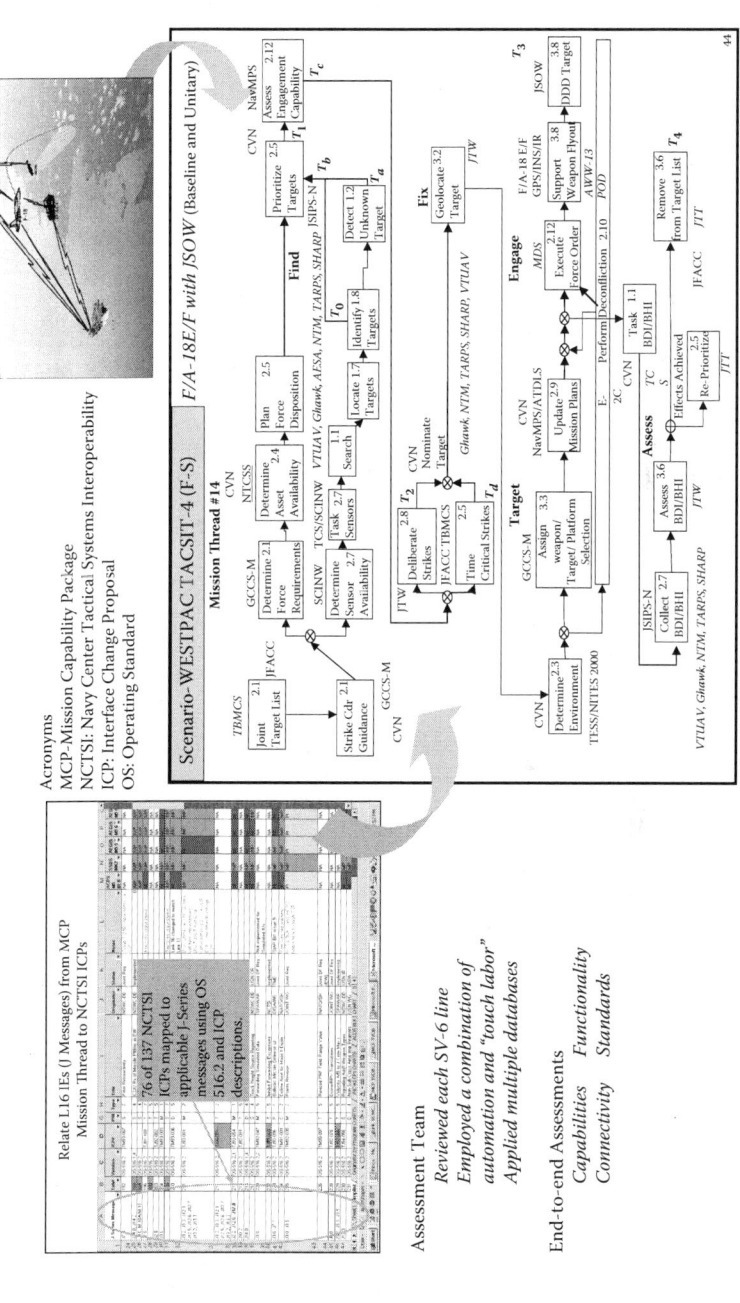

Figure 14.7 Assessment of alternatives (example of Link 16 ICP interoperability trades).

capabilities packages (MCPs) that supported the writing of the Program Objective Memorandum for the 2004 fiscal year (a document referred as POM04). These MCPs also became the subject of the Link 16 trade study.

Analysis of Alternatives for Link 16 ICP Trades

The capability-based trade study found that the most stressing missions involving the 76 ICPs for the J-series messages were found in the MCP for strike missions. This is not surprising because of the increasing attention on time-sensitive targeting, as discussed in Chapter 13 ("Modeling FBE-I with DoDAF"). Therefore, it was decided to use the Strike MCP as the baseline for the capability analysis and trades. In support of the POM04 analysis, thirteen mission threads had been populated and analyzed in three baseline scenarios specified in the POM04 guidelines. This gave a total of 39 strike mission threads, each of which represented an operational capability. The lower right-hand panel in Figure 14.7 displays one such mission thread, which is the same thread that was presented in Figure 13.12 of Chapter 13.

An assessment team was formed by the members of SPAWAR and the Office of the Chief Engineer who had worked on the Strike MCP for POM04. The team employed a combination of automation and "touch labor" to review and update the detailed SV-6 (as reported in Chapter 14 ["Capabilities Assessment"] to have 2,857 lines of connectivity information) using multiple databases to include the NCTSI information. As with the FBE-I and POM04 capability analyses, the assessments were based on an end-to-end analysis of each of the 39 mission threads using the information on functionality, connectivity, and the application of standards to determine how the 76 ICPs for the J-series messages mapped to the interoperability problems in the 39 mission threads.

Relation of the ICPs to Operational Capabilities

Figure 13.14 of Chapter 13 ("Modeling FBE-I with DoDAF") illustrated the seven Link 16 J-series message interoperability problems for Mission Thread #14, which was in the scenario designated as WESTPAC TACSIT-4 (F-S). The mission thread also calls out essential details of the specific FoS in the Strike MCP used in the thread. Figure 14.8 illustrates how the ICPs were mapped against the identified problems. For some problems, multiple ICPs would need to be implemented; for others, one ICP was sufficient, as indicated in the figure.

A *bundle* is a uniquely identified group of the Link 16 ICPs that would address all the identified J-series message interoperability issues for the specified mission thread. In the case of Mission Thread #14 in Figure 14.8, there were seven issues that needed to be resolved to have a resolution of the end-to-end mission capability. A list of ICPs for an end-to-end solution is displayed in the panel in the middle

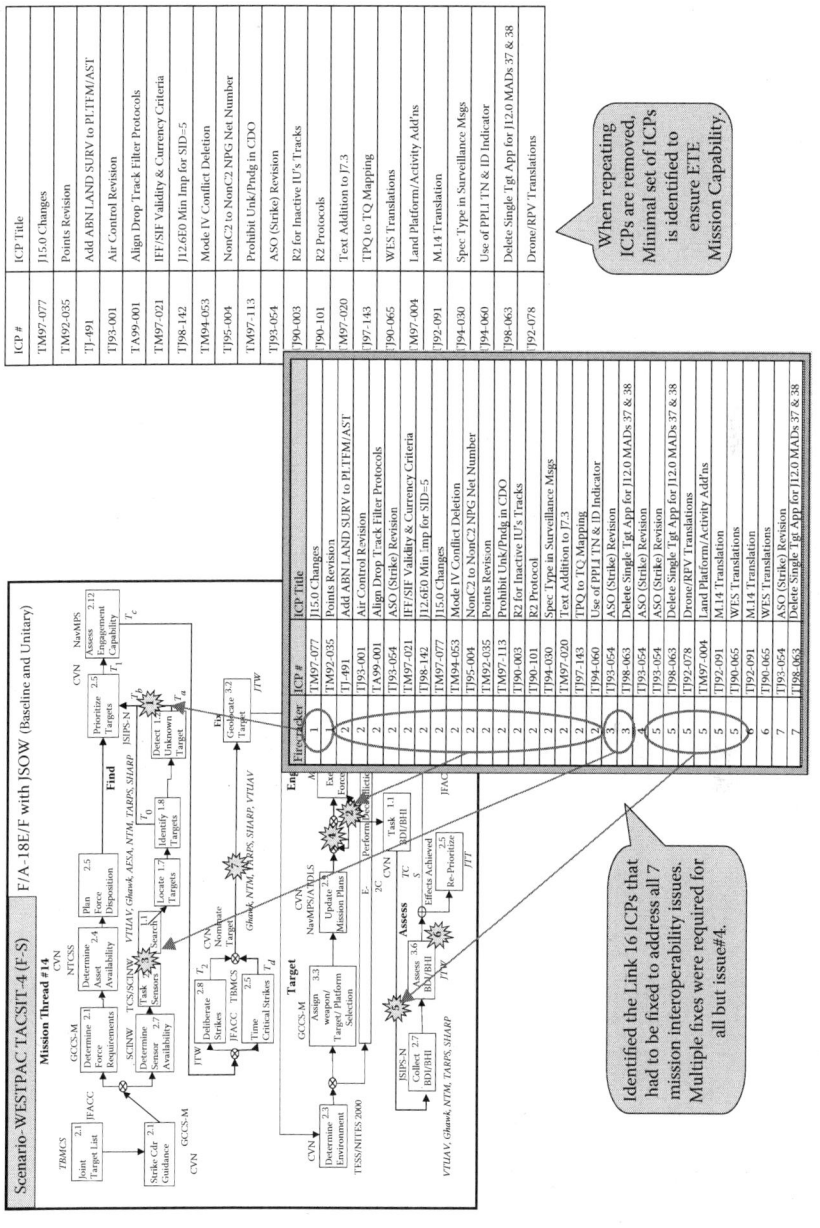

Figure 14.8   Multiple fixes can be required for each gap.

of Figure 14.8. However, a close examination reveals that several of the ICPs are repeated across more than one problem. When the repetition of ICPs is removed, a minimal set of ICPs is uniquely identified to ensure end-to-end operational capability represented by the mission thread. This set is the bundle identified in the upper right-hand corner of Figure 14.8. It contains 22 uniquely identified ICPs. The basic rule here is that the architect and systems engineer should not look at systems or fixes to problems one at a time. To assess operational capabilities, it is necessary look at bundles of solutions.

Choosing a Portfolio of ICPs from the Bundles

Each of the 39 mission threads were examined in this way. The bundle identified in Figure 14.8 was actually bundle #13 in the capability analysis. Each time a bundle was identified, the assessment team applied it to the other 38 mission threads to determine how many threads the bundle enabled end to end. In Figure 14.9, the rows in the table are the 39 mission threads, and the columns are the 15 bundles of ICPs that were identified as end-to-end solutions for a single mission thread. A cell is shaded in the table if a given bundle (column in the table) "fixes" every identified problem in the designated mission thread (row in the table). The white cells in the table indicate partial fixes. The bundles have been sorted and numbered so that the larger bundle numbers generally correspond to being end-to-end solutions for a larger number of the mission threads. However, it must be remembered that this type of ranking is only partially ordered.

The initial capabilities analysis of the Link 16 ICPs considered that the construct in Figure 14.9 might be a decision-making tool for the selection of a bundle or portfolio of bundles to recommend to NCTSI for implementation. However, there are two key attributes missing from this type of approach. The first is cost. It is not impossible to trade cost against the type of ranking illustrated by Figure 14.9. However, the decision-making process would be easier if there was a way to transform the partial ordering of the bundles into a total ordering (i.e., make it completely linear). This can be done if a critical piece of missing information is included: is there a ranking of the capability needs? That is, can the user (in this case the staff of the deployed military forces in conjunction with the strategic planners) offer a ranking to the operational capabilities expressed in the 39 mission threads? Because of the previous work of the Office of the Chief Engineer in FBE-I and the POM04 planning, the mechanisms were still in place to render a ranking on the mission threads. The ranking is a very serious issue. If one of the mission threads happened to be far more important to the user than the other 38, then, without ranking, it could be possible that 97.5 percent of the desired capability sought was achieved (i.e., 38/39 of the mission threads) but the singlemost important capability (represented by the one remaining thread) was not enabled by the choice of solutions.

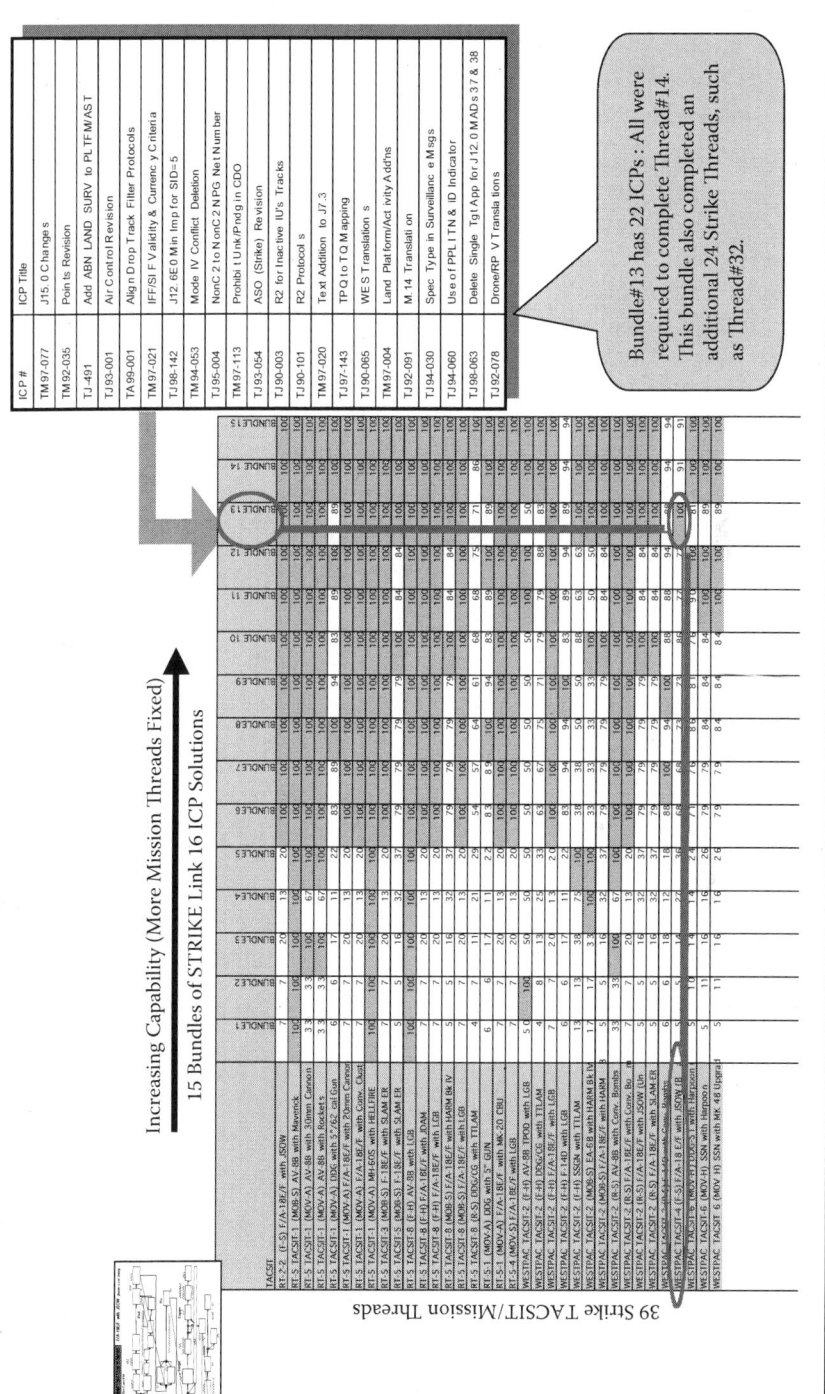

Figure 14.9 Choosing a portfolio from bundles.

The Traditional Cost–Benefit Trade Elevated to a Capability-Based FoS Level

Figure 14.10 is a synopsis of the final reduction of the FoS-level capability trades into a graphical construct that is very similar to a traditional cost–benefit trade-off. The key to "linearizing" the capability metric derives from the ranking exhibited in the upper left-hand corner of the figure. Once the ranking is established for the operational capabilities needed, as represented by the mission threads, the "ranked capability," as measured by the percentage of ranked mission threads enabled (by a bundle), becomes a linear variable that can be traded directly with cost. Each is then assigned two linear variables: one is cost and the other is "ranked capability." The result in the case of the Link 16 capability analysis is the lower left graph displayed in Figure 14.10. In the case of the NCTSI Link 16 study, the 50 percent capability rank corresponded to bundle #7 (ranked by cost), which was comprised of 15 ICPs and estimated to cost $413K to implement, as indicated in the lower right-hand panel of Figure 14.10. This was the recommendation from the capability assessment team. It was integrated with the results of the NCTSI Link 16 trade study to make the final procurement decision.

Summary of AoA at the FoS Level

Architecture can be used to model and assess FoS capabilities. Operational capabilities derive from the individual operation and collective interoperation of systems. FoS performance and interoperability assessments must be in concordance with one another. With proper organization, DoDAF artifacts support an architecture-based approach to an analysis of alternatives at the FoS level. The Link 16 trade study demonstrates how FoS capability assessments can support capabilities-based acquisition (based on JCIDS and DoD 5000 acquisition policies).

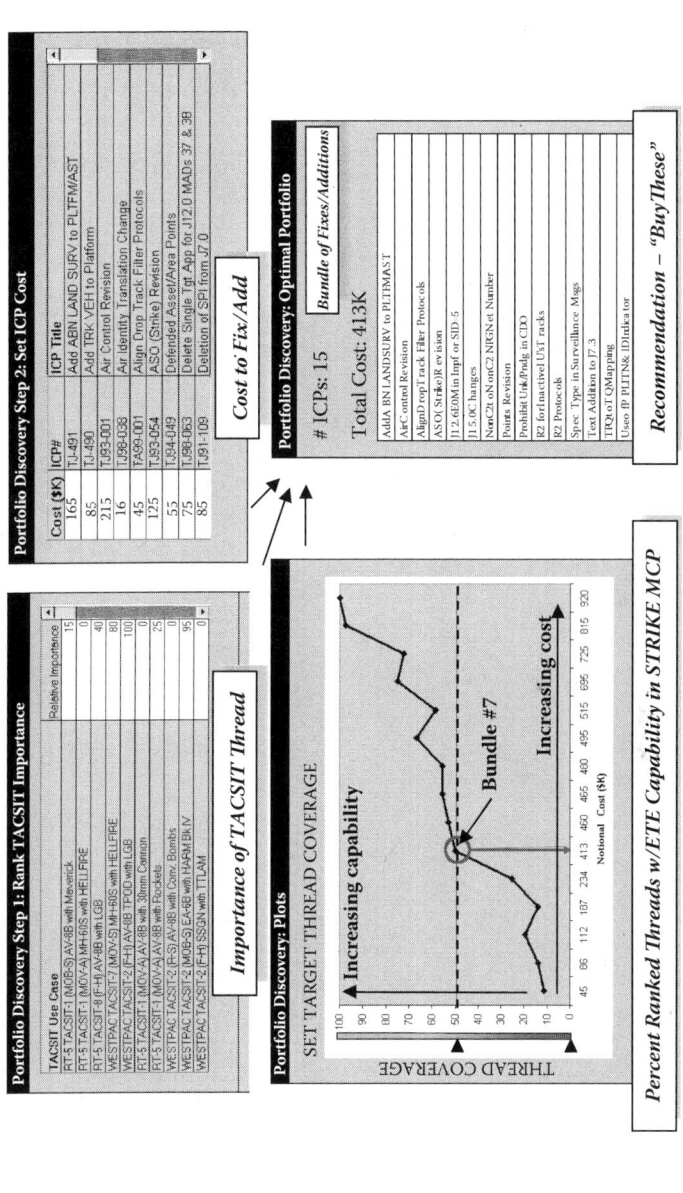

Figure 14.10 Using system bundles for FoS level trades.

References

Blanchard and Fabrycky. 1990. *Systems Engineering and Analysis*, 2nd edition. Englewood Cliffs, NJ: Prentice Hall.

Dickerson, C.E. et al. 21 June 2005. Architecture-based operations analysis. *73rd Annual Military Operations Research Society Symposium*, U.S. Military Academy, West Point.

Chapter 15

Toward Systems of Systems and Network-Enabled Capabilities

This chapter will focus on system of systems (SoS) initiatives and SoS Engineering (SoSE) technical processes being used for the development and acquisition of defense systems of systems. It will begin with the Revolution in Military Affairs (RMA) that led to the concepts of Network-Centric Warfare (NCW), and then show how the concepts of SoSE and network enablement are intertwined. Together, they not only enable mission capabilities, but also are a key to the integration of capabilities. Four SoS initiatives in the U.S. Department of Defense (DoD) will be presented. The application of Model Driven Architecture (MDA™) in one of these initiatives will emphasize the information-intensive nature of SoS and make clear that MDA has an important role to play in SoSE. The chapter will conclude with some of the current activities and emerging research opportunities in SoSE.

Key Concepts

Revolution in Military Affairs
System of systems
Network enablement
Model Driven Architecture

Background

As the defense community has continued to move toward increasingly complex weapon systems that must support joint and coalition operations, the need for SoSE becomes more critical to the achievement of military capabilities. The U.S. DoD and the U.K. Ministry of Defence (MOD) continue to face a critical challenge: the integration of multiple capabilities across developing and often disparate legacy systems that must support multiple warfare areas.

Over the past decade, the need for SoS-enabled mission capabilities has led to changes in policy for the acquisition of systems and the assemblage of battle forces. Defense acquisition structures have been reorganizing to integrate acquisition activities in a way that leads to the achievement of capabilities through an SoS approach rather than from just the performance of individual systems or platforms.

Earlier efforts to meet the challenge faced by the defense community have included solutions using Network-Centric Operations (NCO) in the United States and Network-Enabled Capability (NEC) in the United Kingdom. Although progress has been made, future progress can be expected to depend on the advancement of the practice of systems engineering at the SoS level. SoSE processes and practices must enable the acquisition and implementation of systems of systems that support the integration of multiple capabilities across multiple warfare areas. Recently, there has been extensive debate across a broad community of government, industry, and academic stakeholders about systems engineering processes and practices at the SoS level. However, whatever form SoSE takes for defense systems, the technical processes of SoSE must support the realization of mission capabilities attainable from an SoS that cannot be attained from a single system.

The Move toward Capability Engineering

Defense systems are designed and developed under a defense acquisition policy. Recent changes in the U.S. DoD acquisition policy were motivated by a focus on capabilities-based acquisition. The idea of organizing the defense acquisition strategy around specific military capabilities is traceable to reforms in the British Ministry MOD in the late 1990s. These reforms were a significant departure from the traditional practice of organizing the acquisition strategy around specific threats to be countered by individual systems, platforms, or military components. The concept of acquiring capabilities has been institutionalized in the DoD and across the individual services through the Joint Capability Integration and Development System (JCIDS) process and revisions in the DoD 5000 policies. Capabilities-based planning, including prototyping and JCIDS, is now the emerging overarching concept being used in DoD acquisition. In the United Kingdom, the MOD has continued to evolve its strategy for the acquisition of defense systems, emphasizing that the "whole of their defense acquisition community, including industry," must be "able to make the necessary shifts in behaviors, organizations, and business processes" (Secretary of State for Defence 2005). The basic principles of the MOD's Smart Acquisition are "the primacy of through-life considerations; the coherence of defense spread across research, development, procurement and support; and successful management of acquisition at the Department level."

In line with this policy, the U.K. MOD regards military capability as comprising contributions from across the set of "Defence Lines of Development (DLoD)" (Secretary of State for Defence 2005). One of these is Equipment, the others being Training, Personnel, Infrastructure, Concepts and Doctrine, Organization, Information and Logistics. (The equivalent in the United States is DOTMLPF.) A ninth DLoD, Interoperability, is also defined and is seen as embracing all others. An integrated capability thus requires suitable components from all of the DLoDs to be assembled into a coherent whole, to meet a specific mission need. The United Kingdom has adopted the term *NEC* to reflect the aspiration to ensure that military capability is network enabled; that is, all elements of a force not only communicate effectively with one another, but also enjoy all the benefits of shared awareness and understanding, which lead to better-informed decision making. This is now fundamental to U.K. force development and represents a major contribution to the Interoperability DLoD mentioned here. To do all this effectively requires a broader approach to SoSE, in which all the components of the capability, at all stages of their respective life cycles, are regarded as systems that must be engineered to operate together. Work continues in the United Kingdom to mature this concept and devise practical approaches to its realization; this is beyond the scope of this chapter, which concentrates on developments within the U.S. DoD in this area.

Late 20th-Century SoS History in U.S. Defense Systems

In the last decades of the twentieth century, dramatic changes occurred in both the commercial markets of technology and the traditional role of government-sponsored research and development (R&D) for defense technologies.

Description of Forces and Technologies

The decade of the 1990s was a time of transition for coalition defense systems, and especially so for U.S. defense systems. In the previous decades, the Cold War had set the concept of operations and determined the types of systems needed. A classical symmetric war against a massive ground force was envisioned. Among the advanced technologies that were intended to give coalition forces a significant advantage were stealth aircraft and precision-guided munitions, but information technology (IT) was in its infancy in the defense community. However, by the end of the 1990s, the commercial investment in R&D for IT vastly exceeded that of the U.S. DoD. R&D for the microchip, for example, which had been developed at Texas Instruments in the 1960s by the DoD, was now being driven by the commercial market so strongly that defense acquisition strategies emphasized the need for commercial off-the-shelf (COTS) technologies. By the end of the 1900s, the U.S. Navy had seen the development of its first COTS open architecture combat system, which would be deployed on the Virginia Class submarines shortly after the turn of the century.

A Revolution in Military Affairs

By the 1990s, the power of precision weapons coupled with an effective architecture based on sensor technology and the emergence of IT had been amply demonstrated in the first Persian Gulf War and later in Kosovo. The apparent discontinuous advantage in military capability and the effectiveness of this power were evidence of the viability of concepts originating from Soviet military thought in the 1970s and 1980s, which are known as RMA. This theory of future warfare is intimately related to the concepts of SoS and NCW. Toward the end of the 1990s, this theory was becoming part of the Joint Vision 2010 (Joint Chiefs of Staff 1996) for warfare in what is now the 21st century. The interest in RMA as an organizing concept goes beyond the national boundaries of the Cold War nations, such as the United States and the United Kingdom in a symmetric posture against the communist regime of the Soviet Union in the 1970s and 1980s. India, Singapore, and the People's Republic of China are currently counted among the many nations with an interest in RMA. However, the limiting factor for a country to enter directly into the RMA is perhaps the infrastructure cost of the architecture.

The RMA point of view can be compared with radical changes in military thought that emerged from new concepts of warfare between the first and second world wars, such as Blitzkrieg. This was a revolution in offensive military capabilities based on the integration of weapons and communications technology to achieve new capabilities, and the command structure was designed to exploit both simultaneously. All three elements were essential to the success of this battlefield tactic. And the Nazi war machine enjoyed a discontinuous advantage in military capability and effectiveness comparable to those enjoyed by coalition forces in the Persian Gulf and Kosovo.

A Transition to Network Enablement and SoS

Network Enablement of Naval Forces

At the turn of the decade, in the year 2000, under the auspices of the National Academy of Sciences, the Naval Studies Board released the report, *Network-Centric Naval Forces* (Committee on Network-Centric Naval Forces 2000). This report was motivated in part by a declaration from the Chief of Naval Operations that the navy would be shifting its operational concept from one based on platform-centric warfare to one based on the concepts of network-centric warfare. The concept of NCO was introduced as a new approach to war fighting. In an NCO system, a set of assets, balanced in their design and acquisition, was envisioned to be integrated so as to operate together effectively as one complete system to accomplish a mission. The assets assembled in such a system were envisioned to encompass naval force combat, support, and Command, Control, Communications, Computing, Intelligence, Surveillance, and Reconnaissance (C4ISR) elements and subsystems integrated into an operational and combat network. Although the design and development of the assets in an NCO system would require systems engineering, the assets would be part of an integrated network supported by a common command, control, and information infrastructure.

NCO was envisioned by the Naval Studies Board to derive its power from a geographically dispersed naval force embedded within an information network that would link sensors, shooters, and command and control (C2) nodes to provide enhanced speed of decision making, and rapid synchronization of the force as a whole to meet its desired objectives. The NCO approach would enable naval forces to perform collaborative planning and to achieve rapid, decentralized execution of joint actions. In this way, NCO forces could focus the maneuvers and fires of widely dispersed forces to carry out assigned missions rapidly and with great economy of force. Figure 15.1 illustrates these concepts and how NCO can enable the massing of effects without a massing of forces; that is, in the figure, the weapons effects can be massed against the threat even though the platforms may be geographically dispersed. Figure 15.1 also illustrates that network enablement is as much about the capabilities achieved through the interoperation of systems as it is about networks. The networks simply enable the interoperation (Alberts et al. 1999). The conceptual shift from a platform-centric system architecture to a network-centric system architecture is also illustrated. Platform-centric operations usually involve a sectored Battlespace as a means of weapon systems control and threat engagements (as illustrated in the left-hand panel of Figure 15.1). Platforms carry sensors, processors, and weapons (or combinations thereof), the effectiveness of which may be increased by the integration enabled by a network-centric architecture (as illustrated in the right-hand panel of Figure 15.1). The real-time fusing of multiple sensor outputs is

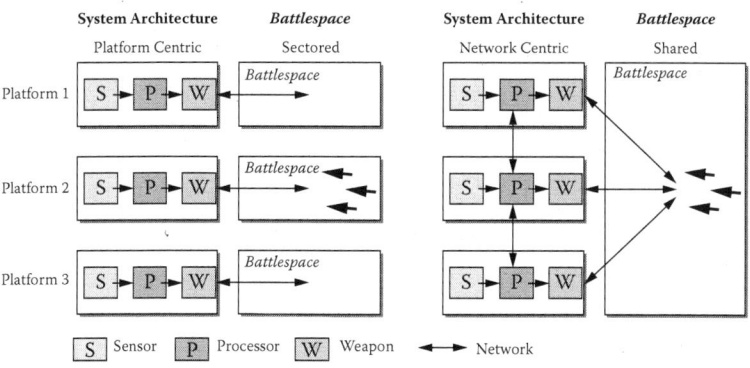

Figure 15.1   System of systems integration through networks.

another driver for the target engagement architecture illustrated in Figure 15.1. Its importance cannot be overemphasized; it is fundamental to NCO.

NCW Report to Congress

By the year 2001, all of the armed services and many of the agencies were actively seeking to become network centric. The U.S. Congress called for a report so as to better understand NCW and consider its impact on defense spending. The Office of the Secretary of Defense (OSD) collected and integrated the inputs from the various services and agencies into the NCW DoD Report to Congress (Office of the Secretary of Defense 2001). The Army, Navy, Air Force, Marine Corps, National Security Agency/Central Security Office, Ballistic Missile Defense Office, National Imagery and Mapping Agency, and Defense Threat Reduction Agency had all adopted visions and implementation plans for NCW.

NCW was now viewed as a maturing approach to warfare that would allow the DoD to achieve Joint Vision 2020 operational capabilities. Furthermore, NCW sought to achieve an asymmetrical information advantage over threat forces. As illustrated in Figure 15.2, this information advantage is achieved, to a large extent, by allowing force access to a previously unreachable region of the information domain—the network-centric region—that is broadly characterized by both increased information richness and increased information reach.

The source of the transformational combat power enabled by NCW concepts was attributed by the report to Congress to be the relationships in warfare that take place simultaneously in and among the *physical,* the *information,* and the *cognitive* domains. Key elements of the transformation included Information Superiority, Decision Superiority, Dominant Maneuver, Precision Engagement, Focused Logistics, and Information Operations.

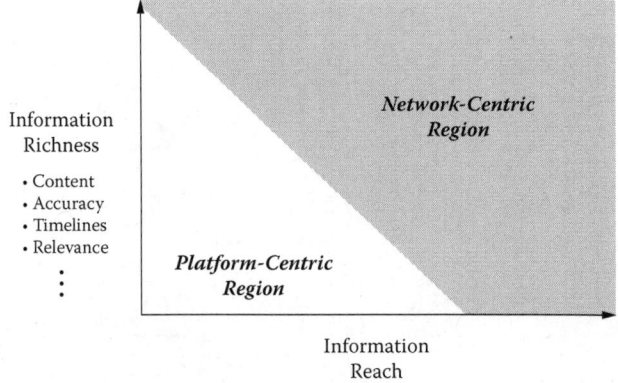

Figure 15.2 Network-centric region of the information domain.

Each service submitted NCW vision statements, which are briefly summarized as follows. In its NCW vision, the army stated that accomplishing its vision was strongly dependent on the potential of linking together and networking geographically dispersed combat elements. The theory behind the army's NCW vision was that by linking sensor networks, C2 networks, and shooter networks, it could achieve new efficiencies in all military operations from the synergy that would be derived by simultaneously sharing information in a common operating environment. In addition, such linkages would allow for the discovery of new concepts of operations both among the army forces and joint forces in theater.

The navy's "Network Centric Operations (NCO)" articulated its path to NCW. It stated that, "In developing NCW systems, a different approach to applying the principles must be taken. NCW requires that technology, tactics, and systems be developed together." The navy submission pointed to three military trends: a shift toward joint effects-based combat, heightened reliance on knowledge superiority, and use of technology by adversaries to rapidly improve capabilities in countering U.S. strengths. It noted that "the power, survivability and effectiveness of the future force will be significantly enhanced through the networking of war fighters."

The air force's NCW vision statement recognized that dominating the information spectrum is just as critical today as controlling air and space or occupying land was in the past. And the time available for collecting information, processing it into knowledge, and using it to support war fighting initiatives is shrinking. Processing, exploiting, and manipulating information have always been essential parts of warfare, but information has evolved beyond its traditional role "Today, information is itself both a weapon and a target." The vision also stated that the key to improving air force capabilities would involve not just improvements to individual sensors, networking of sensors, and C2 for sensors, but also in new ways of thinking about warfare and the integration of U.S. forces.

Although the Marine Corps has not historically used the term NCW, the corps noted in its vision statement that the principles embodied by the term have long been an integral part of Marine Corps operations. The corps acknowledged that their continued capabilities will depend on their ability to capitalize on and expand its networked C2 structure to train and educate future forces in mission-effects-sensitive decision making.

From Network-Centric Enablement to SoS

The link between these two fundamental concepts can be traced back to a perspective on the RMA that emphasizes the integration of advanced weapons technology and IT, and the military organization and doctrine to exploit both simultaneously. An SoS point of view was taken by Admiral William Owens (Owens 1996) when he considered how C4ISR should be organized to support future warfare. Specifically, he considered that the force assets for these key functions should be organized into three systems that must interact with one another:

1. Intelligence, Surveillance, and Reconnaissance
2. Command, Control, Communications, and Intelligence Processing
3. Precision Force

This SoS concept is clearly enabled by the concepts of NCO.

SoSE for Air Force Capability Development

Four years after the submission of the NCW DoD Report to Congress, the U.S. Air Force Scientific Advisory Board delivered an SoSE report to the air force. (Saunders et al. July 2005). While recognizing existent policy and practices for systems engineering in the acquisition of defense systems, but given the emergence of both the concepts and concerns about SoS, the board considered the question of how best to proceed with SoSE. At that time, there was confusion about what constituted the SoSE, and there was little in terms of codified practices that could be adapted for use within the DoD. Also, case studies from the commercial world were not readily applicable to the DoD. Four considerations were emphasized in the study to gain insight into the way ahead: the role of the human in SoS, discovery and application of convergence protocols, motivation issues, and experimentation venues.

The role of the human in an SoS can be part of the overall design, but it could also equally result from a lack of adequate interfaces to support the interactions of the systems. Either can create a challenging environment for the human. The lack of sound Human System Interface designs can exacerbate the challenges. The term *convergence protocol* was used to characterize the election of a single protocol or standard by a group that simplifies connection among different systems. Examples include simple standards such as S-video connections for entertainment products and communications protocols such as the Transmission Control Protocol/Internet Protocol (TCP/IP).

A highly successful convergence protocol for the DoD for SoS applications could enable the use of dynamic discovery technologies for information management among the systems. Motivations within the commercial sector that have led to the successful development of SoS solutions are not readily applicable to DoD. The report recognized that further research was needed to learn how to motivate, for example, DoD program managers to seek SoS solutions. Three experimentation venues were recommended. The first one was for developing concepts of operations, the second was for the evaluation of candidate convergence protocols, and the third was focused on the rapid fielding of SoS-enabled capabilities. The enhancement of infrastructure was also seen as essential for the successful development of SoS capabilities.

SoS or Network Enablement?

The development of military concepts related to the RMA over the last two decades raises the question of whether the RMA is based on SoS or is driven by network enablement.

A Lesson from History

The use of Blitzkrieg as a case study can shed some light on the question of SoS versus network enablement of military capabilities. The Blitzkrieg RMA was based on the integration of weapons and communications technology to achieve military capabilities. Each tank in a Panzer division can be considered an autonomous system, and in fact, this was fundamental to the concept of operations for Blitzkrieg. So, the tanks were a collection of systems, which rightfully could be considered an SoS.

However, the lynchpin of the new tactic was the radio (Hall 2000). Although radios had been available to the military in World War I, they were bulky owing to power supply limitations. During the time period between the world wars, early efforts at miniaturization had reduced power demands, allowing reliable radios to be installed in both tanks and aircraft. Portable radio sets were provided as far down in the military echelons as the platoon. In every tank, there was at least one radio. Although radios and communications links should not be confused with networks, it is a simple extrapolation to apply the SoS concepts for C4ISR and Precision Force advocated by Admiral Owens to the concepts of Blitzkrieg. That is to say, in a more modern time, the SoS (of tanks) would have been network enabled.

Systems Engineering Considerations

Although there are various interpretations of the terms *SoS* and *SoSE*, it still is useful at this point to consider formal definitions of the term *system* and understand its implications for SoS and SoSE. The issue to be resolved is whether an accepted definition of the term *system* can be applied to itself to give a technically consistent definition of the term *system of systems*. Two well-known definitions of *system* will be used as case studies for this purpose. However, it is not the purpose of this chapter to advocate or challenge specific definitions of the terms of systems engineering. Fundamental issues such as this are currently a subject of discussion in the systems engineering community.

The International Council on Systems Engineering (INCOSE) definition of system will be used as a case study for this purpose, primarily because of its simplicity. The *INCOSE Systems Engineering Handbook* (INCOSE 2007) defines the term *system* as follows:

> A system is a combination of interacting elements organized to achieve one or more stated purposes.

A similar definition put forth by Hitchins 2003 emphasizes the nature of emergence:

> A system is an open set of complementary, interacting parts with properties, capabilities, and behaviors emerging both from the parts and their interactions.

The concept of open systems (referred to via open sets in the Hitchins definition) is derived from biological systems, which ingest and interact with their environment, in contrast with the concept of closed systems in physics, where energy is neither gained nor lost (at the system level). The case study for Blitzkrieg showed how radios were the lynchpin for this RMA and the mechanism for the interaction of tanks in a Panzer division. Consequently, when either of the definitions is considered in light of the Panzer SoS, radios and information exchange become key enablers. Will this be generally true of most or all SoSs?

The basis for answering this question and resolving the broader issue of a technically consistent definition of system and SoS can be expected to be resolved by modeling the definitions rather than just relying on arguments based on the natural language definitions. Figure 15.3 provides an example of a model that satisfies both the aforementioned definitions system and aligns well with traditional principles of systems engineering. The diagram in Figure 15.3 uses graphical notations from both the Unified Modeling Language (UML) and the Systems Modeling Language (SysML). The SysML notation makes clear that the *realization* of the system derives from the organized *actions* and *interactions* of the *elements* that comprise the *combination*. This is a commonly accepted principle of systems engineering, referred to as *emergent behavior*.

Can the model presented in Figure 15.3 give a technically consistent definition of system and SoS? It is clear that this model for system can be logically substituted for the keyword elements in Figure 15.3. This could provide a model of the term SoS. Whether this model can be interpreted to reasonably satisfy the various accepted definitions of the term *SoS* across the systems engineering community is a separate question. However, it should be clear from these arguments that the issue

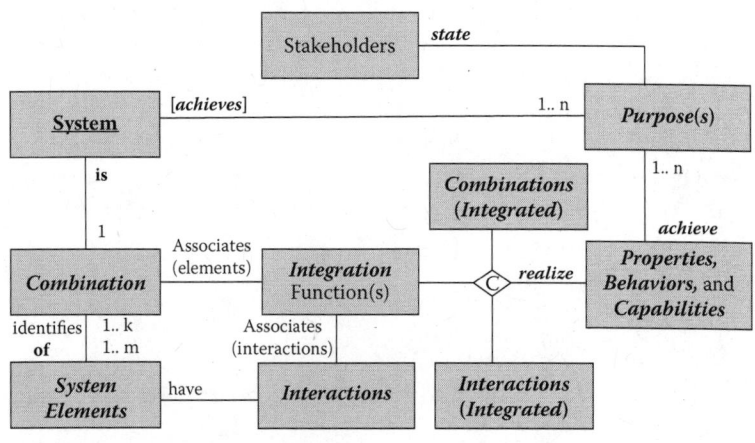

A <u>**system**</u> is a *combination of interacting elements integrated* to *realize properties, behaviors,* and *capabilities* that *achieve* one or more *stated purpose(s)*.

Figure 15.3 A logical model of system.

of a technically consistent definition will not be resolved until methods (such as modeling languages) that are more rigorous than natural language are employed.

With regard to the question of information exchange raised by the case study of the Panzer SoS, it is useful in the model presented in Figure 15.3 to consider what is being transported by the interactions between the elements (of the system). Hatley has considered that *interactions* in human-made systems are described by exactly three types of *transport elements*: energy, material, and information (Hatley et al. 2000). Now, if an SoS is considered to be a system whose elements are other systems, then there are exactly three types of exchanges that can be made between the systems. If these systems are considered individually to have any kind of autonomy, then the exchange of material of energy between the individual systems, although possible or necessary, is not the primary type of exchange. Again, information exchange is the critical and perhaps primary type of exchange between the systems.

The next case study will again emphasize the importance of information exchange between systems in an SoS. It will also introduce the idea that the integration of systems can occur at the capabilities level.

Critical Roles of Information Exchange and Capabilities Integration

This section presents another case study, which illustrates the interplay between the concepts of SoS and network enablement: Fleet Battle Experiment-India (FBE-I) and Time Critical Strike (TCS). See Dickerson et al. (2003). The success of these concepts was demonstrated in actual combat by coalition land forces in Afghanistan after the attacks of September 11, 2001, and again by the Marine Corps in its swift march to Baghdad during the invasion of Iraq from March 18 to May 1, 2003. This was a marked departure from a centuries-long tradition of how the marines had fought. The FBEs were a series of joint experiments initiated by the Chief of Naval Operations (CNO) and designed to investigate NCO as a framework for the development of new doctrine, organizations, technologies, processes, and systems for future war fighting. The Navy Warfare Development Command (NWDC) was the CNO's agent for planning and implementing these experiments in partnership with the numbered fleets. It should be noted that shortly after the time of the FBE-I, the navy adopted the air force (and ultimately joint) terminology of Time Sensitive Targeting (TST) instead of TCS.

Figure 15.4 exhibits the operational concept (more specifically, the DoDAF Operational View-1 [OV-1]) for the TCS experiment. The OV-1 may not look like the kind of operational concept that is more frequently seen, that is, a graphic or cartoon with systems operating in a theater or scenario. This is because the planners for FBE-I were truly focused on NCO, that is, on operations not systems. This is an operator's view of the experiment, exactly as it was intended to be executed.

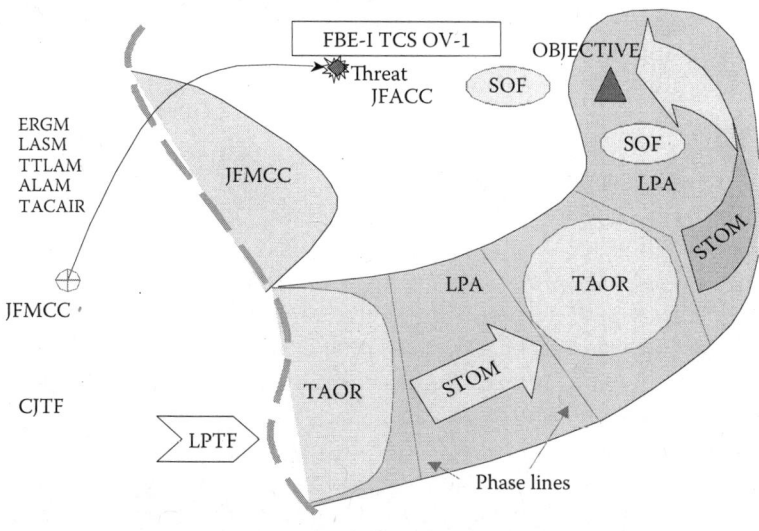

Figure 15.4 Fleet Battle Experiment-India operational concept.

There are three types of entities in Figure 15.4: high-level operational activities depicted in the context of an abstract theater, force elements and the C2 structure that supports them, and an identification of the combination of systems that enables the TCS capabilities sought in the objectives of the experiment:

1. Operational activities:
   a. Ship to Objective Maneuver (STOM)
   b. Time Critical Strike (TCS)
2. Force elements:
   a. Littoral Penetration Task Force (LPTF)
   b. Special Operations Force (SOF)
3. C2 structure:
   a. Commander, Joint Task Force (CJTF)
   b. Joint Force Air Component Commander (JFACC)
   c. Joint Force Maritime Component Commander (JFMCC)
4. Enabling systems:
   a. Extended Range Guided Munition (ERGM)
   b. Land Attack Standard Missile (LASM)
   c. Tactical Tomahawk Land Attack Missile (TTLAM)
   d. Advanced Land Attack Missile (ALAM)
   e. Tactical Aircraft (TACAIR), in this case F/A-18 strike fighters

The theater is divided into two parts by a littoral interface depicted by a dashed line: (1) land (to the right of the line) and (2) sea (to the left). Historically, the marines would "storm the beach," establish a safe beachhead, and then build up supplies (called the *iron mountain*) at the beachhead in preparation for further fighting to secure a larger area. Once the larger area was secure, the army would land and take over inland operations.

In this depiction of the theater, the CJTF leads the operation. The JFMCC both secures a portion of the littoral interface (e.g., with fires) and provides various weapon systems to perform TCS operations. The JFACC controls the air space, so the JFMCC and JFACC must coordinate and collaborate. The joint forces in this OV-1 are geographically dispersed.

What has changed for the marines in FBE-I when they become an LPTF? In this new concept, they are employing classic fire and maneuver tactics to penetrate more deeply and swiftly than ever before. The Tactical Areas of Regard (TAOR in the OV-1) are areas where they expect to fight. The Littoral Penetration Areas (LPAs) are areas of maneuver, where STOM is a maneuver toward the objective. (More than one STOM may be required.) Why have they not done this before? In the past, they only went a short distance and carried a more significant amount of fire power than in the LPTF concept. To penetrate deeper (to a designated objective) and more swiftly, it was necessary to leave fire power behind. So, where will the requisite fire power come from when they need it? It came from TCS in FBE-I, and in current terminology, it will come from TST. They are also supported by SOF.

This experiment embodies both the tenets of NCO and the SoS concepts of Admiral Owens in which three (grand) systems of ISR, C4, and Precision Force are integrated to create emergent behaviors and military capabilities not otherwise achievable. Based on the systems engineering considerations of the previous section, it is clear that the mission-essential interactions between these three systems are based on the exchange of information. This is not to diminish the essential need for energy (in the form of kinetic weapons) and material (in the form of logistic support) to be placed into the Battlespace. However, it is the integration of the ISR, C4, and weapon systems at the mission level that enabled the new Precision Force capabilities (i.e., TCS) sought by FBE-I.

This case study should prompt the systems engineer (and the SoS engineer) to consider two important things: first, the exchange of information is both central and critical for an SoS to enable a military mission, and second, but equally important, mission capabilities enabled by an SoS (or any system for that matter) must be understood through integration of operational capabilities in a mission context.

The concepts of capabilities engineering and integration are currently being explored by the defense community, but will face technical challenges until stricter standards are established for the terminology of systems engineering and SoSE. For example, the INCOSE natural language definition of system is stressed by the concept of capabilities integration, because if interactions are only between the elements of the system, then there is no mechanism (in the definition) for the interactions of the effects caused in the Battlespace. However, the model depicted in Figure 15.3 could support concepts where the system elements interact with elements of the Battlespace (which are not part of the system). Ongoing research in the defense community can be expected to address these types of questions in the years to come.

Examples of U.S. DoD Defense Applications of SoS

The U.S. DoD has defined SoS through the joint staff instruction CJCSI 3170.01E, dated May 11, 2005, as follows:

> A system of systems (SoS) is a set or arrangement of interdependent systems that are related or connected to provide a given capability.

In the DoD concept of SoS, the loss of any part of the SoS will significantly degrade the performance or capabilities of the whole. This definition clearly supports the SoS concept of Admiral Owens. *A priori*, it exhibits no preference for the nature of the interdependence (e.g., does information exchange have primacy for enabling SoS capabilities?). And it may be a candidate definition for a technically consistent definition of system and SoSs.

OSD SoS SE Guide

The writing of the SoS SE Guide was initiated in May 2006, when a task was issued by the Deputy Under Secretary of Defense for Acquisition and Technology (DUSD [A&T]) to develop a guide for SoS SE. The objective was to provide guidelines to defense system program managers for a better understanding of how the system under their management might be part of a larger SoS and how the management of a defense acquisition in an SoS environment might differ from the acquisition of a single system. Government, industry, and academia collaborated under the leadership of DUSD (A&T), and version 0.9 of the guide was released on December 22, 2006. The guide (Office of the Under Secretary of Defense 2006) describes the application of systems engineering at an SoS level, and has served as a mechanism for the broader community to come together, to discuss the wide spectrum of views on this topic, and capture the views that everyone does agree upon. For six months after its release, the OSD used the guide to conduct a pilot study on how SoS SE is being practiced in the DoD. The defense community was then asked to submit its comments as part of the effort to pilot the recommendations in the guide. Over twenty U.S. and international stakeholder organizations from government, industry, and academia have participated in the pilot study.

Emerging insights from the pilot show the following characteristics of an SoS in the DoD today: (1) they tend to be ongoing efforts to satisfy user capability needs through an ensemble of systems, (2) they are not new acquisitions per se, (3) the SoS manager typically does not control the requirements or funding for the individual systems, (4) the SoS is focused on the evolution of capability over time, and (5) a functioning SoS requires start-up time but, in a steady state, seems well suited to routine incremental upgrades.

The comments on the version 0.9 release have been collected and were assembled and reviewed by the working group for the guide. A new version of the guide has been issued for broad review and use. In the original OSD SoS SE Guide, the U.S. DoD services submitted SoS program descriptions, some of which are summarized in the following subsections.

Future Combat System (FCS)

The U.S. Army's FCS is a new development program that is intended to operate as a cohesive SoS in which the whole of its capabilities is to be greater than the sum of its parts. As the key to the army's transformation, the network, and its logistics, and Embedded Training (ET) systems, enable the future force to employ revolutionary operational and organizational concepts. The network enables soldiers to perceive, comprehend, shape, and dominate the future battlefield at unprecedented levels. The FCS network consists of four overarching building blocks: System of Systems Common Operating Environment (SoS COE); Battle Command (BC) software; communications and computers (CC); and intelligence, reconnaissance and surveillance (ISR) systems. The four building blocks synergistically interact, enabling the future force to see first, understand first, act first, and finish decisively.

FCS is an example of an SoS currently under development, with many degrees of freedom from an engineering perspective. A traditional SE approach (top-down analysis, synthesis, and evaluation) has been applied.

Single Integrated Air Picture (SIAP)

The SIAP under development by the Joint SIAP Systems Engineering Organization (JSSEO) consists of common, continual, and unambiguous tracks of airborne objects of interest in a surveillance area. SIAP is derived from fused real-time and near-real-time data, and consists of correlated air object tracks and associated information from multiple sensors. The purpose of an SIAP is to solve a confusing air picture. Real-world operations, exercises, and evaluations highlight joint war fighting shortfalls. Disparate systems within a service as well as across multiple services that share track management data have different views. The views as well as the track identification number may be different, causing confusion in communication among system operators.

The SIAP requirements must provide a solution that is scalable, filterable, and supports situation awareness and battle management. Each airborne object must have one and only one track identifier and associated characteristics. The solution will reduce the risk of fratricide to U.S. and coalition forces caused by incorrect correlation and track identifier association, provide confidence in the air picture, provide point and area defenses the opportunity to engage beyond their self-defense zones, and quickly coalesce to repel asymmetric threats such as cruise and ballistic missiles.

The SIAP will initially be introduced via capability drops on host platforms of the U.S. Navy Aegis, SSDS, E2, and the Air Force E3 AWACS, Battle Control System (BCS), and the RC- 135V/W Rivet Joint. Capability drops are targeted at eliminating specific interoperability issues, providing C4I enhancements, and delivering an executable integrated architecture in a two-year spiral delivery process.

The SIAP operational Joint Battle Management and command and control (BM/C2) reflects new capabilities of networked interoperable heterogeneous systems as opposed to traditional monolithic systems with a single-system orientation. Tactical system networks are dynamic with autonomous mobile nodes and ad hoc membership. The autonomous constituent systems make up the SIAP SoS to fulfill a specific mission that may be presented. This is a hybrid SoS whereby there is some level of C2, and autonomous systems fall under the area commander to fulfill a mission. This is characterized by loosely coupled, independently controlled nodes sharing collaboration rules via centralized guidance.

The SIAP is implemented by applying MDA in developing an executable Peer-to-Peer (P2P) Integrated Architecture Behavior Model to form a Platform Independent Model (PIM). The services receive the PIM for their host platforms and create Platform-Specific Models (PSM) for their hosts, applying adaptive layers to interface with their specific sensors shown.

Naval Integrated Fire Control–Counter Air (NIFC-CA)

NIFC-CA seeks to evolve a family of systems of mixed maturity to extend the Naval Theater Air and Missile Defense Battlespace to distances that are well beyond the existing, stand-alone capability of surface-ship-controlled air defense weapons. NIFC-CA is a capabilities-based program that takes current and emerging technology from Core Pillar Programs (Cooperative Engagement Capability [CEC], Aegis, Standard Missile [SM-6], and E-2D Advanced Hawkeye [AHE]) and other related programs, and integrates them together to form the successful implementation of an SoS capability. The NIFC-CA focus is on Integrated Fire Control for over-the-horizon (beyond visual range) and engage-on-remote capability. NIFC-CA is a key component of Navy Transformation "SEA SHIELD." NIFC-CA is an SoS systems engineering capability designed to define the functional allocation for the pillar elements within NIFC-CA (SM-6, E-2D, CEC, and Aegis). The synchronization of programs across cost/schedule/performance is the key challenge for this SoS.

Commissary Advanced Resale Transaction System (CARTS)

CARTS is the replacement point-of-sale (POS) system used to process customer purchases, capture sales and financial data from those purchases, produce management reports, and provide information to other DeCA business systems. CARTS will be a COTS system with middleware to support the interfaces to the DeCA business systems. It will also provide its customers with services similar to future grocery industry technology advances without disrupting commissary operations.

This program has a development strategy of utilizing COTS to the fullest extent and provides the complete system capability in one step, and software reuse is very common. Challenges with COTS include making the determination whether the COTS architecture can support military requirements. A careful examination of the architecture is required to determine if there is sufficient capability for growth to accommodate the additional military requirements. It is important to conduct a cost as independent variable (CAIV) analysis over the complete product life cycle.

Related Work and Research Opportunities

In addition to the breadth of work on SoSE, NCO, and NEC that is ongoing, the concepts of systems engineering itself are currently an active topic of discussion among the larger community. The International Organization of Standards (ISO) is engaged with issues of architecture concepts through two working groups: JTC1/SC7/WG42 and TC184/SC5/WG1. The INCOSE has various initiatives to include a Special Workshop on Architecture Concepts at the International Workshop 2008 (IW08), which will also support the work of the ISO working groups. The IEEE has also recently initiated the International Council on Systems of Systems (ICSOS). And for the past eight years, the Object Management Group (OMG) has been maturing the practices of MDA, which are being linked to the practice of systems engineering through collaboration with INCOSE.

Research on the foundations of systems engineering as evidenced by INCOSE and ISO should lead to a sharper understanding of SoSE and the role of NCO, NEC, and related topics such as capabilities engineering and integration. IEEE efforts through ICSOS can be expected to further promote a broad community interest in SoSE that will both seek and advance solutions to SoS problems and extend the methods and practices of systems engineering to SoSE. Research on the extension of the OMG MDA to systems and SoSE should also be an increasingly active area of research over the next several years.

References

Alberts, D.S. et al. August 1999. *Network Centric Warfare: Developing and Leveraging Information Superiority.* Washington, D.C.: CCRP Press.

Committee on Network-Centric Naval Forces, Naval Studies Board. 2000. *Network-Centric Naval Forces: A Strategy for Enhancing Operational Capabilities*, Washington, D.C.: National Academy Press.

Dickerson, C.E. et al. August 2003. *Using Architectures for Research, Development, and Acquisition.* Office of the Chief Engineer of the Navy, Assistant Secretary of the Navy. Defense Technical Information Center (www.dtic.mil): ADA427961.

Hall, C., MITRE Corporation. 29 February 2000. Private communication.

Hatley, D.J., Hruschka, P., Pirbhai, I. 2000. *Process for System Architecture and Requirements Engineering.* New York: Dorset House Publishing.

Hitchins, D.K. 2003. *Advanced Systems Thinking, Engineering, and Management.* Boston: Artech House.

International Council on Systems Engineering. August 2007. *INCOSE Systems Engineering Handbook*, v.3.1. Seattle: INCOSE Central Office.

Joint Chiefs of Staff. 1996. *Joint Vision 2010.* Washington, D.C.: Government Printing Office.

Office of the Secretary of Defense. 27 July 2001. *Network Centric Warfare, Department of Defense Report to Congress.* Washington, D.C.

Office of the Under Secretary of Defense, 2006. Director, Systems and Software Engineering, Deputy Under Secretary of Defense (Acquisition and Technology), Office of the Under Secretary of Defense, (Acquisition, Technology and Logistics). December 22, 2006. *System of Systems Systems Engineering Guide, Version 9.*

Owens, W.A. February 1996. *The Emerging U.S. System-of-Systems.* Washington, D.C.: The National Defense University, Institute of National Security Studies, Number 63.

Saunders, T. et al. July 2005. *Systems-of-Systems Engineering for Air Force Capability Development.* Washington, D.C.: United States Air Force Scientific Advisory Board, Report SAB-TR-05-04.

Secretary of State for Defence. December 2005. *Defence Industrial Strategy.* London: Defence White Paper Cm 6697.

Chapter 16

Model Driven Architecture

The level of maturity of Model Driven Architecture (MDA™) can be judged by the fact that key concepts from MDA are beginning to be infused into a model-based approach to architecture and systems engineering. As seen in Chapter 15 ("Toward Systems of Systems and Network-Enabled Capabilities"), MDA can be a key enabler of SoS. It will be seen in this chapter that MDA has demonstrated the utility of shifting from a document-based approach to architecture to a model-based approach. Model transforms will be seen to be at the heart of software system design in MDA. They will also be seen as a key link between the use of architecture models in systems analysis and design, and the systems engineering approach to design. Architecture and networks will be seen as integrating functions for SoS.

Key Concepts

System of Systems (SoS)
Utility of models over documents
Model transforms
Architecture and networks as integrating functions

Motivation

The promise of MDA to the software community by OMG (Miller and Mukerji 2003) has been made clear in their goal statement that MDA will allow definition of machine-readable application and data models, which will allow long-term flexibility of implementation, integration, maintenance, testing, and simulation. Implementation will benefit from the integration and targeting of new infrastructure by existing designs. Data integration bridges will be produced by automation. As system design artifacts become available in machine-readable form, maintenance and configuration control will enjoy new levels of efficiency. Testing and simulation of models can be used to generate code to validate requirements, test infrastructure, and simulate behavior.

The OMG (Object Management Group) has been an international, open membership, nonprofit computer industry consortium since 1989. The OMG's current standards include the Unified Modeling Language (UML), the Systems Modeling Language (SysML), the Meta Object Facility (MOF), and Model Driven Architecture (MDA). UML is a standardized specification language for object modeling. SysML is a general-purpose graphical modeling language for specifying, analyzing, designing, and verifying complex systems. MOF is a standard for model-driven engineering, and is the mandatory modeling foundation for MDA. And MDA is an approach to using models in software development.

What is MDA? It is a significant paradigm shift in software engineering in which OMG made a dramatic move from their Object Management Architecture to models. MDA was initiated in late 2000 and has been public since 2001. It is a trademarked term from OMG. The standards for MDA™, along with a large body of reference material, can be found on the OMG open Web sites at

http://www.omg.org/
http://www.omg.org/mda/

MDA provides an open, vendor-neutral approach to the challenge of business and technology change. It separates business and application logic from underlying platform technology, and seeks to insulate the core of the application.

Why should software and systems engineers be interested in MDA? The first reason is that it has been shown to improve design quality and to better support program management than document-based approaches. The architecture models are a means to preserve the conceptual integrity of the system over time. And configuration management is improved by using models instead of documents. The second reason is that it is supported by the UML, an OMG modeling language that is widely used in software engineering. Recall that in Chapter 2 ("Logical and Scientific Approach"), the question was raised, "What are the formal languages and methods of systems engineering?" It may be argued that SysML is the formal language for systems engineering, whereas UML is the formal language for software engineering, but SysML is a new language, with the standards approved by OMG only last year (2007). And in Chapter 5 ("Architecture Modeling Languages") we see that SysML reuses a substantial part of UML. What benefit should the systems engineer derive from MDA? The methods of MDA can be extended into systems engineering. This was demonstrated in Chapter 4 ("Structure, Analysis, Design, and Models") when model transforms were introduced into structured design. And at the end of this chapter, we shall see how the methods of MDA have direct application to system of systems (SoS) engineering.

The last reason for software and systems engineers to be interested in MDA is that significant benefits and return on investment (ROI) have been realized through the practice of MDA in software development. Life-cycle cost savings for software development have been reported to be in the range of 30 to 60 percent. The ROI

enjoyed by businesses that have instituted MDA is a reflection of the financial benefits of the first two reasons given earlier. The ROI is driven by demonstrated greater
efficiency and lower labor costs, increased reuse of computer code, and reduced cost
of poor quality. The following OMG Web site has numerous reports of success:

http://www.omg.org/mda/products_success.htm

These reports come from companies that have instituted MDA in their business
pursuits and from MDA tool vendors, and as such are not peer reviewed. However,
they give a clear indication of the success that has been enjoyed by MDA.

The success story from the Codagen testimonial on MDA benefits and ROI is
a good model of how and when cost savings are realized. Figure 16.1 is the graph
of level of effort (measured in dollars) plotted against time, with basic systems
engineering program phases called out. Using MDA the Codagen software development costs for a program called CGI was $5.475 million. Based on their his-
torical experience of developing software without MDA, they estimated the same
work would have cost $8.073 million without using MDA. The savings realized
was estimated at $2.598 million (32 percent). The region of cost savings indicated
in Figure 16.1 show that as the project progressed from the analysis and design
phase to the maintenance phase, deep cost savings with a long tail were realized.
Experience shows that this occurs because, around the peak of the analysis and
design phase, large numbers of highly skilled technical professionals are involved.
In a model-based approach such as MDA, this is the point in system development where the models have been finished. Because the models are captured using
a formal language such as UML in an electronic engineering environment, the

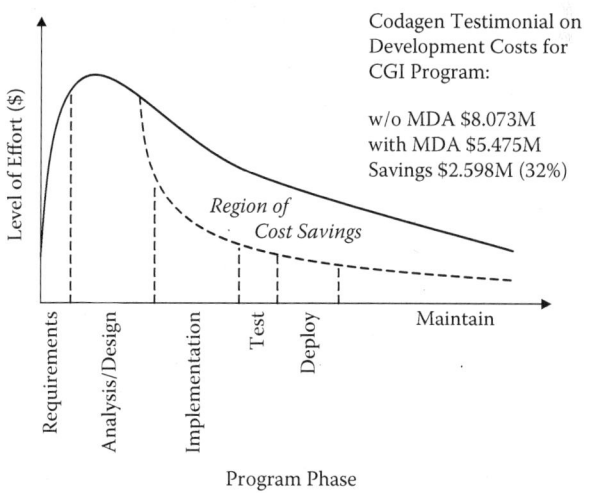

Figure 16.1   MDA™ cost savings over program life: Codagen testimonial to OMG.

maintenance and configuration management requires far less labor. In a project the size of the CGI program that Codagen worked on, it can even be the case that a single system architect is all that is needed after deployment.

MDA Concepts and Terminology

This section will provide a summary of the OMG concept of MDA based on open source material in the MDA Guide v1.0.1 (Miller and Mukerji 2003). The systems engineer should keep in mind that the OMG viewpoint is from software development.

MDA is considered an approach to system development. In MDA, the model of a system is a description or specification of that system and its environment for some certain purpose.

The MDA approach is model driven because it provides a means for using models to direct the course of understanding, design, construction, development, operation, maintenance, and modification [of a software system].

In MDA the term *system architecture* is defined as follows:

> The architecture of a system is a specification of the parts and connectors of the system and the rules for the interactions of the parts using the connectors.

Practitioners of architecture who understand MDA will recognize that more general concepts of architecture have been carefully tailored to the specific purpose and methods of MDA for software development. A logical model of the MDA definition of architecture was provided in Chapter 3 ("Concepts, Standards, and Terminology").

MDA prescribes certain kinds of models to be used and how those models may be prepared. It also prescribes the relationships of the different kind of models.

MDA specifies three viewpoints on a system. In MDA terminology the terms *view* and *viewpoint* are used in the following way. A viewpoint on a system is a technique for abstraction using a selected set of architectural concepts and structuring rules to focus on particular concerns within that system. The term *abstraction* is used to mean the process of suppressing selected detail to establish a simplified model. A *viewpoint model* or *view* of a system is a representation of that system from the perspective of a particular viewpoint.

MDA uses the terminology *platform* and *independence* in a specialized way. A platform is a set of subsystems and technologies that provide a coherent set of functionality through interfaces and specified usage patterns. (The definition in the MDA Guide v1.0.1 goes on to require that any application supported by the program can use the subsystems and technologies without concern for the details of how the functionality provided by the platform is implemented.) Platform independence is a quality that a model may exhibit, where the model is independent of the features of the platform.

MDA has three key viewpoints: the Computational-Independent Viewpoint, Platform-Independent Viewpoint, and Platform-Specific Viewpoint. The Computational-Independent Viewpoint focuses on the environment of the system and requirements for the system. The details of the structure and processing of the system are hidden or as yet undetermined. The Platform-Independent Viewpoint focuses on the operation of a system while hiding the details necessary for a particular platform. It shows that part of the complete specification that does not change from one platform to another. The Platform-Specific Viewpoint combines the Platform- Independent Viewpoint with an additional focus on the detail of the use of a specific platform.

From these three viewpoints, MDA specifies three views (models): the Computational-Independent Model (CIM), Platform-Independent Model (PIM), and Platform-Specific Model (PSM). The CIM is a view of the system from the Computational-Independent Viewpoint. It does not show details of the structure of the system. The PIM is a view of the system from the Platform-Independent Viewpoint. It exhibits a specified degree of platform independence so as to be useful for use with a number of different platforms of similar type. The PSM is a view of the system from the Platform-Specific Viewpoint. It combines the specifications of the PIM with the details that specify how that system uses a particular type of platform.

In MDA, system specification is accomplished by means of the Platform Model (PM), which is a set of technical concepts representing the different kinds of parts that make up a platform and the services provided by that platform. The PM also specifies requirements on the connection and use of the parts of the platform, and the connections of an application to the platform. A generic platform model can amount to the specification of a particular architectural style.

Implementation is a specification that provides all the information needed to construct a system and to put it into operation.

Model transformation is at the heart of MDA. It is the process of converting one model to another model of the same system. An MDA mapping provides specifications for transformation of a PIM into a PSM for a particular platform. The PM will determine the nature of the mapping. Model transformation can be thought of as part of the design and development process. See, for example, Raistrick et al. 2004.

The graphic in Figure 16.2 summarizes the MDA approach described earlier. This figure is adapted from Jones et al. (2007). Numerous graphs like this are popular in the MDA community. They all convey the same concept of the centrality of model transforms in MDA to generate the PSM from the PIM and the PM from the PSM. When the OMG promises to allow definition of machine-readable application and data models that can be used in software system design and development, this is the method that makes MDA viable. The software community can

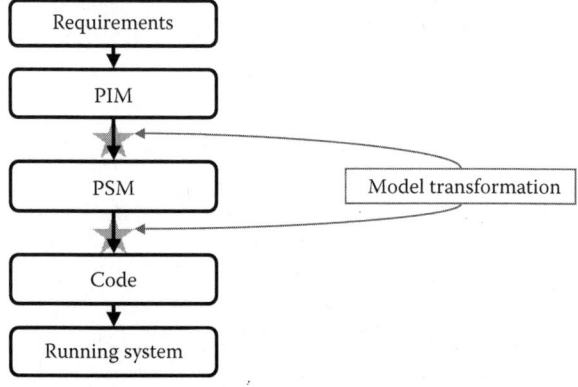

Figure 16.2　A simple view of the MDA™ approach. (Adapted from Jones, V., van Halteren, A., Konstantas, D., Widya, I., and Bults, R. [2007]. *International Journal of Business Process Integration and Management,* **2(3):215–229).**

reach for this kind of automation in system design and development because the technologies used to implement software solutions are well defined and bounded. These technologies comprise things like computer languages, operating systems, and computing platforms. Although the technology space of the systems engineering community is much greater, if not unbounded, the approach and methods of MDA do have extensions into systems engineering.

For example, the general pattern set forth by Richard Soley, the chief operating officer of the OMG, will be seen in the next section to be precisely the pattern implemented by logical modeling and structured analysis and design. Soley's general pattern is to

Discover multiple syntaxes for a single semantic
Derive and design a model which underlies that semantic
Develop transformations between those models (Soley 2004)

Transformation Example

The following simple example of model transformation and its relation to Richard Soley's general MDA pattern will serve to make specific all of the abstractions in the previous section.

Consider the simple narrative requirements,

Problem: Employees need to be paid.
Design a payroll system to pay the employees

Assume that agreement has been made between employees and management on the pay rate for time worked. And assume that no employees are hired or leave during the pay period.

The graphical tools from Part II of this book will be used for model transformation in a structured design approach that emulates MDA. The following procedure parallels the general MDA pattern:

Discover the syntax in the problem statement for the system by writing a sentence summarizing requirements.

Derive a logical model of the sentence.

Design an implementation-free behavioral model by modeling flow in the logical model.

Use a relation-preserving transformation of the DFD graphical model to specify an implementation model.

This pattern captures one part of the model-based approach to architecture and systems engineering that has been threaded through this book.

Using the problem statement for the system, the first step is to write a sentence summarizing requirements of what the payroll system must do. Conciseness is paramount here because this sentence will be interpreted into a logical model. The summary of requirements will be written as follows:

Employees are paid a salary calculated based on the time they have worked and a pre-agreed pay rate.

Precondition:

Agreement has been made between employee and management on the pay rate for time worked.

Figure 16.3 is a logical model of the requirements summary. Because of the emphasis on what the system must do, rather than any interactions with the proposed payroll system, both the requirements summary and the logical model make no reference to a payroll system (at this stage of the analysis). The core concept is that employees are paid.

The logical model gives rise to the data flow diagram (DFD) for the system in a natural and repeatable way. The logical model has two constants: (1) the employees and (2) the pay rates.

It also has two variables: (1) the time worked (independent variable) and (2) the salary paid (dependent variable) to the employee. Recall that the DFD approach seeks to identify what processes are needed for system definition. Here is the key concept for identifying processes from a logical model:

Statement: <u>**Employees**</u> are *paid* a *salary calculated based on* the *time* they have *worked* and a pre-agreed *pay rate.*

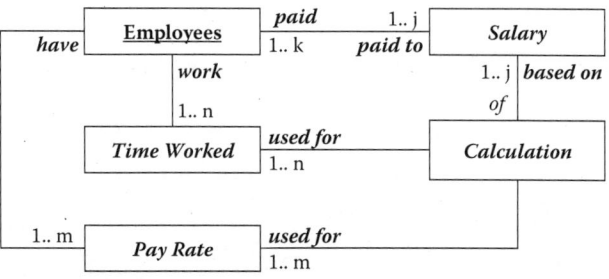

Figure 16.3   Logical model of what the payroll system must do.

> A relationship between an independent and a dependent variable implies that a process must exist that will transform the independent variable into the dependent variable.

This is one reason why preservation of relationships is so important in model transforms. In this case, the preservation of relationships assures the conceptual integrity of the DFD (or IDEF0). When graphical modeling tools such as UML, DFD, and IDEF0 are used, the relationships are captured in the structure of the graph. This visualization technique makes it easy to understand the relationships conveyed by a given model and to make comparisons between models.

An elaboration of this simple idea may help make clear its significance. It is important to realize that this gives a way of identifying the processes required of a system (i.e., the required system functions) with absolutely no reference to the system. This identification is performed strictly based on the flow, as is the intent of DFD analysis. This was made possible by properly writing the requirements summary to focus on what the system must do, which is a subject of system transport or flow. System interactions (external to the system) can be treated through use cases.

The requisite modeling of the flow derives from an analysis of the logical model. The keywords in the logical model are the logical constants and variables that will be used to define the system processes. Each instance of a variable having logical relationship with another keyword identifies a potential system process. In this approach, the logical model becomes the collection of identified or defined relationships that will be the basis for the definition and decomposition of the system.

Figure 16.4 displays the resulting DFD for the requirements summary for the payroll system. The logical variables, time worked and salary, are displayed on the flow·lines in accordance with DFD graphical notation. Note that the EMPLOYEES Data Store in Figure 16.4 contains the logical constants, Employee ID and Pay Rate. As specified by the logical model, both these constants are needed in order

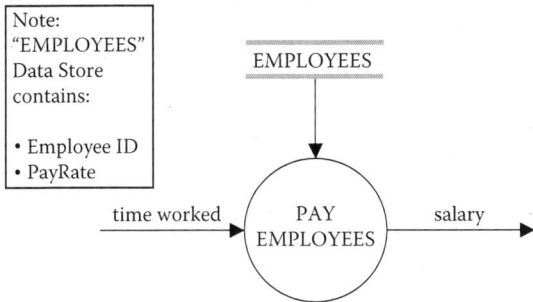

Figure 16.4 A DFD model of what the payroll system must do.

to pay an employee but as represented in the DFD, they are not part of the system flow. The DFD model in Figure 16.4 is implementation independent and is an essential model in structured analysis. In MDA, it is a PIM.

Model transformation will now be examined in detail as a design method that transforms "what" into "how" for the system. The Essential Model for the Payroll System gives us an implementation-independent model of what the system must do to satisfy the requirements summary. The transformation of the essential model into how it is implemented is the specification of the system solution (the implementation model). In MDA, this transformation is from the PIM into the PSM. The relationships determined by the essential model must be preserved under the model transformation in order to maintain conceptual integrity of the system.

The implementation will be the specification of how the employee receives his salary. This raises a design issue: which implementation should be done first: the process or the flows? Specifying the process first may limit the permissible data flows and data attributes. Similarly, specifying the data flows first may limit the permissible processes and process attributes. Figure 16.5 illustrates the trade space. But it is important to realize that in either case, as illustrated in the figure, the

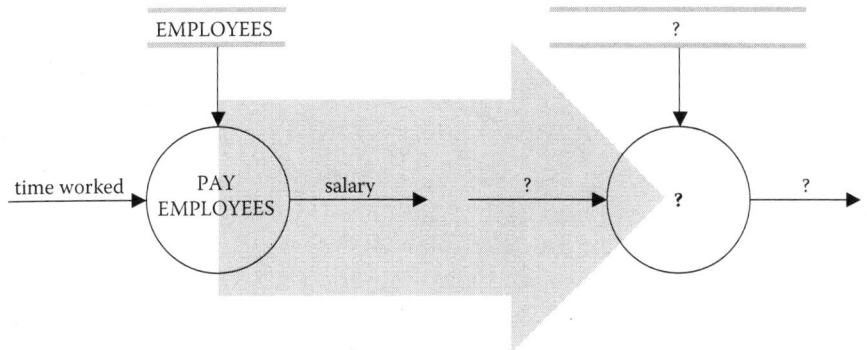

Figure 16.5 Implementation issue for the payroll system.

structure of the graphical relationships of the requirements model is preserved in the graphical relationships of the implementation model. In this particular case, the structures of the graphs are identical, and is called *isomorphism* (i.e., identical form), but in general, the graphs need not be identical; rather, the structure is preserved in some way by the model transformation.

Suppose that the process is implemented first. Perhaps the customer is a small business owner and prefers to have the employees paid in person. Maybe the business owner uses this as a mechanism to ensure that the employees do not leave early on Friday afternoons or to ensure that they actually come to work on Fridays so that the week's work can be finished. In this case, the model transform might be to specify the mechanism to implement PAY EMPLOYEES in Figure 16.5 to be a payroll clerk. This design decision immediately impacts the system flow. The payment to the employee will be paper based. Payment might be in the form of cash or it might be in the form of a check. In either case, payment will use some form of paper as the transport of payment from employer to employee. This may also impact the data store EMPLOYEES and even the reporting of the logical variable Time Worked. If the payment is paper based, then the business owner may choose to make the entire system paper based, which is a design decision. Employee data may be stored in something as simple as file folders. In this case, employees' work time may be reported on timesheets or even punch cards. Specific paper-based flows may be caused by the specification of PAYROLL CLERK.

Suppose that the flow is implemented first. Perhaps the customer is a large business and the employees prefer to be paid in person by electronic bank deposit. Specifying this type of flow necessarily limits permissible processes. No matter what approach is taken in the process, the ultimate output is in electronic form, and this will put pressure on the other aspects of the process of paying the employees to also be electronic. Time worked may be tracked using electronic CSV files (comma-separated values, a file format). A clearing house automated payment system (CHAPS) may be used to realize the electronic deposits of salary. And the EMPLOYEE data store may become a database implemented by a commercial technology like MySQL.

Details of MDA Model Transformation

The example of the payroll system can also be used to illustrate MDA model transforms (or *mappings* as they are also called). An MDA transform most commonly is used to provide specifications for the transformation of a PIM into a PSM using a particular (computing) platform. There are two distinct approaches used but most mappings will consist of some combination of them: (1) type mappings and (2) instance mappings.

Model type mappings specify mapping from any model using types specified in the PIM language to models expressed using types from a PSM language. Type

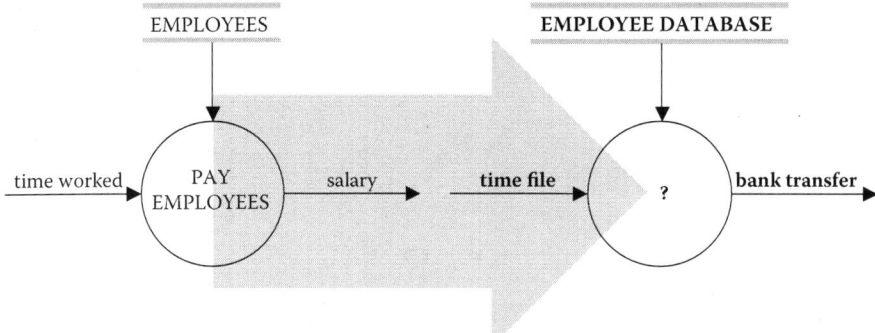

Specify Flow by the rule: use electronic media.

Figure 16.6   Example of a type mapping in a flow transform.

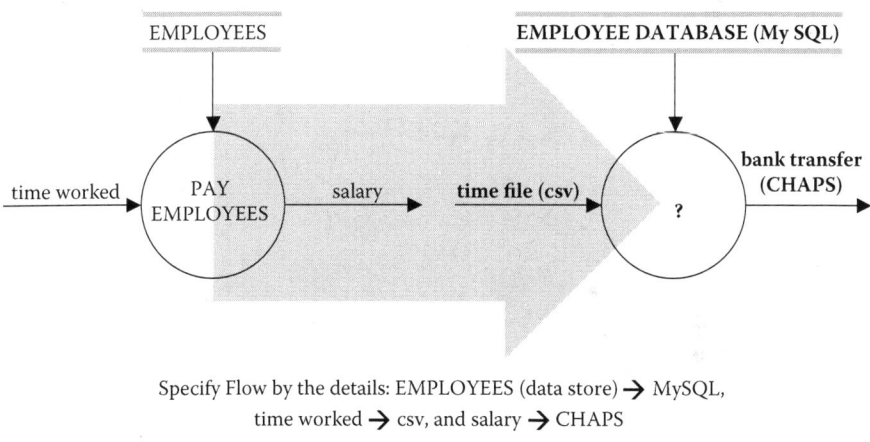

Specify Flow by the details: EMPLOYEES (data store) ➔ MySQL,
time worked ➔ csv, and salary ➔ CHAPS

Figure 16.7   Example of an instance mapping in a flow transform.

mappings are stated using rules. Model instance mappings identify model elements in the PIM by defining marks, which should be transformed in a particular way, given the choice of a specific platform for the PSM.

Figures 16.6 and 16.7 give an example of the distinction between type and instance mappings in MDA. In type mappings, a rule is specified. An example for the payroll system would be to specify system flow by the rule: use electronic media. A more formal definition of type mappings would be to say that type mappings preserve the relations of the model (i.e., the relational structure). Recall from Chapter 3 ("Concepts, Standards, and Terminology") that a relation in mathematical logic is a totality of relationships between entities and a relationship is an instance of a relation.

Model instance mappings will define marks. A mark represents a concept in the PSM, and is applied to an element of the PIM to indicate how that element is to be transformed. The marks, being platform specific, are not part of the PIM. In model instance mapping, the architect marks elements of the PIM to indicate the mappings to be used to transform that PIM into a PSM.

Figure 16.7 illustrates an instance mapping in a flow transform for the payroll system. In this case, the flows are specified by the individual model elements rather than by a rule. For example, the data store EMPLOYEES is marked to become an (electronic) database, which will be implemented by MySQL. The logical variable Time Worked is marked to be recorded in an electronic file, which will use CSV. And the employee Salary is marked for payment, which is implemented by CHAPS. This is the "point-to-point" nature of instance mappings and markings.

In practice, type mappings are very challenging to define. Within the scope of the system under development, they are like design paradigms: very general patterns of system design. Instance mappings are, in a sense, point-to-point transformation, that is, at the level of the details of the system. However, instance mappings are more commonly used because it is difficult to find type mappings.

Relation of MDA to Systems Engineering

The OMG MDA is a realization of the principles of structured analysis and design. Separation of problem specification from the solution is accomplished by the logical separation of the CIM and PIM in MDA. This was clearly illustrated in the example of the modeling and analysis of the Payroll System. This shows that MDA adheres to one of the principles of structured analysis. Logical modeling was seen to have a key role in the interpretation of the CIM into the PIM. The logical model of the requirements summary could be regarded as the UML model of the CIM. The processes identified for system definition were established exclusively by the analysis of the logical constants and variables in the logical model of the requirements summary. These processes were used in the DFD, which was a graphical representation of the PIM. The two types of system models (essential and implementation) are realized through the PIM and PSM. Therefore, MDA adheres to the modeling approach of structured analysis.

MDA also practices the principles of structured design because MDA domain models are highly cohesive and loosely coupled, although based on a different type of cohesion than the traditional approach to structured design, which focuses on cohesion because of interactions communication or system function exchanges. The MDA approach for software applications prescribes cohesion by subject type (i.e., domains of information). The underlying assumption is that domains require information sharing between each other far less than a domain requires information sharing within itself.

Also, relation preserving model transforms from the PIM to the PSM were central to system design. Because the interpretation of the logical model for the CIM was seen to be relation preserving and MDA model transforms should be relation preserving, the PIM, PSM, and the PM should adhere to the principle of structured analysis that the structure of the problem should be reflected in the structure of the solution.

Relationships between the concepts and terminology of systems engineering and MDA have not been firmly established, much less agreed upon by the software and systems engineering communities. However, the previous discussion should make clear that there is a solid foundation upon which to establish these relationships.

There are some obvious starting points from which to consider the possibility of relationships between MDA and the model-based approached to systems engineering. One is that in the model-based definition of *system* that we have developed, combinations of system elements are integrated in some way. In the physical architecture of a system, the combinations of elements are usually integrated to become entities called *components, subsystems* (which are joined by interfaces), and *assemblies*. Parts are usually considered to a low-level integration of elements, if not the lowest (or atomic) level of integration in systems engineering. MDA, on the other hand, typically groups elements into "parts", which are UML Packages of Classes, which become the components of the software system. In UML and MDA, connectors join the parts. Parts can be joined by connectors in systems engineering. However, the term *component* also has a different meaning in UML and MDA.

Relation of MDA to SoS Engineering

Our viewpoint SoS will be to apply the model-based definition of system to itself as in Chapter 3 ("Concepts, Standards, and Terminology"). This gave

> A system of systems is a combination of interacting systems (i.e., elements of the SoS) integrated to realize properties, behaviors, and capabilities that achieve one or more stated purpose(s).

The logical model for system was readily applied to itself to produce the model of SoS given in Chapter 3. However, the definition did not indicate whether all of the elements were capable of directly achieving the stated purpose of the SoS. Each of the "interacting systems" has properties, behaviors, and capabilities that when integrated achieve one or more of the stated purposes. But some of the "interactions" must play the role of integrating or supporting functions.

There were three key properties of the SoS that gave rise to this subtlety. First, each of the systems of interest (SOI) was independently capable of achieving the stated purpose, in that systems in the SOI were considered, in some way, peers.

However, as a system, the SoS had emergent behavior that was realized through the integrating functions of the SoS. The "other elements" of the SoS were necessarily infrastructure that either provided connectivity, integration, or support to the SOI.

When architecting an SoS, most of the physical systems have already been built to preexisting requirements. The mission capabilities of the SoS are primarily enabled by information exchange, as discussed in the Chapter 15 ("Toward Systems of Systems and Network-Enabled Capabilities"). Therefore, details of the system models are less important than details of the exchange of actionable information between the systems and how that information contributes to mission capabilities. Architecting an SoS will be different than architecting a single physical system. Because of the focus on information in the architecture and engineering of SoS, the emphasis can be expected to be more on developing software applications that enable mission capabilities. There will be other problems to be solved at the SoS level such as concept of operations, command and control, etc. but the systems engineering problem will likely settle into software engineering problems.

This was discussed in Chapter 15, which pointed to the Single Integrated Air Picture (SIAP) as a case study. SIAP consists of common, continual, and unambiguous tracks of airborne objects of interest in a surveillance area. Disparate systems within a Service as well as across multiple Services that share track management data have different views.

The SIAP was implemented by applying Model Driven Architecture (MDA) to develop an executable Peer-to-Peer (P2P) Integrated Architecture Behavior Model to form a Platform-Independent Model (PIM). The Services received the PIM for their host platforms and created Platform-Specific Models (PSM) for their hosts applying adaptive layers to interface with their specific sensors shown.

Why did SIAP use MDA? SIAP, similar to many examples of SoS, primarily is about transport and integration of information. Although the systems of an SoS may interact by means of all three types of transport, information is the type primary exchange. This leads to a fundamental premise that the two primary integrating functions are the architecting and the networking of the SoS.

Summary

MDA is a significant paradigm shift in software engineering that uses models to separate application logic from underlying platform technology. Significant return on investment is realized through increased reuse, and configuration management is improved using models instead of documents. Architecture models are a means to preserve the conceptual integrity of the system over time.

MDA is a realization of the principles of structured analysis and design. The MDA concept of architecture supports the model-based definition of system. System design is implemented through model transformations (type and instance mappings). MDA can support SoS engineering, as evidenced by the Single Integrated Air Picture (SIAP). However, relationships between the concepts and terminology of systems engineering and MDA have not yet been firmly established. This relationship is a subject of research by the systems community.

References

Jones, V., van Halteren, A., Konstantas, D., Widya, I., and Bults, R. (2007). An application of augmented MDA for the extended healthcare enterprise, *International Journal of Business Process Integration and Management*, 2(3):215–229.

Miller, J. and Mukerji, J. 2003. *MDA guide, version 1.0.1*. OMG. http://www.omg.org/docs/omg/03-06-01.pdf (accessed 8 August 2008).

Raistrick, C., Francis, P., Wright, J., Carter, C., and Wilkie, I. 2004. *Model-Driven Architecture with Executable UML*. New York: Cambridge University Press.

Soley, R.M. March 2004. Leveraging Model-Driven Architecture° for Model-Driven Systems Engineering. http://www.omg.org/mda/presentations.htm.

AEROSPACE AND DEFENSE SYSTEMS ENGINEERING

4

Chapter 17

Fundamentals of Systems Engineering

Systems engineering is an essential part of the foundation and theoretical basis of this book. This chapter provides an overview of systems engineering, starting with its most fundamental concepts and building blocks. Then, the elemental question, What is systems engineering? is addressed, first by noting that consensus on a single answer is absent but that a set of characteristics is commonly observed across most of the proposed definitions. These characteristics, along with the basic concepts and principles, comprise the fundamentals of systems engineering, which are explained and illustrated by observing a selection of established system engineering processes. This chapter introduces these fundamental concepts and illustrates how they are implemented and embodied in a variety of widely known and accepted systems engineering processes.

Key Concepts

Requirements analysis
Functional decomposition
Synthesis
System Integration

Fundamental Concepts in Systems Engineering

Before proceeding with a definition of systems engineering and the examination of representative systems engineering processes, it is paramount to state some of the fundamental concepts and principles that lay at the foundation of this practice. This section does not intend to cover a comprehensive list of all the concepts in systems engineering, but rather focuses on those that represent the most basic and crucial conceptual building blocks. The starting point is, quite appropriately, the definition of "system." Though a number of accepted definitions exist (see, e.g.,

271

the definitions in Chapter 3 ["Concepts, Standards, and Terminology"]), all of them refer to "a collection of elements." This suggests that in its most basic form, a system is defined by the elements that comprise it, and implies that there are some criteria that differentiate these elements from others outside the system. It also suggests that there is a *hierarchical relationship* between a system and its elements, namely, that an element is a part or member of a system. Following the approach in Chapter 2 ("Logical and Scientific Approach"), combinations of system elements can be identified. Certain combinations of the elements in a system may be viewed as systems, which are regarded as subsystems of the larger system. This leads to the concept of a *hierarchical structure*. In a hierarchical structure, a system exists at a certain level of abstraction, and the elements that are part of that system are at a lower level of abstraction, or at a lower level in the hierarchy. These are referred to as *system levels*, or simply *levels*. Thus, the hierarchical structure states what combinations of elements define systems across multiple levels (Kossiakoff and Sweet 2003; 31–44). As a system is described in terms of its elements, and of the groupings of elements that comprise them, a process of *decomposition* is said to take place.

These fundamental concepts are readily illustrated by a simple example. An aircraft is a system composed of numerous elements such as the wings, fuselage, and engines, among many others. Each of those elements may in itself be considered a system and decomposed into its elements. The jet engine, for instance, can be decomposed into the inlet, compressor, combustor, turbine, etc. The process may be extended further, for instance, decomposing the compressor into its multiple stages, each stage into rotors and stators, and each rotor into blades. However, decomposition is not a concept limited to physical descriptions of systems. Decomposition can be *functional* in nature as well: a function may be decomposed into a set of functions that comprise it. Continuing with the aircraft example, the function of the aircraft is to fly a mission, which can be decomposed into the different phases of the mission such as take-off, climb, cruise, etc. Cruise can be decomposed into functions such as providing lift, generating thrust, maintaining stable control, etc. At this point it is crucial to recognize that whenever a *functional decomposition* and a *physical decomposition* of a system take place, there should be (in the vast majority of cases) an association between physical elements and the functions they perform at the same level in the system hierarchy structure. In the present example, the function "providing lift" is performed (primarily) by the wings of the aircraft, and the function "generating thrust" is performed by the engines.

The counterpart of decomposition, namely *integration*, is another fundamental concept in systems engineering. It can be conceived as the reversal of the decomposition process within the same hierarchical relationship between systems and elements. Thus, just as a system may be decomposed into its multiple elements, a series of elements (or systems) may be integrated into a system. As a final comment in this section, it is important to note that the definition of a system from its elements is strongly dependent on the point of view. The example of the aircraft assumes a perspective where the aircraft is the system at the top level of the hierarchy structure. This may be the case for an aircraft designer, for instance, but not for an air traffic controller, for whom one aircraft is a single element in a collection of aircraft operating within an airspace sector. Conversely, an engine may be viewed as the top-level system by a propulsion mechanic.

What Is Systems Engineering?

With the fundamental building blocks of systems engineering explicitly stated, there are still a multitude of definitions for *systems engineering*. Most of them describe systems engineering in terms of the processes and activities involved in the development of a system. It is not surprising then that no single systems engineering process is currently recognized as the only accepted approach, and thus consensus on the answer to this elemental question remains elusive (SMC 2004). It is therefore necessary, if not inevitable, to address a number of these definitions in hopes of identifying a common characterization. A large number of definitions are available in the literature but in the interest of conciseness, only a small subset of them is presented here. Systems engineering can be defined as follows:

> An interdisciplinary approach that encompasses the entire technical effort, and evolves into and verifies an integrated and life cycle balanced set of system people, products, and process solutions that satisfy customer needs. (EIA 1994)(DoD 201).

> A logical sequence of activities and decisions that transforms an operational need into a description of system performance parameters and a preferred system configuration (DoD 1974; DoD 201).

> An iterative process of top-down synthesis, development, and operation of a real-world system that satisfies, in a near optimal manner, the full range of requirements for the system (Eisner 2002; INCOSE 2003).

> An interdisciplinary approach and means to enable the realization of successful systems. It focuses on defining customer needs and required functionality early in the development cycle, documenting requirements, and then proceeding with design synthesis and system validation while considering the complete problem (INCOSE 2003).

Despite the obvious differences among definitions, there are key concepts visible across all of them that provide an adequate contextual construct for the remainder of this chapter. First, systems engineering encompasses a collection of activities, processes, and techniques. Second, its purpose is to enable the generation of a system that represents a design solution and meets certain needs or requirements.

Systems engineering therefore enables the transformation of information from requirements to system definition. Third, the processes are characterized as being iterative, recursive, and interdisciplinary, and take a multitude of aspects of the system into consideration. Because these processes are based on a structured, logical approach to problem solving, systems engineering offers a means to manage the complexity of systems development, revealing a strong incentive to implement its practices and techniques. The following sections introduce well-established systems engineering processes that embody and clearly illustrate these and other fundamental concepts.

The "Vee" Model

One of the most well-known set of systems engineering processes for systems development is the "Vee" model, shown in Figure 17.1. The "Vee" is a graphical representation "portraying the technical aspects of the project cycle [that] clarifies the role and responsibility of system engineering to a project" (Forsberg and Mooz 1991). This model was developed by Forsberg and Mooz based on an earlier version by NASA in its participatory role within the Software Management and Assurance Program (SMAP). Forsberg and Mooz also expanded the basic "Vee," elaborating on the project cycle and the role of systems engineering, thus developing a comprehensive version of the model. In this chapter the model is explained with the basic chart shown in Figure 17.1 for simplicity. However, as per Forsberg and Mooz's comments, the core activities in the expanded "Vee" can be directly mapped to those of its basic version (Forsberg and Mooz 1991).

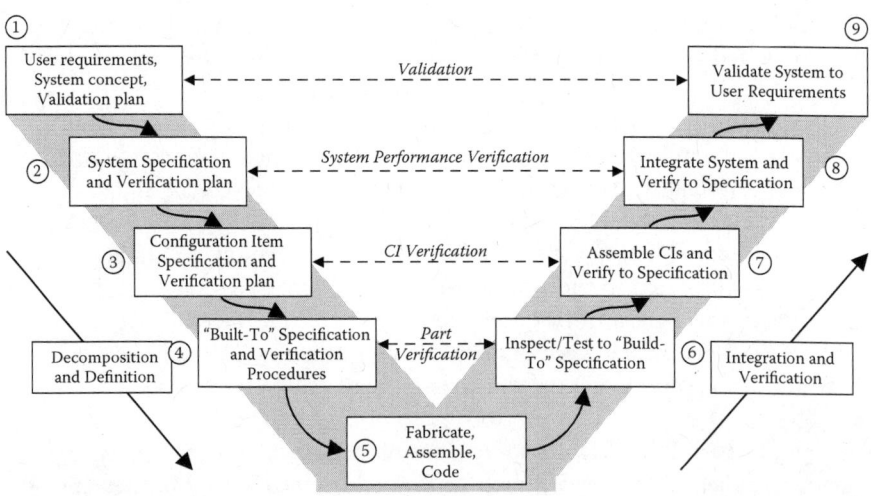

Figure 17.1 The Systems Engineering "Vee." (Adapted from Forsberg and Mooz "vee" model.)

The sequence of steps in the "Vee" starts at the top left with the user needs and ends at the top right with the user-validated system. Time and project maturity flow from left to right, whereas the level of system decomposition increases from top to bottom, and then reverses as a level of integration from bottom to top. The "Vee" is divided into two main parts. The downward steps on the left half correspond to the "Decomposition and Definition" portion of the model. Decomposition and definition are a key principle in systems engineering: starting with a highly abstracted concept, a greater level of system definition is attained by decomposing the system into deeper levels of detail and providing sufficient specification for that level of detail. It follows that the outcome of this half of the model at the bottom of the "Vee" is a fully defined system, specified at the lowest level of detail needed for its construction via a performance/design-to/build-to evolution of the specification stream. The bottom, or vertex, of the "Vee" is the actual fabrication of the system components as per build-to specification, and constitutes the highest level of decomposition. The second half of the model is Integration and Verification, in which all parts of the system are produced and progressively integrated into larger assemblies that will finally comprise the complete system. Verification at each level of integration provides a means to check whether the parts or assemblies meet the specifications stated earlier at that level, and can be explained in terms of answering the question "Was the system built right?"

Because the verification task at each level of integration is dependent on the development of specifications at the corresponding level of decomposition, there is a direct correspondence between steps on the right and left halves of the "Vee" (Forsberg and Mooz 1991).

Note that at the top level, verification is replaced with validation, a process by which compliance of the completed system with established needs and requirements is demonstrated, and often explained in terms of answering the question "Was the right system built?" Validation and verification are two different, clearly distinguishable tasks; in fact, validation demonstrates that user needs are met irrespective of the system specification. However, they are often considered side by side in systems engineering process planning and referred to as "V&V." Also, these tasks are performed on the integration portion of the "Vee" model, but their plans are formulated in Decomposition and Definition, where requirements and specifications are generated (INCOSE 2003; Forsberg and Mooz 1991). With these concepts in mind, a step-by-step description of the basic "Vee" model is presented in the remainder of this section.

In the first step, the future users of the system and stakeholders are queried to establish their needs and translate them into system requirements. Various notional system concepts are developed at this stage in conjunction with a validation plan, which establishes validation criteria based on needs and requirements and identifies the tests that will help answer the validation question. In the second step, further concepts development take place, followed by concept trades and selection. These tasks are run in parallel with functional allocation to ultimately generate a system performance specification. In this step, system requirements have been decomposed into performance specifications, thus providing a higher level of definition. This step also formulates a system-performance verification plan that, analogous to its validation counterpart, establishes criteria based on system specifications and defines actions to answer the verification question. The third step decomposes the system into Configuration Items (CIs), and expands the previous system-level performance specification into a "design-to" specification for all CIs. A CI is simply one of multiple elements that make up the system at that level of decomposition/integration. This step represents a system decomposition and definition at the CI level and includes the generation of a corresponding CI verification plan. Though not shown in Figure 17.1, Forsberg and Mooz note that further decomposition of the CIs into elements or parts may be necessary, thus generating even more detailed system definitions. For each case, a "design-to" specification is generated along with a corresponding verification plan. When the lowest level of decomposition and detail has been reached, the fourth step evolves the parts' "design-to" specifications to "build-to" specifications. This step usually involves detailed design and manufacturing considerations, and provides the corresponding verification plan, which usually represents a manufacturing inspection procedure. The fifth step at the bottom of the "Vee" is the fabrication and assembly of parts at the lowest level of integration (Forsberg and Mooz 1991).

The sixth step, first on the integration side, involves the execution of the part verification plan using "build-to" specifications. Step seven assembles parts and elements into CIs, and performs the CI verification plan based on "design-to" specifications. If, as noted in the previous paragraph, multiple CI decomposition steps were performed in the first half of the "Vee," an equal and corresponding number of CI integration and verification steps should exist on the second half. The eighth step integrates all top-level CIs into the system, followed by the verification process of system performance specifications. Finally, the fully realized system is validated relative to user needs as per the validation plan in the ninth and last step (Forsberg and Mooz 1991).

The DoD Systems Engineering Process

Another popular and widely accepted systems engineering process is that of the U.S. Department of Defense (DoD), which offers an interesting alternative to the "Vee" model, while capturing the same key concepts of decomposition, definition, integration, and verification as enabling functions for the "structured but flexible [...] transformation of requirements into specifications, architectures, and configuration baselines" (DoD 2001). The DoD process is formulated in the context of Systems Engineering Management, defined as the collection of three primary activities: development phasing, the systems engineering process, and life-cycle integration. As suggested by its name, development phasing divides the overall effort into phases and controls it by developing relevant baselines at each level. Additionally, it serves as a bridge between the technical effort and acquisition management by delivering milestones for viability assessments. The systems engineering process, which will be the focus of the remainder of this section, is the central activity in the DoD Systems Engineering Management formulation, and is performed one or more times in each phase of the development cycle. Finally, life-cycle integration introduces planning considerations across the entire life of the system into the development endeavor to ensure its viability throughout (DoD 2001). The systems engineering process, shown in Figure 17.2, is best described as a "top-down comprehensive, iterative, and recursive problem-solving process, applied sequentially through all stages of development" (DoD 2001). With each instantiation of the process, user needs and requirements are transformed into specifications, baselines, and architectures, thus injecting detail and enhancing system definition into the subsequent phases. Given the architecture focus on this book, it is important to note that DoD Systems Engineering Management utilizes three primary types of architectures to describe and define different aspects of the system under development: (1) the *functional architecture*, which embodies the structure of allocated performance and functional requirements; (2) the *physical architecture*, which provides the breakdown structure of the physical system into various levels of subsystems, components, and parts; and (3) the *system architecture*, which identifies all products and processes required to support the system across its entire life cycle. The genera-

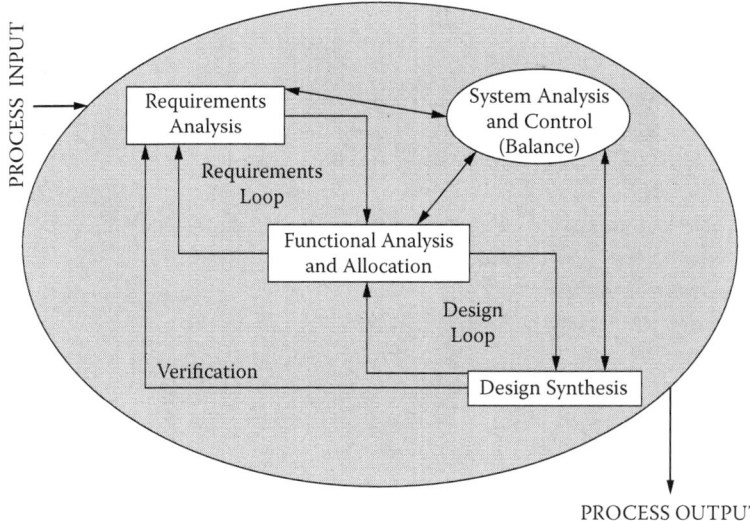

Figure 17.2 The U.S. Department of Defense systems engineering process. (From Department of Defense (DoD), United States. 2001. *Systems Engineering Fundamentals.*)

tion of said architectures throughout the different elements of the process, and the most relevant details therein, are hereby explained (DoD 2001).

The inputs to the DoD systems engineering process are customer needs, requirements, and constraints, which constitute contextual information that provides initial guidance into the definition of a system. For this process, the customers are the people who perform the eight primary life-cycle functions of the system as defined by the DoD: development, production, and construction; deployment; operation; support; disposal; training; and verification. The "users" of the system, as identified by the "Vee" process, are the customers involved in the operation function. Examples of customer needs, requirements, and constraints include missions or measures of effectiveness, availability of resources from previous development programs, budgetary and other programmatic constraints, assumptions about entry into service dates and the associated technology base, and applicable standards, among others (DoD 2001).

Requirements analysis, the first step in the DoD process, consolidates the process inputs into a cohesive set of customer requirements, and then translates them into functional and performance requirements. Customer requirements, often focusing on the operator, *state expectations* about the system in terms of measures of effectiveness, mission objectives, operational environment, and design constraints. The latter are limitations to development flexibility, such as environmental limits or regulatory standards. Functional requirements *state the required actions* that must be performed by the system to meet the aforementioned expectations, and embody top-level functions used in the functional analysis step. In turn, performance requirements *state the extent to which actions must be performed* to meet expectations, and are expressed in terms of quality, quantity, scope, and availability. Together, functional and performance requirements constitute the core of the Requirements Analysis output, effectively defining *what* the system must do and *how* well it must do it, respectively. In addition, this process yields two more important sets of information based on the consolidation of customer requirements. First, an operational context for the system is defined, thus describing the environment in which the system is expected to operate. Second, a range of design constraints and limitations is identified by the customers that effectively bounds the design space in which the system must evolve. These outputs are documented in a *decision database*, and are expressed in terms of three coordinated views: *functional, physical,* and *operational.* The functional view addresses what the system must do to attain a prescribed operational behavior, and couples functional, performance, and physical requirements, which shape design specifications downstream in the systems engineering process. The physical view addresses the construct of the system by establishing the configuration of operator–system interfaces, characterizing the different users and establishing physical limitations. The operational view, embodied in an operational concept document, describes the context in which the system is expected to operate (DoD 2001).

There is no unique way of performing requirements analysis. A variety of approaches have been developed and evolved over time, each featuring unique strengths and weaknesses in particular applications. The DoD, however, recognizes this process to be one of inquiry and resolution where the systems engineer asks questions of the customers and consequently generates customer requirements based on the answers received. Additionally it provides a "Procedure for Requirements Analysis" as a supplement in reference (DoD 2001).

Core outputs of Requirements Analysis, namely the functional (what) and performance (how much) requirements, are passed as inputs to Functional Analysis and Allocation. In this step, the top-level functions comprising the functional requirements are decomposed and flowed down, or *allocated*, into lower-level functions. The performance requirements associated with the top-level functions are then allocated to the lower-level function. This procedure of functional decomposition and performance requirement allocation is then successively repeated to define progressively lower-level functions. This analysis embodies the principles of decomposition and definition, and it therefore presents a fundamental parallelism with the first half of the "Vee." This process is enabled by a series of tools and techniques such as functional flow diagrams, IDEF0 models, and functional allocation sheets. Once a sufficiently detailed level of decomposition has been attained, the resulting description of the system explains what the system does and how well it must do it at each level. This structured description of the system's functionality is the *functional architecture* of the system. In fact, with the completion of every decomposition and allocation cycle, the architecture is said to be defined at a greater level of detail. Figure 17.3 presents a notional functional decomposition and architecture for an

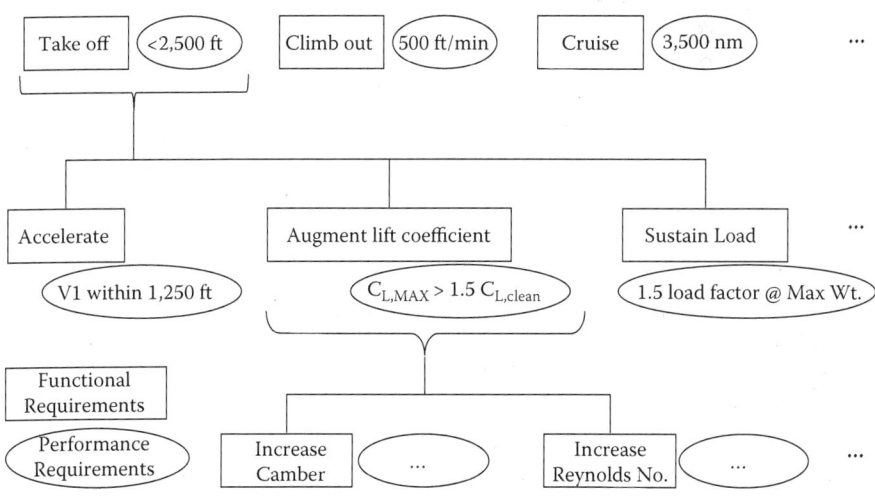

Figure 17.3 **Notional functional decomposition and allocation for an aircraft.**

aircraft whose functional and performance requirements have been defined and are allocated to lower levels of the architecture. Beyond the functional architecture as a core output of this step, significant insight is attained in terms of what the system must do, how well it has to do it, the conflicts between functions, and the trades and compromises that will have to take place downstream (DoD 2001).

On the basis of this insight, the systems engineer will often find issues, such as incompatible requirements or inadequate standards, that demand the reexamination of top-level or higher-level functional requirements. Even if that is not the case, performing Functional Analysis and Allocation always yields a deeper understanding of the requirements, given that all functions at the lowest level should be traceable back to a requirement. Consequently, the Requirements Analysis step is often revisited, leading to the iterative execution of these first two steps, referred to as the Requirements Loop. When iterated upon, this loop allows systems engineers to better define functional requirements based on the voice of the customer, improve the definition of corresponding performance requirements, and consequently produce better functional architectures (DoD 2001).

The functional architecture is then passed as an input to the Design Synthesis step. The objective of Design Synthesis is to generate a *physical architecture* of the system that performs all the functions with the prescribed required performance as established in the functional architecture. The physical architecture is the core output of Design Synthesis, which is manifested in different ways, based on the current development phase. In early development phases, Design Synthesis is embodied by conceptual design and the physical architecture is limited to system concepts and the basic structure of subsystems. Later phases involve preliminary and detailed design where correspondingly more elaborate descriptions of subsystems and their interrelationships are defined. Figure 17.4 presents a notional physical decomposition for an aircraft design at the conceptual level. For simplicity, only one element is

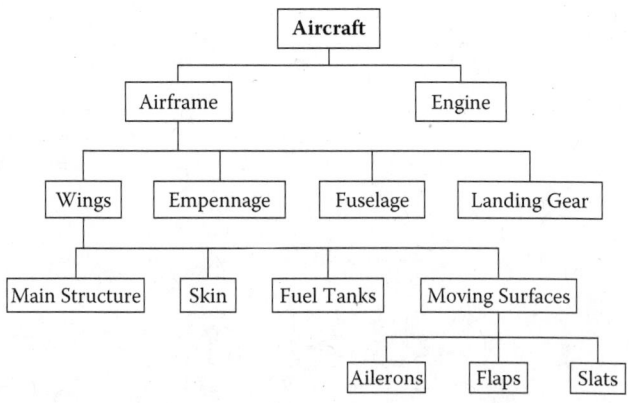

Figure 17.4 Notional physical decomposition for an aircraft.

expanded at the next lower level. Regardless of the development phase, the physical architecture ultimately represents the basis for design specifications, baseline definitions, and other design documentation (DoD 2001).

The word *synthesis* in the name of this step refers to the composition, combination, or bringing together of parts to form a coherent whole (Merriam-Webster Online Dictionary). It is thus suggestive of a fundamental principle in systems engineering, namely integration. Analogous to the second half of the "Vee," Design Synthesis brings together parts of the system, starting from the lowest level of decomposition, and integrates them into a whole. As with Requirements Analysis, there is not one single or correct way of performing this step, but rather there are a number of process features that have been widely recognized as important or desirable. For instance, Design Synthesis is an inherently creative activity, and thus is strongly dependent on the resourcefulness, ingenuity, and experience of the engineer. Development of physical architectures requires that all components contribute to meeting one or more functional requirements. In this regard, a modular approach is often an attractive alternative, allowing for certain assemblies to be responsible for single and independent functions so that entire segments of the systems are independently testable and can be readily replaced by upgraded versions. Finally, selection among a number of architectures requires that the decision be justified and supported by adequate trades and analyses (DoD 2001).

In terms of techniques and methods for Design Synthesis, a single or correct approach is also absent, and instead a large number of methods are featured that are rendered more or less applicable, depending on the specifics of the application. Modeling techniques are of particular importance, allowing the systems engineer to evaluate the performance of a given functional architecture and capture the behavior of the physical product. Furthermore, models enable optimization techniques and leverage trade studies at various levels. Chapter 19 ("Systems Design") of this book presents a number of established design synthesis techniques and tools. More advanced design methodologies, representative of the state of the art, are then introduced in Chapter 20 ("Advanced Design Methods").

The Design Loop is the iteration between Design Synthesis and Functional Analysis and Allocation, whose objective is the evaluation of consistency between the physical architecture and the functional and performance requirements. This loop involves a mapping of elements between physical and functional architectures, and very often provides valuable insight about conflicting or countervailing design trends. This broadening of knowledge about the system can prompt the reevaluation of the functional architecture and even of top-level requirements, but can also help refocus efforts toward promising solutions and design opportunities (DoD 2001).

Once a physical architecture has been fully developed, or chosen among a number of candidates, the Verification process allows the systems engineer to confirm that this architecture satisfies system requirements. In early phases, modeling tools are used to determine the values of metrics of interest that are then compared with system requirements at various levels of decomposition. Later phases require progressively higher-fidelity tools to provide more accurate assessments of the physical architecture. Before the production of the system begins, final prototyping phases require testing and evaluation of the different physical components starting at the lowest level and progressively moving to larger components and assemblies. Testing and evaluation criteria are generated from design specifications and requirements documentation resulting from the physical architecture definition (DoD 2001).

Throughout the DoD systems engineering process, the engineer is required to perform a number of *systems analysis* activities to support the evaluation of alternatives and the decision-making processes. Examples include trade-off studies and effectiveness analyses. Likewise, the engineer must respond to technical management responsibilities related to the allocation of the program resources at different levels. To do so, several *control* activities such as progress assessment and risk management must be executed. The collection of these activities is uniformly applicable across all elements in the systems engineering process, and is grouped in Systems Analysis and Control. These activities add value to the entire process by ensuring that technical decisions are adequately supported and consider impacts on the multiple aspects of the system. They also enable traceability of the entire process by facilitating the documentation of system design evolution as well as most resource allocation decisions (DoD 2001).

The specific outputs of the systems engineering process vary, based on the development phase in which the process has been executed. In general, however, they consist of all "the documents that define the system requirements and design solution" (DoD 2001). In their most basic form, outputs are embodied by functional and performance requirements, the functional architecture, and the physical architecture, all defined for a level of detail and definition. In the context of DoD systems engineering management, the level of technical detail in the systems engineering process progressively increases with development phases and is manifested in the degree of system definition observed in process outputs. Moreover, the outputs of any instance of the process become inputs for the instance that follows immediately after (DoD 2001).

This sequence of DoD systems engineering process instances, as shown in Figure 17.5, as well as the relationship between their inputs and outputs, are best understood by examining the role of the different types of outputs. A *system architecture* is a description of the entire system based on the *physical architecture* and the definition of supporting products and services across all phases of the life cycle. *Specifications* are based on the *functional architecture*; they describe what the system must do by documenting functional requirements, describe how well it must do it by documenting performance requirements, and describe how to verify that these requirements are met by documenting the verification process. A *baseline* is the complete documentation of the system at some level of detail and design definition, and serves as reference for subsequent development phases. Systems developed

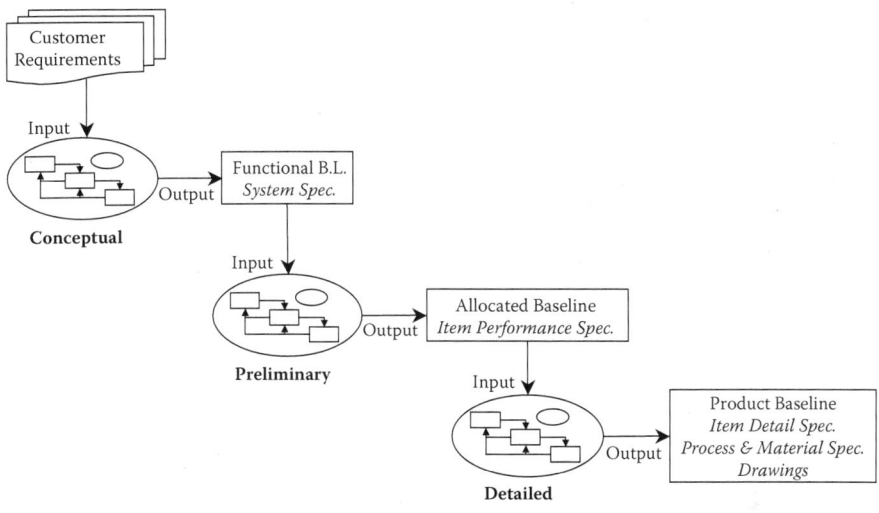

Figure 17.5 Inputs and outputs of DoD systems engineering process sequence.

under the DoD systems engineering management framework usually incorporate three baselines, collectively known as *configuration baselines,* which define the system at different levels within the system hierarchy and are generated during different phases. The *functional baseline,* created in the conceptual phase of design, is a system-level technical description based on top-level system concept functions and performance requirements, which in turn are derived from customer requirements. These elements comprise the *system specification,* which documents the functional baseline. The resulting system architecture in this phase is limited to the higher levels of the system hierarchy. In the preliminary phase, system requirements are allocated to items at the next level and used to generate *item performance specifications.* An *allocated baseline,* also called a *"design-to" baseline,* is documented with these specifications, item interfaces, and other design documents. At this stage the system architecture has been expanded to the item level of specification. The detailed design phase uses the allocated baseline as a reference and defines the system completely, down to the component level of detail. The outputs at this phase include *item detail specifications, process and material specifications,* and a full physical architecture. All these outputs collectively document the *product baseline,* also referred to as the *build-to baseline.* At this phase the system architecture describes the system to the greatest degree of detail (DoD 2001).

In brief, the DoD process features key principles and fundamental concepts of systems engineering and applies them in a structured way to enable the definition of a system from customer requirements. A summary of the DoD system engineering process is provided graphically in Figure 17.6, highlighting the key items in each element of the process.

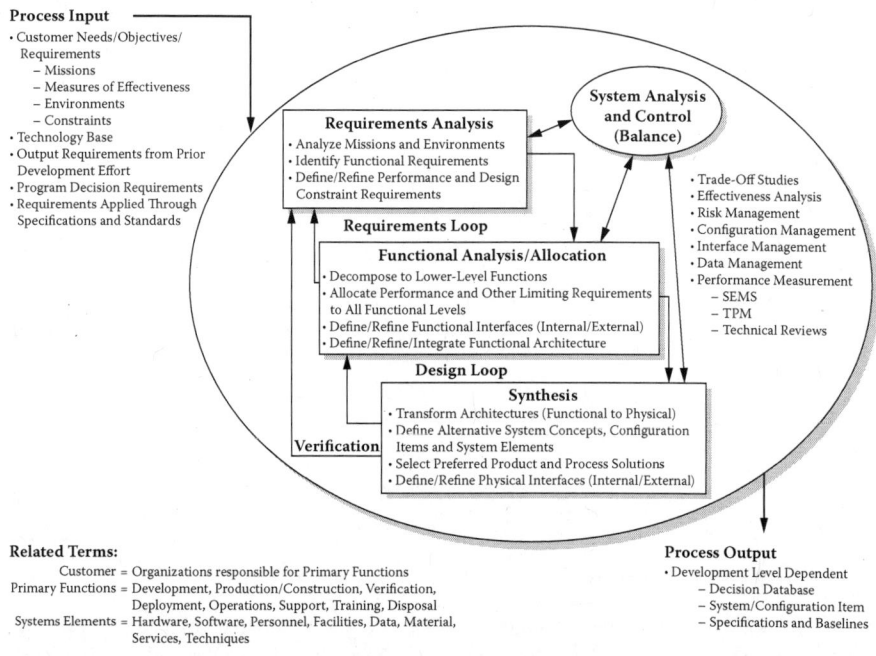

Process Input
- Customer Needs/Objectives/
  Requirements
  - Missions
  - Measures of Effectiveness
  - Environments
  - Constraints
- Technology Base
- Output Requirements from Prior
  Development Effort
- Program Decision Requirements
- Requirements Applied Through
  Specifications and Standards

Requirements Analysis
- Analyze Missions and Environments
- Identify Functional Requirements
- Define/Refine Performance and Design
  Constraint Requirements

System Analysis and Control (Balance)
- Trade-Off Studies
- Effectiveness Analysis
- Risk Management
- Configuration Management
- Interface Management
- Data Management
- Performance Measurement
  - SEMS
  - TPM
  - Technical Reviews

Requirements Loop

Functional Analysis/Allocation
- Decompose to Lower-Level Functions
- Allocate Performance and Other Limiting Requirements
  to All Functional Levels
- Define/Refine Functional Interfaces (Internal/External)
- Define/Refine/Integrate Functional Architecture

Design Loop

Synthesis
- Transform Architectures (Functional to Physical)
- Define Alternative System Concepts, Configuration
  Items and System Elements
- Select Preferred Product and Process Solutions
- Define/Refine Physical Interfaces (Internal/External)

Verification

Related Terms:
Customer = Organizations responsible for Primary Functions
Primary Functions = Development, Production/Construction, Verification,
     Deployment, Operations, Support, Training, Disposal
Systems Elements = Hardware, Software, Personnel, Facilities, Data, Material,
     Services, Techniques

Process Output
- Development Level Dependent
  - Decision Database
  - System/Configuration Item
  - Specifications and Baselines

Figure 17.6 Summary diagram of the DoD systems engineering process. (From Department of Defense (DoD), United States. 2001. *Systems Engineering Fundamentals.***)**

Other Systems Engineering Processes

The "Vee" and the DoD systems engineering process are well-accepted practices that lend themselves very well for the illustration of the key principles and fundamental concepts of systems engineering. There are, however, multiple other frameworks that feature analogous practices aimed at the design and development of systems. In most cases it is relatively easy to identify parallelisms between specific practices and even map corresponding elements between them. Most important, though, it is possible for the systems engineer to identify where principles and concepts are embodied and how they are applied in each instance. A few notable examples will be examined in this section, providing a good starting point in the task of researching and understanding how the fundamentals of systems engineering are realized in other accepted practices.

The "waterfall" model, shown in Figure 17.7, is the basis for the Decomposition and Definition portion of the "Vee." It was originally proposed by Royce (Royce 1970) as a series of steps to be followed for large software development programs. As suggested by its name, the model naturally flows down to subsequent steps once

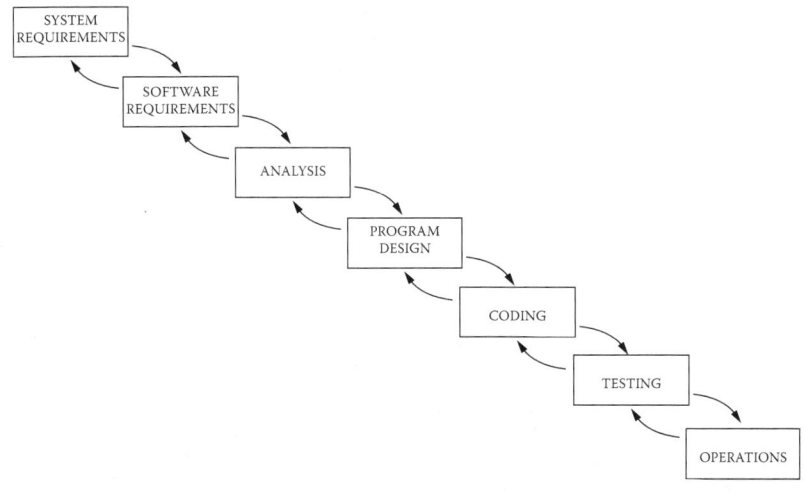

Figure 17.7   Waterfall model for large software development programs. (From Royce, Winston. 1970. Managing the development of large software systems. *Proceedings of the IEEE WESCON*: **1–9.)**

the current one has been completed and fully implemented. Royce incorporates the iterative nature of most systems engineering practices by adding arrows that communicate downstream activities with preceding ones. This concept is directly analogous to the Requirements Loop and the Design Loop in the DoD process. In this regard, and somewhat alluding to the notion of system definition through decomposition, Royce notes that "as each step progresses and the design is further detailed, there is an iteration with the preceding and succeeding steps but rarely with the more remote steps in the sequence" (Royce 1970).

The "spiral" model, shown in Figure 17.8, also finds its origins in the software development field. Given the widespread use of the "waterfall" approach, this "model [became] the basis for most software acquisition standards" (Boehm 1988). The "spiral" model is no exception, and moreover captures most previous models as special cases in the context of specific implementations. In this approach each loop of the spiral represents an instance of the same progression of generic steps: (1) determination of objectives, alternatives, and constraints; (2) evaluation of alternatives; (3) prototyping/simulation for risk analysis; (4) specification; (5) validation/ verification; (6) planning for next cycle; and (7) review. More notably, with each cycle the system is defined at a more detailed level of elaboration, applying the principle of definition by decomposition. Moreover, the output of every cycle becomes the input to the next one, a structure that evokes the recursive use of the DoD systems engineering process through multiple phases of development, using outputs at one phase as inputs to the next one. The "spiral" model also captures the costs

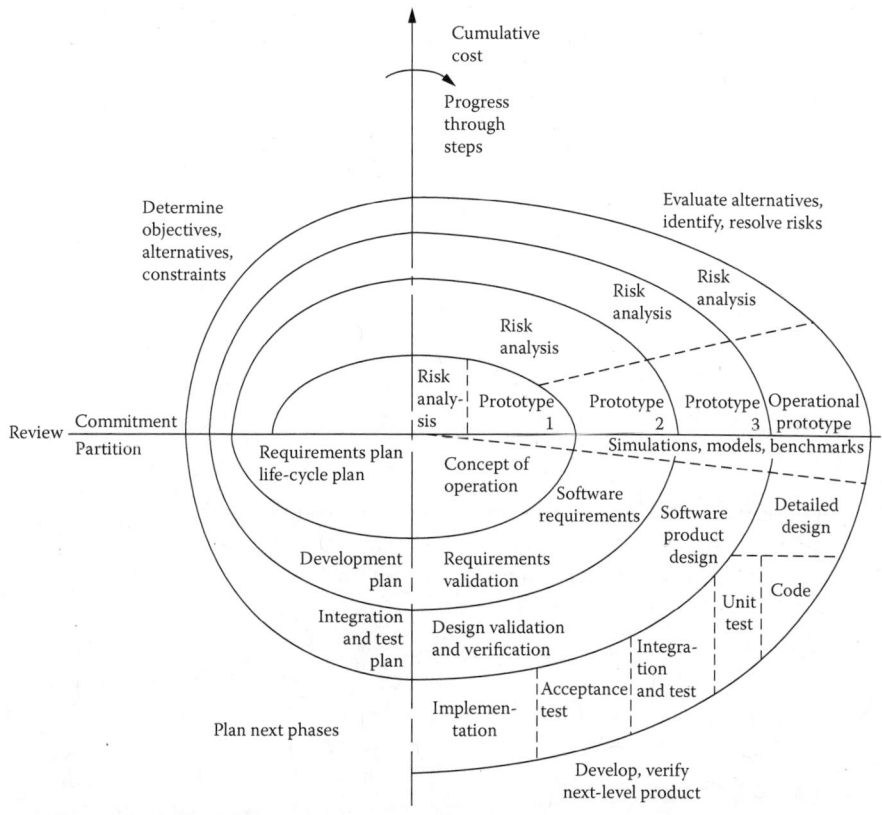

Figure 17.8 Spiral development model. (Boehm, B.W. 1988. A spiral model of software development and enhancement. *Computer* **21(5):61–72.)**

committed in the total development project with the radial distance of the spiral. Finally, the model accommodates a prototype-driven approach where evolutionary progressions are made to minimize risk (Boehm 1988).

A more elaborate and extensive set of systems engineering practices is found in the *INCOSE Systems Engineering Handbook* (v. 3; INCOSE 2003). This handbook documents a number of key processes and activities performed by systems engineers based on the standard *ISO/IEC 15288:2002(E)—Systems engineering—system life cycle processes.* The breadth, depth, and structure in the documentation of relevant processes within a cohesive framework is evocative of the DoD's Systems Engineering Fundamentals rather than of a model such as the "waterfall" or the "spiral." It features a set of Technical Processes that follow the life-cycle stages of a system and that can be related to portions of the "Vee" and the DoD process. These processes are Stakeholder Requirements Definition, Requirements Analysis, Architectural Design, Implementation, Integration, Verification, Transition, and Validation.

Processes for subsequent life-cycle phases, namely Operation, Maintenance, and Disposal, are also provided. Additionally there are a number of Project Processes directly related to the technical management of a project, and are analogous to those in Systems Analysis and Control of the DoD process (INCOSE 2003).

The last set of practices presented is documented in the *NASA Systems Engineering Handbook* (NASA 2007), which also represents an extensive collection of practices structured within a major framework. Much like the INCOSE handbook, this document presents a model for project life cycle and explains the various systems engineering processes that directly address each phase as well as those that support the process as a whole from a technical management perspective. The NASA program and project life cycle is composed of the following phases: concept studies, concept and technology development, preliminary design and technology completion, final design and fabrication, system assembly integration and test, operations sustainment, and closeout. Design activities initiate the cycle of product development, starting with stakeholder expectation definitions and their transformation into technical requirements. The latter represent the basis of a logical decomposition process, which then leads to the design solution definition. During product realization activities, the various product elements are "produced, acquired, reused, or coded; integrated into higher-level assemblies; verified against design specifications; validated against stakeholder expectations; and transitioned to the next level of the system" (NASA 2007).

Summary

Absent a consensus on a single definition of systems engineering, there is a plethora of processes used to develop a system from customer requirements. In turn, no single approach is recognized as the single accepted practice; rather, a number of processes are available for systems engineering to choose from and adapt to their particular needs. However, a number of fundamental concepts and principles are commonly found across the more well-known processes and models among whose elements parallelisms are often observed. In general, systems engineering processes, inherently *iterative* in nature, start with customer needs and requirements that are then processed in a systematic way to begin describing the solution system in terms of what it must do and how it must do it. *Definition of the system through recursive decomposition cycles* leads to greater levels of detail and specification. *Synthesis and integration* provide the means of describing and constructing the system in ever larger items and assemblies starting from the most detailed level of definition. To ensure that the solution is in fact meeting requirements, a sequence of *verification* steps assess the system relative to system definitions and specifications in progressively higher levels of integration. Various model processes and general approaches that embody these fundamental principles were presented, collectively offering a basic overview of the most representative and accepted systems engineering practices.

References

Boehm, B.W. 1988. A spiral model of software development and enhancement. *Computer* 21(5):61–72.

Department of Defense (DoD), United States. 2001. *Systems Engineering Fundamentals.*

Department of Defense (DoD), United States. 1974. *MIL-STD- 499A, Engineering Management.* May 1.

Electronics Industry Alliance (EIA). 1994. *EIA Standard IS-632 Systems Engineering.*

Eisner, H. 2002. *Essentials of Project and Systems Engineering Management.* New York: John Wiley & Sons.

Forsberg, K. and Mooz, H. 1991. The relationship of systems engineering to the project cycle. Paper presented at the *1st Annual Symposium of the National Council On Systems Engineering (NCOSE),* October 21–23, in Chattanooga, TN, USA.

INCOSE (International Council on Systems Engineering). 2003. *INCOSE Systems Engineering Handbook v.3.*

Kossiakoff, A. and Sweet, W.N. 2003. *Systems Engineering Principles and Practice.* Hoboken: John Wiley & Sons Inc.

Merriam-Webster Online Dictionary. Synthesis. http://www.merriam-webster.com/dictionary/synthesis. (accessed June 18, 2008).

NASA (National Aeronautics and Space Administration), Office of the Chief Engineer. 2007. *NASA Systems Engineering Handbook, NASA/SP-2007-6105 Rev. 1.*

Royce, W. 1970. Managing the development of large software systems. *Proceedings of the IEEE WESCON:* 1–9.

United States Air Force Space and Missile Systems Center (SMC). 2004. *SMC Systems Engineering Primer and Handbook—Concepts, Processes and Techniques,* 2nd ed., January 15.

Chapter 18

Capability-Based Acquisition Policy

The Cold War era after World War II shaped the defense system acquisition policies of the Western democracies until the late 1990s. The underlying philosophy of that era was to build systems to counter well-defined threats to national security. Acquisition decision making had become threat driven and technology based. A technology was selected because of its contribution to meet a specified set of technical requirements to defeat the threat. Stealth technology is such an example. The requirements in a threat-driven and technology-based acquisition environment are often very specific in nature and focused on maturing a technology rather than achieving broader operational capability. But over the past ten years there has been an increasing focus on capabilities-based acquisition for national defense. The idea of organizing the acquisition strategy around specific military capabilities delivered by families of systems is traceable to reforms in the U.K. Ministry of Defense (MOD) in the late 1990s. These reforms were a significant departure from the traditional practice of organizing the acquisition strategy around specific threats to be countered by individual systems, platforms, or military components (Dickerson et al. 2003). The U.S. Department of Defense (DoD) has also shifted toward a capability-based approach to the acquisition of defense systems. However, this shift imparts a tension to government acquisition structures that have been highly tuned to the delivery of tightly specified systems rather than broadly defined military capability.

Key Concepts

Acquisition policy
Requirements
Capability
Functional viewpoint
Through-Life Management

U.S. DoD Acquisition Policy

The paradigm shift to capabilities-based acquisition within the DoD is causing a fundamental shift in the way defense-related systems are both engineered and purchased. New mission needs and technological advancements have led to new directives that are pushing defense acquisition toward a capability-based approach. Government customers such as the DoD must acquire products and services through a formal process. The DoD Directives 5000.1 and 5000.2 (Department of Defense 2003a,b) describe the operation of the U.S. Defense Acquisition System, which mandates how the DoD acquires systems. In CJCSI 3170.01F (CJCS 2007a), the Joint Capabilities Integration Development System (JCIDS) provides extensive guidance on acquisition practices. With regard to system architectures, the Clinger–Cohen Act (U.S. Congress 1996) was designed to improve the way the U.S. Government acquires and manages information technology. It mandates the use of an Enterprise Architecture for all federal agencies. The DoD Architecture Framework (DoDAF; Department of Defense 2007a), formerly the C4ISR Architecture Framework, facilitates compliance with the Clinger–Cohen Act and allows depiction of total system solutions using multiple integrated "views."

Before one can really understand capability-based acquisition, it is important to understand what a capability is. According to CJCSI 3170, a capability is "the ability to achieve a desired effect under specified standards and conditions through combinations of ways and means to perform a set of tasks" (CJCS 2005). A simpler definition provided in the DoD dictionary (Department of Defense 2002) was "The ability to execute a specified course of action." New products and services must be acquired when a gap exists in capabilities. A capability gap exists when a required or desired capability cannot be or is not being performed with current policies or technologies. Of particular interest are using those synergistic resources (DOTMLPF, discussed later) that are unavailable but potentially attainable to the operational user for effective task execution in order to fill capability gaps. (See CJCSI 3170.01F [CJCS 2007a]).

The DoD 5000

The DoD 5000 provides overall acquisition policy, including how acquisition decisions should be made. There are five key policies emphasized in the DoD 5000.1: flexibility, responsiveness, innovation, discipline, and streamlined and effective management. *Flexibility* refers to the need to structure each acquisition program according to the set of strategies, documentation, reviews, and phases that make sense for that program. Not every acquisition program must be structured in the same way. *Responsiveness* refers to the need to react quickly to new and advanced technologies. Time phasing of capability needs and evolutionary acquisition strategies designed to leverage future technologies as they become available help increase responsiveness. *Innovation* is required to constantly improve and streamline acquisition practices. In particular, electronic solutions and best commercial practices are encouraged wherever possible. Overall, the goal is to reduce cycle time and cost while improving teamwork. All acquisition programs are to be managed with *discipline*. This means ensuring compliance with regulations; establishing goals for cost, schedule, and performance for the life cycle of the program; and effectively managing deviations from those goals as well baseline parameters and exit criteria. *Streamlined and effective management* refers to the decentralization of acquisition responsibility, giving a single individual with knowledge about the program full responsibility and accountability for cost, schedule, and performance reporting, and authority to accomplish development, production, and sustainment objectives.

Joint Capabilities Integration and Development System (JCIDS)

The Joint Capabilities Integration and Development System (JCIDS) is an enabler of this paradigm shift to capability-based acquisition. The shift in acquisition policy from a systems focus to a capability focus impacts requirements analysis and requirements flow-down. The JCIDS is a key component of the defense acquisition process. It aids in the transition between the Capability-Based Assessments (where the gap analysis is done) and the actual procurement and production of a new system or technology. The JCIDS dictates several documents that, when approved, provide the basis for transition between key acquisition milestones.

The JCIDS was developed in 2004 to replace the Requirements Generation System (RGS). The RGS used a bottom-up approach, leaving acquisition up to the individual services. As the complexity of defense systems has increased, this strategy became ineffective at acquiring capabilities that maximized interoperability between the services. This was problematic, as a war is not fought with individual services but rather with a joint and integrated effort. To address this problem, the DoD created the JCIDS to move toward a top-down, global capability-based approach that derives requirements from high-level warfighting needs and addresses gaps across the entire military, not just the individual services.

In the overall DoD acquisition strategy, the JCIDS fits in between Capabilities-Based Assessment (CBA) and the procurement and production of new systems. The overall acquisition process is depicted in Figure 18.1 (CJCS 2007a). Capabilities-Based Assessment has three pieces: the Functional Area Analysis (FAA), the Functional Needs Analysis (FNA), and the Functional Solutions Analysis (FSA). The FAA identifies what capabilities are required for a military unit and the criteria for measuring success of those capabilities. The FNA maps existing solutions to the needs identified in the FNA and identifies gaps between current and desired capabilities. The FSA looks at potential materiel and nonmateriel solutions for filling or mitigating those gaps (CJCS 2007a).

The JCIDS has three key milestones that must be met before the decision is made to produce a new system. Milestone A determines if the operational risk associated with looking into fulfilling an existing gap is acceptable. This is supported

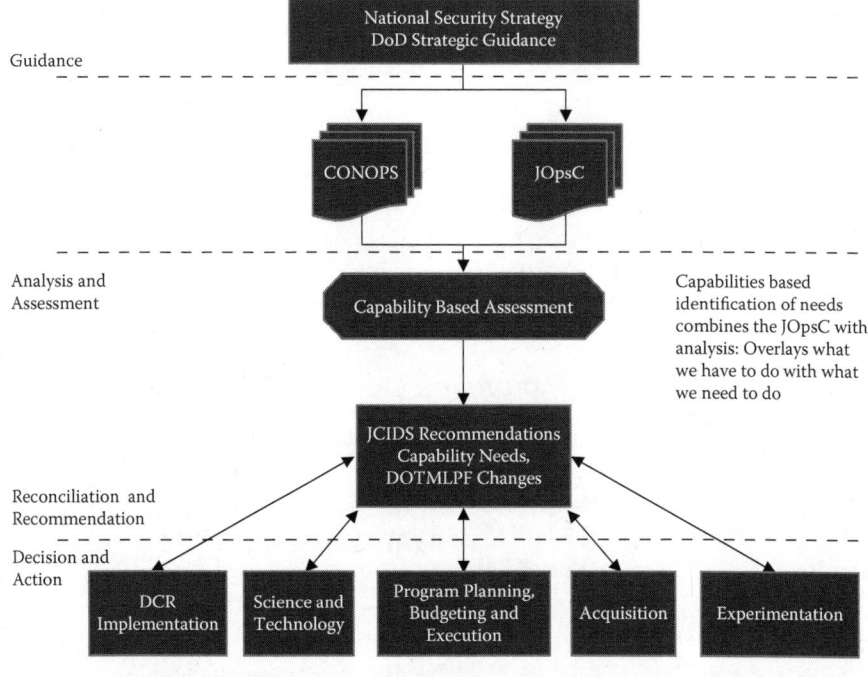

Figure 18.1   DoD's acquisition strategy. (Adapted from JCIDS 2007.)

by two JCIDS documents, the Joint Capability Document (JCD) and the Initial Capability Document (ICD). The JCD describes the gaps identified by the FAA and FNA, and the relative priority or importance of those gaps. The ICD evaluates Doctrine, Organization, Training, Materiel, Leadership and Education, Personnel and Facilities (DOTMLPF) approaches. If no DOTMLPF-based solution is found, the ICD describes the analysis of alternatives and basic concept refinement. The ultimate goal is to identify if there are potential feasible and affordable solutions to filling the identified gaps. The JCD/ICD are then reviewed by the Joint Requirements Oversight Council (JROC). There are three possible outcomes of this review. The first is that DOTMLPF change is recommended. This results in the creation of a DOTMLPF Change Recommendation (DCR) document and the JCIDS process in exited. This typically occurs when a new materiel solution is not needed. The second possibility is that the JCD/ICD is approved, in which case the process proceeds to the next phase. If the JCD/ICD is not approved, the JROC has determined that there are no feasible or affordable solutions, and no further action is taken at that time. The JCIDS process through Milestone A is shown in Figure 18.2 (CJCS 2007a).

Once a JCD or ICD is approved at Milestone A, analysis by a service or agency is performed to identify the best technical approach. The Capability Development

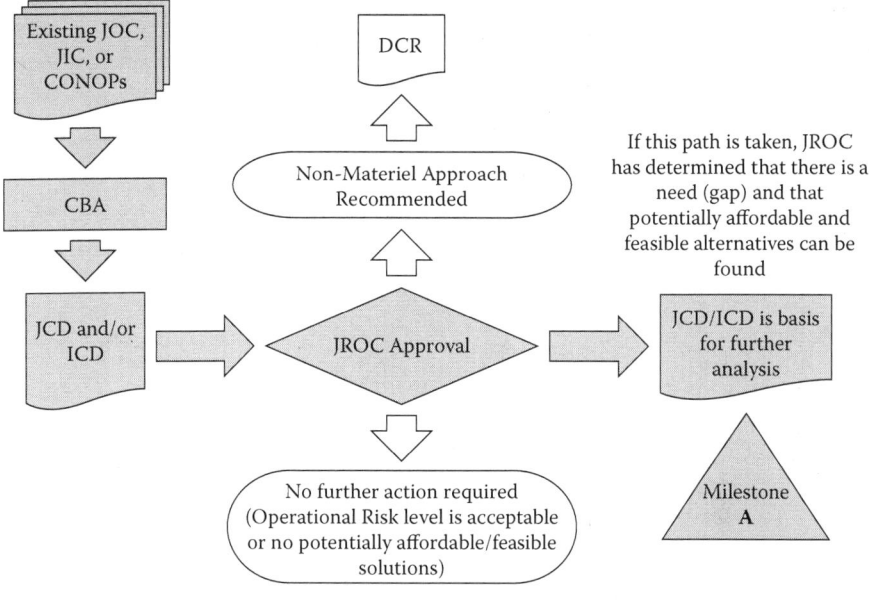

Figure 18.2 JCIDS process through Milestone A.

Document (CDD) summarizes the proposed approach, and then estimates performance attributes of the proposed solution. The CDD is then reviewed by the JROC. If the JROC approves the CDD, the Milestone Decision Authority (MDA) will determine whether or not to initiate development of the system. The JROC may not approve a solution if the proposed approach is deemed unaffordable or if the Key Performance Parameters (KPPs) summarized in the CDD are believed to be unreasonable. This is shown in Figure 18.3.

Once development is initiated, the system development process begins. In this phase, the system goes through detailed design and testing, resulting in the creation of a Capability Production Document (CPD) that details the production attributes of the system. The CPD is then reviewed by the JROC. If the system does not meet the KPPs to acceptable levels or is too costly, the CPD will not be approved. If the system meets the needs originally defined in the ICD/JCD, and performance and cost are in line with CDD predictions, then the JROC will approve the CPD and the MDA will determine if they want to initiate a production program. This finishes Milestone C, the final milestone of the JCIDS process. The transition between Milestones B and C is depicted in Figure 18.4.

An important observation can be made from review of the JCIDS documentation. JCIDS provides a detailed checklist of what should be addressed in an acquisition, but little guidance on where or how to obtain the information required to support development of these pieces.

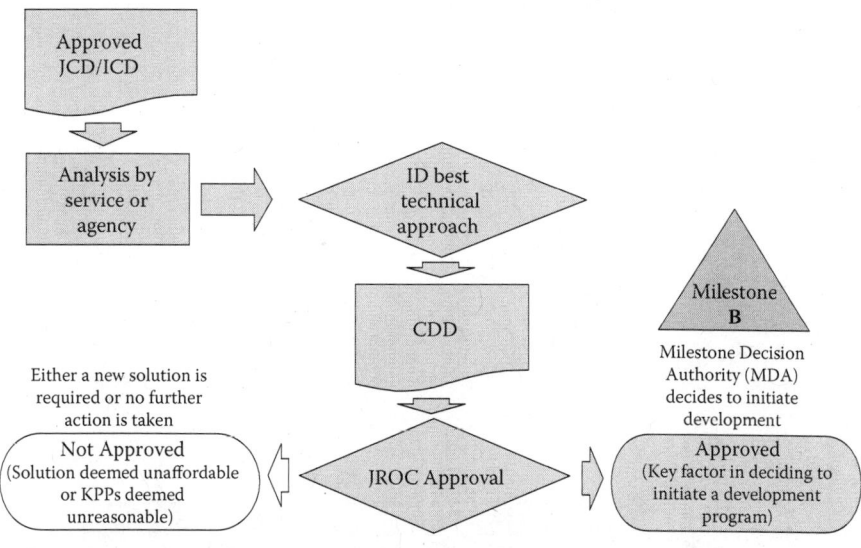

Figure 18.3 JCIDS process from Milestone A to Milestone B.

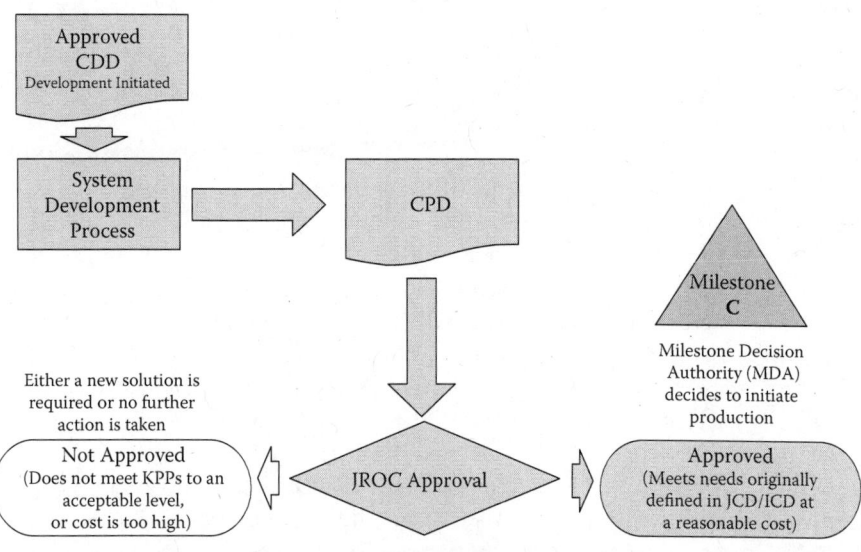

Figure 18.4 JCIDS process from Milestone B to Milestone C.

JCIDS is just one part of a comprehensive acquisition framework. JCIDS acts as a "front end" to the acquisition process and preconceptual design phase. This is depicted in Figure 18.5.

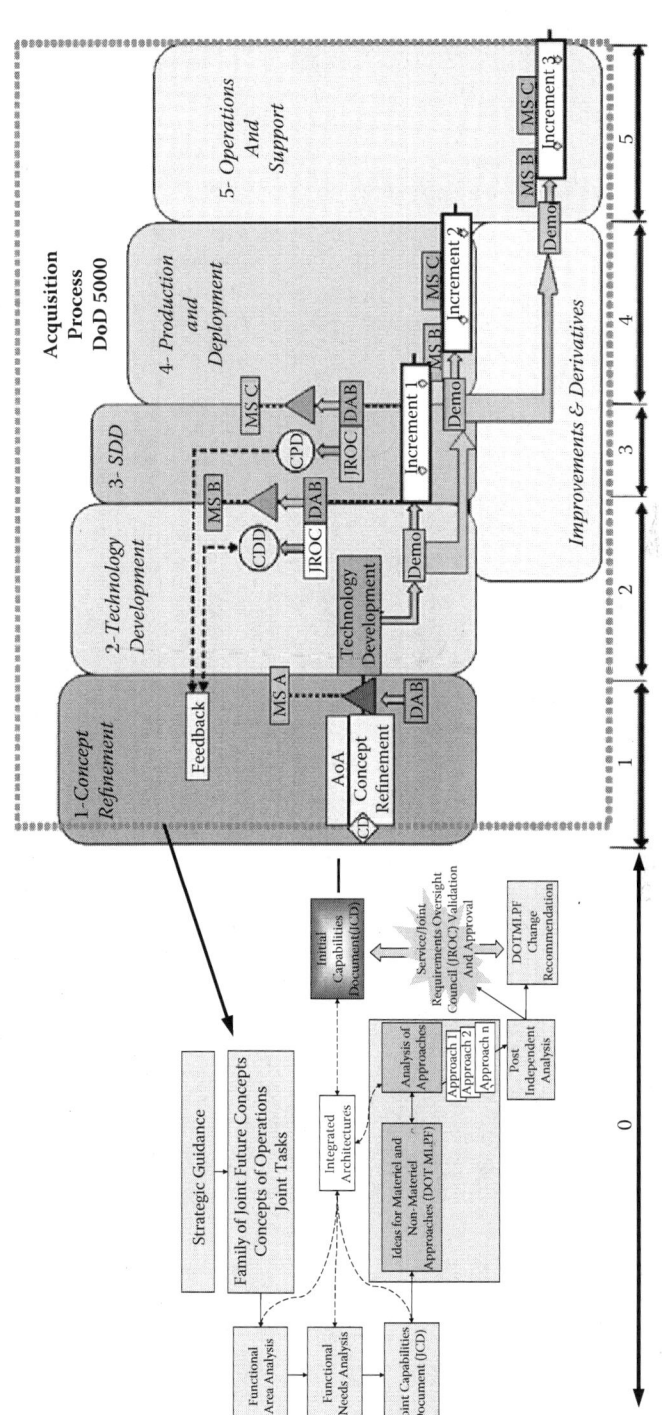

Figure 18.5   Summary of the DoD acquisition process.

The Role of Systems Engineering in Acquisition

One of the primary challenges of using a capability-based approach is that it is very easy to compare competing systems based on specified and measurable metrics such as speed. It is much more difficult to compare systems based on more nebulous capability need statements. It is also much more difficult to create capability requirements, and to determine when these requirements are met. A top-down systems engineering method to evaluate competing solutions based on how well they are predicted to meet capability needs can help address this challenge. A more complete discussion of systems engineering is contained in Chapter 17 ("Fundamentals of Systems Engineering") and will not be presented again here. This method should maintain clear traceability in decision making and show a clear link between design and acquisition decisions, and customer needs. In particular, applying systems engineering in the earliest phases of acquisition can help clarify requirements and goals, and provide credibility when the most important design decisions are being made. If implemented well, early-phase systems engineering can help reduce cost and risk, and improve the timeliness and performance of acquisition programs and the resulting procurements.

Each piece of the CBA and JCIDS process will have enabling systems engineering techniques. These will be highlighted here. For a more complete discussion of these methods, refer to Chapter 17 of this book, or to the many published systems engineering textbooks and manuals.

The FAA, which aims to identify operational tasks, conditions, and standards needed to accomplish military objectives, can be supported using functional decomposition and scenario development. The FNA, which assesses the ability of current and programmed capabilities to accomplish the tasks outlined in the FAA, required a solid analysis of alternatives and system effectiveness, and decomposition and recomposition. The FSA, which focuses on operational-based assessment of doctrine, organization, training, materiel, leadership/education, personnel, and facilities (DOTMLPF) approaches to solving capability gaps identified in the FNA, can benefit from using a matrix of alternatives and analysis coupled with multiattribute decision-making techniques (such as TOPSIS or AHP), and modeling and simulation where possible. Independent analysis of the results is then needed to identify a high-priority capability gap (or set of gaps) that is addressable in an affordable, effective, and time-efficient manner. Pulling together the techniques discussed earlier into a parametric, on-the-fly digital environment can allow decision makers to vary the scenario (which perhaps includes current defense priorities or current available budget) to see how this affects which capabilities become more important or more affordable. This will result in the information required to produce the ICD. Other systems engineering techniques can be used to support the other elements of the JCIDS process; however, these are outside the scope of this text.

In order to translate these capabilities into top-level requirements, there needs to be an understanding of the customer's objectives, the customer's mode of operations, and the functions and capabilities needed to achieve objectives. In the

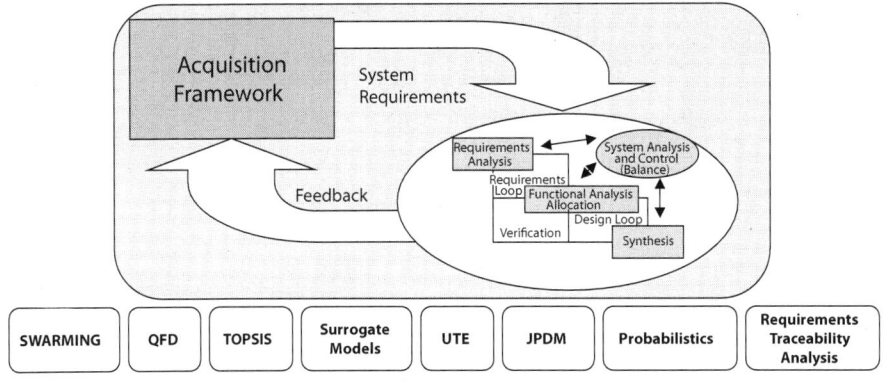

Figure 18.6   Systems engineering, requirements analysis, and defense acquisition.

context of Requirements Engineering, systems engineering provides a framework that aids in decomposing top-level capabilities into system and subsystem requirements. Enabling systems engineering techniques facilitate each step of the requirements analysis process and are consistent with the military acquisition process, as shown in Figure 18.6. SWARMing and QFD can help support requirements identification, decomposition, and prioritization. TOPSIS and surrogate models, combined with probabilistic decision making can help understand the impact of changing requirements on the solution space. Joint Probabilistic Decision Making (JPDM), a Unified Trade-off Environment (UTE), and requirements traceability analysis can help examine the multiple dimensions of a complex problem space simultaneously and understand the sensitivity of the entire design space to changing requirements, as well as understanding the coupling between requirements.

MOD Acquisition Policy

The current approach to capability-based acquisition in the U.K., MOD reflects an evolution of thinking over the past decade. Great emphasis is placed on Through Life Capability Management, which is governed by the MOD Defence Acquisition Operating Framework (AOF). Through Life Capability Management (TLCM) has been succinctly described as

> "… an approach to the acquisition and in-service management of military capability in which every aspect of new and existing military capability is planned and managed coherently across all Defence Lines of Development from cradle to grave" [McKane 2006].

Through the Defence Lines of Development (DLoDs), TLCM shares a common philosophy with JCIDS. The DLoDs are the MOD equivalent of the DoD

DOTMLPF. Specifically, the DLoDs are Training, Equipment, Personnel, Information, Doctrine and Concepts, Organisation, Infrastructure, and Logistics. The acronym for the DLoDs is TEPIDOIL, which is pronounced "Tepid Oil." The acquisition of military capability draws from across the DLoDs.

Defence Acquisition Operating Framework (AOF)

The AOF is an authoritative source of policy and good practice for all members of MOD and their industry partners concerned with acquisition. As illustrated in Figure 18.7, TLCM translates the requirements of MOD policy into an approved program that delivers the required capabilities, through-life, across all DLoDs. The Scope of Capability Management can be illustrated as a series of transitions from the analysis of Defence Policy through to the award of an industrial contract. Further details can be found at the U.K. MOD AOF Web site, www.aof.mod.uk.

Transitions in TLCM from National Capability to Lines of Development

As illustrated in Figure 18.8, TLCM begins at the level of National Capability based on the Defence Strategic Guidance and flows down through Capability Change Plans, which become governed by Capability Management Strategy and Plans, which are flowed into the Lines of Development governed by Through Life Management Plans. At each transition in Capability Management, trades must be made that may affect the requirement through the performance, cost, and delivery timescale perspectives of a program, across all DLoDs.

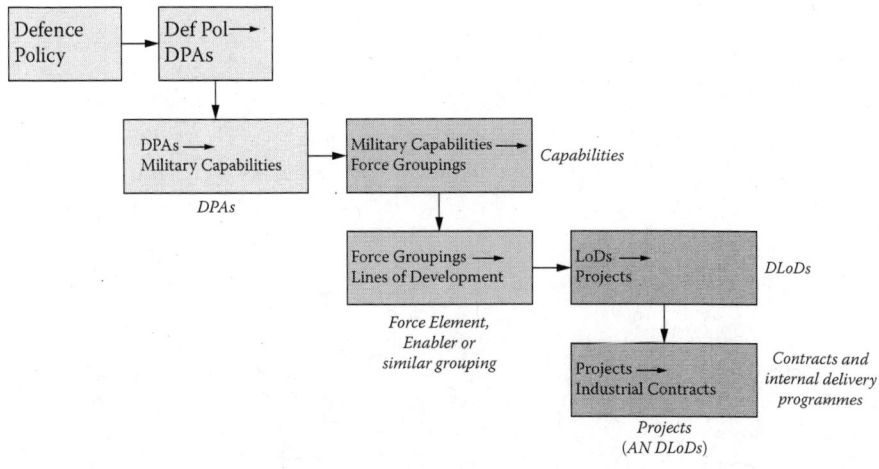

Figure 18.7 Defence Acquisition Operating Framework (AOF).

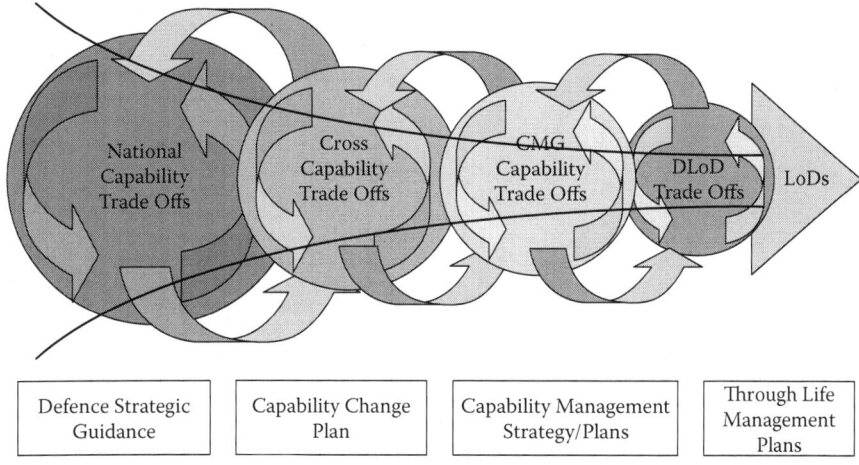

National Capability Trade Offs

Cross Capability Trade Offs

CMG Capability Trade Offs

DLoD Trade Offs

LoDs

| Defence Strategic Guidance | Capability Change Plan | Capability Management Strategy/Plans | Through Life Management Plans |

Figure 18.8 Through Life Capability Management.

References

Commander, Joint Chiefs of Staff (CJCS). 11 May 2005. CJCSI 3170.01B. Washington, D.C.

Commander, Joint Chiefs of Staff (CJCS). May 2007a. CJCSI 3170.01F: *Joint Capabilities Integration and Development System*. Washington, D.C.

Commander, Joint Chiefs of Staff (CJCS). May 2007b. CJCSM 3170.01C: *Joint Capabilities Integration and Development System Manual*. Washington, D.C.

Department of Defense. 2002. *DoD Dictionary of Military Terms* (JP 1-02). Washington, D.C.

Department of Defense. May 12, 2003a. *Department of Defense Directive, DODD 5000.1: The Defense Acquisition System*. Washington, D.C.

Department of Defense. May 12, 2003b. *Department of Defense Directive, DODD 5000.2: Operation of the Defense Acquisition System*. Washington, D.C.

Department of Defense. 2007a. *Department of Defense Architecture Framework (DoDAF), Volume I: Definitions and Guidelines*. Washington, D.C.

Department of Defense. 2007b. *Department of Defense Architecture Framework (DoDAF), Volume II: Product Descriptions*. Washington, D.C.

Dickerson, C.E. et al. August 2003. *Using Architectures for Research, Development, and Acquisition*. Office of the Chief Engineer of the Navy, Assistant Secretary of the Navy. Defense Technical Information Center (www.dtic.mil): ADA427961.

McKane, T. 2006. *Enabling Acquisition Change*. London: Ministry of Defence.

U.S. Congress. 1996. *Information Technology Management Report Act (Clinger-Cohen Act of 1996)*. Washington, D.C.

Chapter 19

Systems Design

Dmitri Mavris and Hernando Jimenez

Systems design brings together the tools and methods enabling systems architecting while embodying its fundamental concepts. These can be easily appreciated by observing how design is conducted and recognizing it as a cognitive activity geared toward the purposeful transformation of information.

Key Concepts

Transformation of information
Quality Function Deployment
Phased approach
System life cycle

Introduction

The first part of this book presented the foundations of architecture and systems engineering, thus providing a conceptual context. The second part presented the tools and frameworks with which systems architecting is enabled. The stage is now set to revisit the design process so as to study it in more detail and illustrate its most crucial features through relevant examples. That is the purpose of this chapter. In Chapter 17 ("Fundamentals of Systems Engineering"), it was stated that no single systems engineering process is regarded as the one correct alternative, and, rather, that a variety of processes, techniques, and methods existed from which the systems engineer could draw upon, based on the specific needs of the task at hand. Design is most certainly characterized the same way, in part because of its profound connection with systems engineering. It is critical to note that this chapter is not intended to document "the correct" design process, but rather *a* design process that adequately fits the theme, scope, and purpose of the entire book.

This chapter begins with a characterization of design where relevant definitions, features, and general paradigms in the structure of accepted design processes are presented. The formulation of the design process described in the sections that follow is heavily reliant on the fundamental concepts presented in earlier material of this book, particularly those presented in Part 1 and in Chapter 17. Thus, a brief overview of these fundamental concepts is provided for convenience. After describing the design process in detail, the final section introduces Quality Function Deployment (QFD), a valuable technique that supports the design process by revealing key attributes and design trade-offs, and documenting relational information that follows the approach of model transforms presented in Chapter 4 ("Structure, Analysis, Design, and Models").

Every effort has been made to strike a balance between generality and specificity throughout this entire chapter—that is, to provide sufficient detail in process instantiations and specific applications so as to adequately illustrate design principles without loss of generality or applicability. To facilitate this objective, a representative example is used to illustrate the application of the design process while showing relevant enabling tools and techniques.

A Characterization of Design

Design is a fascinating field that has elicited vast amounts of research across its numerous aspects and applications. It is perhaps for this reason that such a large number of definitions for "design" exists, though the vast majority of them consistently characterize this practice with the same features. A survey of published definitions (Jimenez and Mavris 2007) suggests that the most widely accepted ones describe design as

- A decision-making process
- A problem-solving process
- A cognitive activity (i.e., relies on the construction of knowledge)
- Involving transformation of information
- Involving considerations of the entire life cycle of the product
- Inherently multidisciplinary

These features of design are inherently consistent with the concepts of relational structures and model transforms presented in Chapter 4 ("Structure, Analysis, Design, and Models"). Knowledge about a system lies primarily in the relational structure of system parameters. This knowledge is further developed throughout the design process by adequately transforming relational information from one system model into another. This idea is captured by Nahm and Ishikawa who define design as "transformation process from a set of functional specifications and requirements into a complete description of a physical product or system which meets those specifications" (Nahm and Ishikawa 2004). This characterization of design, notably similar to those for systems engineering presented in Chapters 3 ("Concepts,

Standards, and Terminology") and 17 ("Fundamentals of Systems Engineering"), confirms the profound connection and overlap between the two fields.

Other treatments of design highlight the fundamental role of logic and logical modeling, presented in Chapters 2 ("Logical and Scientific Approach") and 4 ("Structure, Analysis, Design, and Models"), by drawing similarities between the basic structure of the design process and the scientific method. Steps of need identification, conceptualization, and feasibility analysis in design stand in one-to-one correspondence with steps of scientific curiosity, hypothesis formulation, and logical analysis, respectively (Dieter 1999, p. 8).

Design as a Phased Process

A *phased approach* of design, where the overall process is broken down into smaller, more manageable related segments, represents the current paradigm across multiple fields. The *conceptual design phase* is the first of these segments, and includes the initial identification of needs, definition of the problem, collection of information for the generation of a relevant context, and the definition of a *design concept* usually requiring concept generation and concept selection. A design concept provides a definition of the system meeting the identified needs at the highest level of abstraction in the system structure hierarchy. The *preliminary design phase* involves the *decomposition and definition* of the system through lower levels in the hierarchy structure, producing descriptions embodied by the *functional and physical architectures*. The *detailed design phase* provides the final stage of decomposition and definition, and produces a complete engineering description of the system ready for production. (DoD 2001; Dieter 1999) These represent the three main segments in the phased design paradigm, though different names or additional phases may be featured at times to explicitly account for planning of all system life-cycle phases.

Systems engineering processes presented in Chapter 17 ("Fundamentals of Systems Engineering") depict obvious similarities to the phased design process. Initial steps gather *stakeholder needs* and consolidate them into a set of *customer requirements*. These are then transformed into *system requirements*, which are recursively decomposed within a structure that embodies the *functional architecture* of the solution system. Elements of functionality are then associated with physical components that comprise the *physical architecture* of the system and define it at that level of decomposition. The process of decomposition and definition is not done all at once, but rather in cycles. The reason is that with each level of decomposition, the design must be assessed against requirements and specification. Specifications directly result from the functional decomposition of the previous cycle. This assessment process, referred to as *verification*, usually requires some type of analysis to support the decisions made in the definition of a new level of the physical architecture. Once verified, a design is inherently defined with a greater degree of detail and can be considered a *baseline*, which is the core output of the current design cycle and serves as the input of the next cycle.

Problem Definition for Design Context

The design process is, in the vast majority of cases, invoked after a need is identified. Some notable exceptions can be entertained: it may be argued that conquering flight was not a "need" in the early 20th century but rather a challenge whose fascination yielded the design of the first powered flight aircraft. Exceptions of this type aside, the exact way in which a need is identified or stated can vary widely. A practitioner may notice that some task is particularly difficult, cumbersome, or inefficient, and may note that the design of a tool can resolve this situation. Consider, for example, cash counting and the design of automated machines that perform the same task that a human teller would but with increased accuracy and in a fraction of the time. Private entities such as businesses can identify needs in their operations and ask for the design of a solution, such as a software system or a piece of machinery. A social need may be collectively identified and lead to inventions such as the telephone or the automobile. In some cases, specific entities such as government institutions are charged in part with the responsibility of identifying societal needs, issuing a call for a solution, and supporting the ensuing efforts. Examples include the National Aeronautics and Space Administration (NASA) and the Department of Defense (DoD).

Such is the case of the notional example used throughout this chapter to illustrate the design process, namely that of a high-speed commercial aircraft. This example is based to some extent on NASA's High-Speed Civil Transport (HSCT) study, which investigated the technical feasibility, economic viability, and environmental compatibility of supersonic transport concepts and aimed to identify key high-payoff technologies. The study, which spanned between 1990 and 1999, was a joint effort with industry responding to recommendations by the White House Office of Science and Technology Policy (OST) in 1985 and 1987 (National Research Council 1992, p. 61). Though partially based on an actual design research study, the example used in this chapter is notional, and in no way attempts to characterize, describe, or document the HSCT study. Details about the HSCT study can be found in published literature. With this caveat in mind, the design example begins with a government entity that identifies supersonic transport and technology development needs, and states them as development goals in a recommendations document.

Regardless of the mechanism, the identification of a need not only calls upon the design process, but also represents the first piece of information comprising the context within which design will take place. Because design attempts to respond to a need, it does not take place in a vacuum but rather is primarily driven by the contextual conditions in which it is invoked. It is therefore imperative that designers make every attempt to understand as much as possible this context before proceeding logically with the generation of a solution. Designers recognize the criticality of this initial step, commonly referred to as *problem definition*, noting that "the ability to define the problem is the most important and difficult task in engineering" (Suh 1990). This observation is particularly evident in the design of complex and unconventional systems whose interactions with other systems and the environment is difficult to understand, or for systems with limited or no empirical history.

Due to its importance, a structured approach that helps address the natural difficulty for problem definition is highly desirable. An example of such an approach is the System-Wide Assessment, and Research Method (SWARM) (Soban et al. 2005), a comprehensive and rigorous data collection, analysis, and knowledge management method that identifies the stakeholder needs and contract specifications to enable the decomposition and allocation of requirements, and leverages the identification of additional resources from which to derive requirements. SWARM overlaps with requirements analysis, discussed next, but extends beyond by truly generating an information and knowledge contextual basis for design. In its basic formulation, SWARM identifies important statements in the customer need, and involves the stakeholders as much as possible if there is need for clarification. It then guides the process of identifying and exhausting available resources containing relevant information associated with the stakeholders' needs. Information gaps are identified and addressed either by approaching the stakeholders or referring to alternative resources.

In support of requirements analysis, SWARM identifies the key performance parameters to be validated through subsequent methods, and helps avoid potential misunderstandings with the customer by differentiating between hard requirements, regulatory constraints, and design preferences or desires. Beyond requirements, SWARM highlights the value of past experience and points at resources such as previous contractual agreements between stakeholders/customers and designers, case studies, and subject matter experts. The context of design therefore observes the history of relevant design efforts critical in establishing how solutions were previously developed based on requirements. This information points at the identity of designers, which then become a potential source of information. The lessons learned from said efforts also provide insight about design choices that proved to be successful or failures, and suggest what the physical phenomena and technical challenges are that will likely be most important. SWARM also highlights the value of industry standards and policies, government regulation, and institutional codes as valuable sources. They reveal what the relevant regulatory bodies are and aid in the development of key design constraints and derived requirements. This knowledge base helps strengthen the credibility of the designer and the communication with relevant parties by pushing the designer to learn the lingo, thus ensuring a consistent set of definitions and terms with stakeholders. As will be shown in later sections, the definition of a problem context (such as would be generated through SWARM) supports all subsequent steps in the design process.

For the supersonic aircraft example, problem definition includes a variety of sources documenting previous design efforts of supersonic aircraft, particularly of commercial transports such as the Concorde or the Tupolev 144, as well as the multiple U.S.-led supersonic programs. It identifies Federal Aviation Regulations (FAR) and airport compatibility reports as applicable standards. Technical issues such aerodynamic, propulsive, controls, or material considerations of supersonic aircraft performance are highlighted. Entities such as the Federal Aviation Administration (FAA) and NASA are identified as key authoritative entities, along with those responsible for successful supersonic transport designs of the past.

Building context is a task that is continuously executed throughout the design process, although an important part of it takes place in the early phases. However, when the problem is sufficiently defined, the process is cleared to continue to the next steps.

Requirements Analysis

With a sufficiently complete design context at hand, the next step of design calls for customer needs to be collected into customer requirements and transformed into top-level system requirements. Systems engineering processes introduced in Chapter 18 ("Capability-Based Acquisition Policy") state that system requirements are composed of functional requirements, which state *what* the system must do, and performance requirements, which indicate to what *extent* it must be done. The nature of requirements analysis is one of "inquiry and resolution" (DoD 2001), and thus the process relies heavily on the knowledge gathered in the problem definition, particularly past experience and communications with stakeholders.

In the current example, a government entity identified a need for a commercial air transport vehicle that dramatically reduces long-range travel time, is environmentally friendly, and represents a favorable business case from the economic perspective of various relevant stakeholders. The responding party, referred to as *the designer*, consolidates the need into customer requirements, stated as a commercial supersonic transport that is technically feasible, economically viable, and environmentally compatible. The designer also performs the initial requirements analysis to establish key system requirements such as design mission range, seat-class (approximate number of passengers), the maximum takeoff field length, and a minimum supersonic cruise Mach number. Sometimes a design study is large enough that it is divided into tasks and awarded to contractors who inherit system requirements and may perform additional requirements analyses. Otherwise, all the system requirements must be developed by the designer.

Requirements analysis for the notional supersonic transport is supported by problem definition and identifies FAR environmental constraints that prohibit supersonic flight over land (Part 91, section 91.817), as well as limits for noise (Part 36) and emissions (Parts 33 and 34), and transform them into corresponding system requirements. Other FARs represent performance requirements, such as climb rates or speed limits for certain operations. Similarly, airport compatibility documentation limits the airports at which the aircraft will be able to operate, based on its available runway length, allowable aircraft weights, allowable aircraft dimensions, and local noise exposure regulations (FAR Part 150), among other factors. Takeoff and landing performance, and noise and weight limitations thus represent additional performance constraints. Communication with aircraft manufacturers and airlines reveals the economic criteria and thresholds representative of an economically viable design. Examples include program cost, cash flow features, and return on investment for the manufacturer, and acquisition cost, operating costs, and return on investment for the airlines. These are a representative set of system requirements, indicating *what* must be done (functional requirement) and the *extent* to which it must be done (performance requirement).

System Decomposition

In the next step of design, system-level functional requirements are decomposed and performance requirements are allocated at the new level of decomposition. As mentioned earlier, this process is done in cycles so as to allow for recursive verification activities. In the first cycle, the system-level attributes are decomposed to the subsystem level. Though this task is intrinsically functional in nature, the designer may implicitly carry through assumptions about the expected physical architecture and perform the functional decomposition with a notion of what subsystems are likely to enable system- and subsystem-level requirements. Problem definition offers information about previous systems and their architectures that can introduce this tacit handling of physical considerations, trivial as they may be.

For instance, consider the system requirement "cruise at a minimum Mach number of 2" for the notional supersonic transport example. The decomposition process yields subsystem-level requirements such as "generate lift for level flight," "provide necessary thrust for cruise speed," and "maintain stability and control in cruise conditions." Though seemingly trivial, the designer implicitly carried the notion of a subsystem that provides lift through this decomposition activity, and in all likelihood expects that subsystem to be a wing. The same can be said about engines providing thrust and control surfaces, and an empennage of some kind to provide stability and control in cruise.

Solution Definition

When the designer is satisfied with the functional decomposition at the current level, the physical architecture is then formally stated. This collectively constitutes a functional and physical description of the system at the current level of decomposition, and is for all practical purposes a solution definition. Notions of the physical architecture have been implicitly carried, so now a more explicit set of analyses and decisions must take place to formalize physical architecture at this level of system definition.

Continuing with the notional supersonic transport example, the designer expects that a wing will be the subsystem that will generate lift in cruise (as well as in all other segments of flight, of course). The question that naturally follows is "What type of wing is it?" Analogous questions are stated for engine types, engine placement, seat arrangement for the given passenger capacity (affecting fuselage geometry), and empennage configurations, among others. Analysis of Alternatives (AoA) helps the designer answer these questions by providing a structured and rigorous approach for identifying configuration alternatives, leveraging relevant analyses to support design decisions and formalize said decisions to define the system. These decisions are not only important for design purposes, but also because the type of subsystems selected will have implications on the choice of adequate Modeling and Simulation (M&S) tools in later design cycles. This technique is primarily based on morphological analysis, a field originally developed in the mid-1960s by Swiss-American astrophysicist Fritz Zwicky (1898–1974). Morphological analysis is a method that facilitates the rigorous investigation of large multidimensional spaces by enumerating and studying the possible alternatives for system elements via a *morphological matrix*, and the relationships among them through a *cross-consistency matrix*. (Zwicky 1967, 1969) In analysis of alternatives, the morphological matrix, sometimes referred to as a *matrix of alternatives*, lists all the alternatives for all relevant system elements. A complete system alternative is generated by selecting one element alternative from all elements under consideration. Thus, this matrix contains all possible configuration alternatives for the assumed configuration space. A sample matrix of alternatives for the notional example is shown in Figure 19.1, where only a small fraction of physical architecture elements is shown for simplicity.

The total number of system alternatives, estimated as the product of the number of alternatives over all elements, grows quickly and can become unmanageable for the designer. For instance, the matrix in Figure 19.1 is overly simplified, and yet it contains a total of 576 alternatives ($4 \times 4 \times 3 \times 4 \times 3$). A cross-consistency matrix captures the interrelationship between elements, and thus defines the feasible combinations according to the specified compatibility rules. These compatibility considerations are supported by expert opinion and technical considerations in the problem definition. As a result, this matrix (referred to as a *compatibility matrix* in this context) effectively reduces the number of combinations to a smaller, feasible subset. The compatibility matrix corresponding to the ongoing example is shown in Figure 19.2.

Elements	Alternatives			
Wing Type	Delta	Double Delta	Variable Geometry	Multi-sectioned
Engine Type	Turbojet	Turbojet + Afterburner (AB)	Low Bypass Turbofan	Low BP Turbofan + AB
Engine Placement	Under Wing	Aft Fuselage	Fuselage Immersed	
Empennage Configuration	Conventional	T-tail	Vertical Fin Only	Canard
Internal Seat Arrangement	2–2	2–3	3–3	

Figure 19.1 Matrix of alternatives for notional supersonic transport.

	Delta	Double Delta	Variable Geometry	Multi-Sectioned	Turbojet	Low BP Turbofan	Low BP Turbofan	Low BP Turbofan + AB	Under Wing	Aft Fuselage	Fuselage Immersed	Conventional	T-tail	Vertical Fin Only	Canard	2–2	2–3	3–3
Delta	■	X	X	X														
Double Delta	X	■	X	X														
Variable Geometry	X	X	■	X										X	X			
Multi-Sectioned	X	X	X	■														
Turbojet					■	X	X	X										
Turbojet + AB				X		■	X	X										
Low BP Turrbofan					X	X	■	X										
Low BP Turrbofan + AB					X	X	X	■										
Under Wing									■	X	X							
Aft Fuselage									X	■	X							
Fuselage Immersed									X	X	■							X
Conventional											X	■	X	X	X			
T-tail												X	■	X	X			
Vertical Fin Only		X										X	X	■				
Canard		X										X	X	X	■			
2–2																■	X	X
2–3																X	■	X
3–3												X				X	X	■

Figure 19.2 Compatibility matrix for notional supersonic transport subsystems.

This technique for design builds upon some of the fundamental concepts and definitions introduced in Chapters 3 ("Concepts, Standards, and Terminology") and 4 ("Structure, Analysis, Design, and Models") regarding relations and relational structures, and is further developed in Chapter 20 ("Advanced Design

Methods"). First, note that compatibility is a symmetric relationship, meaning that if X is incompatible with Y, then Y is incompatible with X. It follows that the structure of the compatibility matrix features symmetry along the main diagonal. Because the entries are essentially duplicated, one of the halves is oftentimes not shown. However, other types of cross-consistency matrices may not capture symmetric relationships, leading to an asymmetric matrix. It is also common to assume the alternatives within elements are incompatible among themselves, so that if the designer selects a delta wing, for instance, the choices of all other alternatives for the wing element are no longer available. This explains the incompatibility entries, denoted with an "X," along the matrix diagonal. Additionally for this example, designers have considered that a variable geometry will require a traditional or T-tail empennage, and consequently established incompatibility between a variable geometry wing and alternatives for vertical fin only and canard. Similarly, considerations for airframe geometry indicate that the wider fuselage cross-section resulting from a 3-3 seat configuration is incompatible with fuselage-immersed engines. Though the example used in this chapter is notional, a morphological approach to analysis of alternatives has been successfully implemented for concept generation and selection of an Efficient Multi-Mach Aircraft similar to the one used in this chapter (Jimenez and Mavris 2005).

The analysis of alternatives then calls for the designer to make the selection of a system alternative. To support this selection, the designer must analyze the implications of making certain choices relative to its impact on other elements and aspects of the system. Said implications have already been partially captured by the compatibility matrix. Chapter 20 will introduce the Interactive Reconfigurable Matrix of Alternatives (IRMA), which enables the designer to embed more contextual information in a software instantiation of the matrix and enhance the dynamic process of alternative selection. Design is known to be a practice of compromise, meaning that improvement in one aspect of the system will oftentimes degrade or impose more stringent design requirements on other aspects. It follows that the selection of an alternative to improve a certain aspect of the system may very well degrade multiple others. As a result, the designer is forced to perform some trades. A trade-off analysis is commonplace in design and systems engineering practice. When performing trade-offs, a designer defines an objective, identifies alternatives to address the objective, quantifies the sensitivity of objective criteria to alternatives, and compares alternatives against each other to perform a final selection (Kossiakoff and Sweet 2003, p. 430–432).

For instance, the supersonic transport is forced to operate at subsonic speed when flying over land, but can accelerate to a supersonic cruise over water. Because a significant portion of the design mission may take place at subsonic speeds, the delta and multi-sectioned configurations, which are geared to predominantly supersonic missions, are tentatively eliminated. To obtain the most favorable aerodynamic performance in both speed regimes, the best option for a wing is one with variable geometry, like the symmetric swing wing depicted in Figure 19.1, and featured in aircraft such as the B-1B bomber and the F-14 Tomcat. This alternative provides good aerodynamic properties for low-speed flight, as in takeoff, landing, and subsonic cruise, and can reconfigure to provide similarly favorable aerodynamic properties in supersonic cruise. However, the penalty associated with this choice is added structural complexity, which translates to more weight (about 20 percent increase of the wing element), more reliability considerations, added maintenance, and requirements on the power plant to power the actuation system, all of which translates to increased costs (Beissner et al. 1984; Raymer 1999). A double delta, whose design also attempts to capture good subsonic and supersonic characteristics, provides somewhat lower aerodynamic performance in both regimes relative to the variable geometry alternative, but does not induce the steep weight and cost penalties on the rest of the system. Trades like this can be easily extended beyond the physical architecture, and may include general features about the mission of the system. For instance, choices of supersonic cruise Mach number and altitude have implications on aerodynamic, propulsive, and structural performance, and require the study of trades between physical and functional architecture elements.

In general, trades analyses at this *conceptual phase* are conducted via subject matter experts and supported by problem-definition context. When all relevant trades have been completed, the designer can make informed decisions about the physical architecture at this level of decomposition. The resulting design is then checked against customer requirements, though this verification assessment may be difficult to complete, given that no explicit assessment of system performance exists yet. However, the designer can estimate, based on the trades that have taken place, whether or not the chosen system alternative meets the customer requirements. In later iterations, quantitative data from calculations and execution of M&S tools will support this verification at deeper levels of decomposition and system definition.

If the verification assessment against customer requirements is favorable, the result is a *system-level specification* and the system is thus defined by its functional and physical architectures. This system definition is documented as a *baseline* at the conceptual level, also referred to as the *functional baseline* in the DoD systems engineering framework. Moreover, the design steps described thus far effectively constitute a transformation process from customer requirements to a system definition (at the subsystem level of decomposition).

Decomposition and Definition: Second Cycle

In the second cycle of the design process, the existing baseline is used as a starting point to further decompose the system, thus progressing from subsystem concept selection to subsystem attributes definition. These subsystem attributes are chosen so as to meet system requirements in the verification process and represent a performance specification at that level of decomposition. The DoD systems engineering process refers to it as the *item performance specification*, as noted in Chapter 17 ("Fundamentals of Systems Engineering").

Once again, past experience and the problem definition context support the decomposition of subsystems into attributes. In the present example, a double delta configuration was chosen for the wing. Key attributes relevant to that wing type include airfoil selection and planform geometric parameters such as aspect ratio, and inboard and outboard sweep angles. Similarly, relevant engine attributes would have to be determined for the chosen engine type, such as the engine's bypass ratio and overall pressure ratio (OPR). Physical architecture considerations, though sometimes trivial, are also tacitly carried through the functional decomposition phase in the second cycle. For instance, the designer expects that the engine inlet, fan, and compressors will change the pressure of the flow moving through the engine, and has thus assumed that these components will be charged with achieving a prescribed OPR.

In the previous cycle, the trade studies performed relied primarily on expert opinion; this cycle incorporates quantitative estimates produced by first-order M&S tools chosen to fit the subsystem types selected for the functional baseline in the first design cycle. In this design phase, trade studies enable the designer to determine subsystem attribute values that strike a compromise among countervailing design trends, and that also meet system-level requirements. This necessitates a tighter integration of multiple disciplines within a cohesive analysis relative to the previous design cycle, and thus M&S tools used in this phase often feature the synthesis of various discipline-specific modules. For instance, a first-order aircraft mission performance tool will likely include a propulsion module where subsystem attributes such as bypass ratio and OPR can be dialed in by the designer. Initial system sizing also takes place in this phase as enabled by the aforementioned M&S tools, which yields values of metrics of interest corresponding to system-level performance requirements.

For example, trade studies enabled by a performance and sizing tool for the supersonic transport reveal the sensitivity of takeoff field length (a system requirement) to variations in the wing geometry parameters. It also allows the designer to determine the amount of thrust that must be generated by the engines during takeoff for the prescribed wing attributes, and in turn to size the engines and observe the sensitivity of takeoff field length to variations in engine attributes. The designer can now perform trades between engine and wing attributes to strike a compromise solution and, ideally, meet system requirements. Similar trades can be made for other subsystem attributes relative to a given system requirement. Advanced techniques for quantifying and visualizing sensitivities are presented in Chapter 20 ("Advanced Design Methods").

After all relevant trades have been performed and subsystem attribute values are determined, the system is verified against system requirements. If the design is verified, the functional and physical architecture at the current level of decomposition represents a new system definition that is documented in a new baseline, referred to as the *allocated baseline* in the context of DoD systems engineering. The item performance specifications that also result from this iteration are documented and used in the next design cycle.

Additional Iterations

The design process is successively executed in additional iterations, each time performing a functionally driven decomposition, associating functions with elements in the physical architecture, performing supported trade studies in the definition of element attributes, verifying against specifications resulting from the previous cycle, and documenting the system definition in a baseline. In the supersonic transport example, the third design cycle involves the definition of the engine inlet, fan, and compressors based on the decomposition of the engine subsystem into its components. Attributes of these engine components, such as fan pressure ratio, are traced through the functional architecture to the engine attribute OPR. The sensitivity of fan pressure ratio and compressor pressure ratio to OPR are quantified so that these fan and compressor attributes can be traded against each other. Definition at this level is verified against the item performance specifications. An analogous process is performed with the wing where airfoil attributes are traced to wing attributes, traded against each other, and verified. Eventually, the designer attains a completely detailed system definition and is able to document the design in the production baseline.

It is possible that at some point a baseline cannot be "closed," that is, that system requirements cannot be met at some level of decomposition, which implies that the design cannot be verified against the relevant specification. If the design process is performed rigorously and is properly documented, problematic requirements identified at a low level can be traced up through the functional architecture to system-level requirements and customer requirements. At this point, either customer requirements need to be reassessed or recommendations need to be made about technology infusion at different levels in the system architecture. For instance, phase II of NASA's HSCT study focused on the technology improvements to the baseline produced during phase I necessary to achieve economic viability. It was determined that up to a 40 percent weight reduction was needed on the airframe (Haggerty 1996, p. 38–41). Two methodologies that address requirements trade-offs, and technology portfolio evaluation and selection, respectively, are presented in Chapter 20 ("Advanced Design Methods"), along with methods and techniques to incorporate uncertainty into the design and technology infusion processes, or to design for robustness relative to uncertain operational environments.

Supporting Information Transformation in Design with Quality Function Deployment

Quality Function Deployment (QFD) is a well-established, systematic approach used to support requirements prioritization and relate stakeholder needs to technical requirements (Breyfogle 2003, p. 349–356; Dieter 1999, p. 69–75). This qualitative technique is relevant to systems design because it uses expert opinion to facilitate and document relevant information transformations, namely, the transformation of customer requirements ("What's") into product attributes ("How's"). Moreover, this transformation mechanism can be extended and deployed successively across multiple levels of system decomposition and definition, hence supporting a phased design process. Most important, however, is the fact that QFD places particular emphasis on the relational structure of system parameters, and provides guidance in terms of their relative importance across multiple iterations of information transformation.

The information developed in QFD is documented in a matrix-type document referred to as the *house of quality*, shown in Figure 19.3 with its main components. Oftentimes, these components are referred to with names of rooms, following the analogy between the QFD matrix and a house. The *customer requirements*, consolidated by the problem definition task, are listed on the left of the matrix. Depending

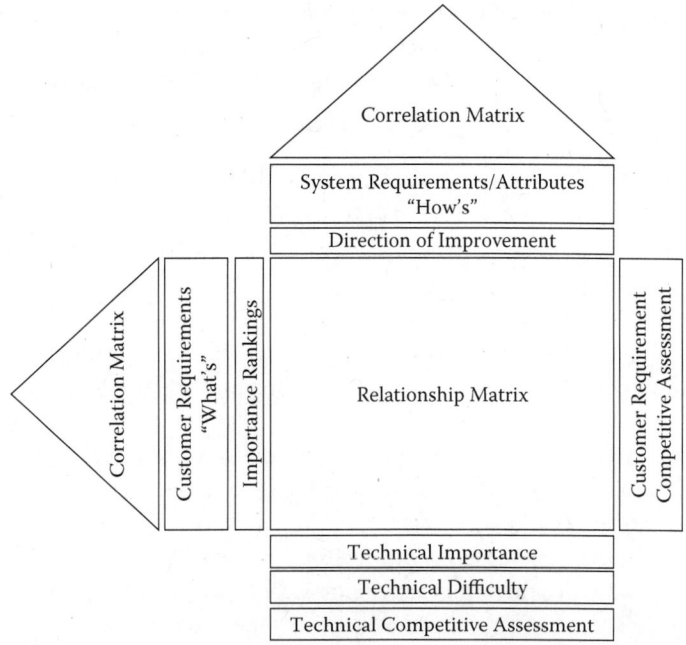

Figure 19.3 Main elements of the QFD house of quality.

on the form in which requirements are issued to the designer, a top-down decomposition or bottom-up consolidation technique may be needed to identify the relational structure of requirements across various hierarchical levels, or tiers. For instance, if the requirements for the notional supersonic transport example are issued in terms of "technical feasibility, economic viability, and environmental compatibility," then a top-down approach allows the designer to decompose these three requirements into more detailed and concrete ones, thus creating second- and third-tier customer requirements as needed. It also facilitates the identification of redundant or derived requirements. A top-down (decomposition) tool such as a *tree diagram*, shown in Figure 19.4, can be used for this purpose. Conversely, if requirements are issued in terms of "minimum takeoff field length, minimum cruise Mach number, minimum range, etc.," then a bottom-up technique such as an *affinity diagram* allows the designer to identify higher-level requirements that motivate the design. When using an affinity diagram, designers list out customer requirements and brainstorm any additional ones that may be relevant. Designers then assemble the requirements into groups based on their mutual conceptual similarity, which in turn reveals a higher level issue or requirement (Breyfogle 2003, p. 349–356; Dieter 1999, p. 69–75). This process is illustrated in Figure 19.4. Traceability is a key feature and objective in QFD, so it is desirable to capture the structure of customer requirements in more than one tier.

To enable the prioritization of requirements, the designer needs to establish *customer requirement importance ratings* to characterize how some are more important

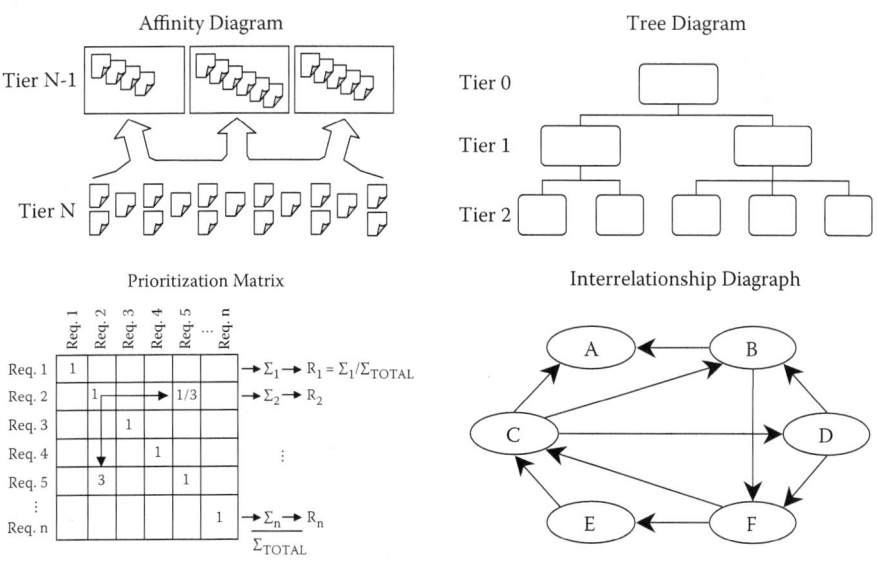

Figure 19.4   Supporting tools for Quality Function Deployment.

than others. There are numerous ways in which these rankings can be generated, like customer inquiries or surveys. A *prioritization matrix* can support this task by providing a structured approach of comparing all requirements against each other and synthesizing ranking values. This matrix documents pair-wise relative importance values using a ratio scale. For example, if requirement i is much more important than j, then the entry (i,j) has a value of 9 and the entry (j,i) has a value of 1/9. Analogous entries are made for other relative measures of importance—say, more important, {3, 1/3}, and equally important {1}. The values of the matrix are summed across the row and then normalized to get an importance rating. The prioritization matrix and the estimation of importance ratings are illustrated in Figure 19.4. This approach clearly explores the relational structure of customer requirements in a form consistent with basic mathematical formulation provided in Chapter 4 ("Structure, Analysis, Design, and Models"), and relies on subject matter opinion about the strength of each pairwise relation. However, it is worth noting that the prioritization matrix does not make use of binary entries as was shown in Chapter 4 (tick mark, no tick mark) but rather uses a set of multiple possible values. Additionally, it is inverse-symmetric, meaning that its entries in the top diagonal are the inverse of the corresponding ones in the lower diagonal.

The top-level *system attributes*, as developed via the requirements analysis and supported by the problem definition task, are listed on the top of the matrix. System attributes, sometimes called *engineering characteristics*, address *how* the customer requirements will be addressed and in turn represent *system-level requirements*. As with the customer requirements, the hierarchy structure of the system attributes is important to the execution of QFD. In this case, a bottom-up approach is most appropriate to thematically group system attributes. A top-down approach would imply decomposition beyond the scope of the current house of quality matrix. The direction of improvement of each system attribute is also noted: an upward arrow indicates "higher is better," a downward arrow indicates "lower is better," and a zero denotes a target value. For instance, system weight and fuel consumption of the notional supersonic transport represent improvements when reduced (downward arrow), and the lift-to-drag ratio represents an improvement when increased (upward arrow).

The relationships between system attributes are captured in the *correlation matrix*, also referred to as the *roof of the house*. The entries in this matrix indicate whether an increase in a system attribute k is accompanied by an increase (positive correlation), no change (zero correlation), or a decrease (negative correlation) on another system attribute l (Breyfogle 2003, p. 349–356; Dieter 1999, p. 69–75). A simple {positive-zero-negative} correlation notation can be used, or extended to incorporate the extent of the change with schemes such as {strong positive, weak positive, zero, weak negative, strong negative}. In the case of a supersonic transport, for instance, designers may note that weight is positively correlated with cost (heavier aircraft are more expensive).

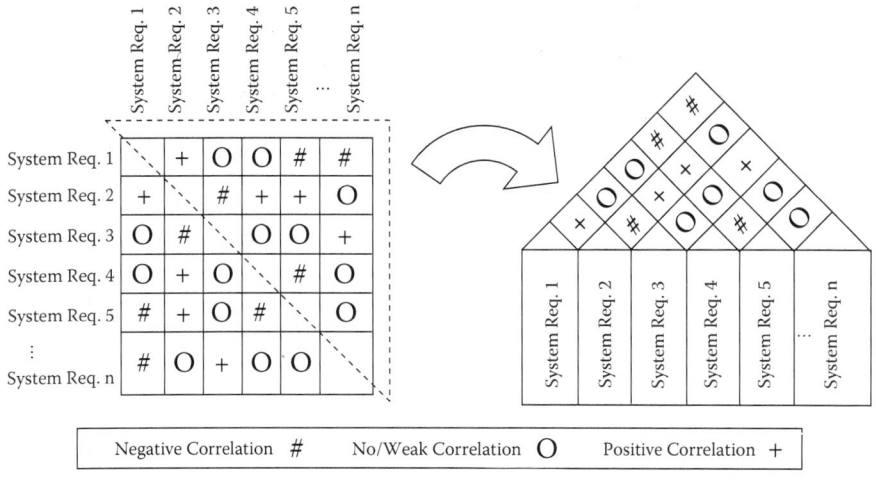

Figure 19.5   Sample correlation matrix.

The correlation matrix is another element of the house of quality that explicitly explores and documents a relational structure, in this case correlation among all system requirements. Correlation is a symmetric relation, and thus this matrix is also symmetric. It follows that the roof of the house is the single diagonal half of the correlation matrix needed to document the relational structure for system requirements, as shown in Figure 19.5. Moreover, it is worth noting that the role of the correlation matrix at the roof of the house is directly analogous to that of the $\underline{M}$ matrix in the mathematical formulation of model transforms presented previously in Chapter 4. The information documented in the correlation matrix supports the execution of trade-off studies in the design process by revealing how changes in a system attribute aimed at satisfying a given requirement can be accompanied by change on another attribute that may have a negative impact on the same or another requirement. It also serves to identify where conflict will arise between two system attributes. More specifically, those that have negative correlation and the same direction of improvement will represent a design conflict most likely requiring a compromise. The same will occur with those that have positive correlation and different direction of improvement.

The generation of the correlation matrix is mostly supported by expert opinion, but may make use of tools such as the *interrelationship digraph* or the cause/effect diagram. If using any of the two aforementioned tools for this purpose the designer must proceed with caution in later stages of design, recognizing that a cause/effect relationship between k and l is sufficient to establish correlation between the two, but correlation between k and l is *not* a sufficient indicator of a cause/effect relationship. An interrelationship digraph, which denotes stronger influence between two elements by the origination of an arrow connecting them (Breyfogle 2003, p. 349–356; Dieter 1999, p. 69–75), is notionally shown in Figure 19.4.

As with top-level system attributes, the relational structure among customer requirements can also be explored and documented to support qualitative trade-off analyses and help designers identify conflicting requirements. This relational structure among customer requirements is documented in a second triangular matrix on the left side of the house of quality, commonly referred to as the *greenhouse*. The function of the correlation matrix in the greenhouse is directly analogous to that of the $\underline{N}$ matrix in the mathematical formulation for model transforms presented in Chapter 4.

The cells in the relationship matrix at the center of the house of quality, or main room, indicate the importance of each system attribute relative to attaining the different customer requirements; that is, more importance denotes a stronger relationship. This matrix captures the relational structure between two sets of system attributes, that is, customer requirements and system attributes, thus embodying the model transform functionality of matrix Q introduced in Chapter 4. In the main QFD matrix, a discrete scale of numerical values is used to map the different degrees of relationship strength, usually following a given factor rate and including the zero value for null relationships (e.g., $\{9, 3, 1, 0\}$).

The *technical importance* of each system attribute can be estimated as a weighted sum of all customer requirement relationship values for that attribute, where the weights are the importance ratings for the customer requirements. Technical importance values are recorded at the bottom of the house of quality, and can then be normalized for convenience. The *technical difficulty* of each system attribute, interpreted as design requirement, is assessed and documented with a discrete scale in a row at the bottom of the matrix (Breyfogle 2003, p. 349–356; Dieter 1999, p. 69–75). The combination of technical importance and difficulty provides a measure of *technical risk*. For example, takeoff field length is an important constraint for the success of the supersonic transport program in the notional example of this chapter, but also a difficult one to meet with respect to other mission constraints. This increases the technical risk of this system requirement and reveals a key design item that may require special attention or elaborate trade studies. The designer can now begin to assess where and how to focus design efforts.

There are additional pieces of information that can be documented in the house of quality and used to enhance the supporting role of QFD in design. For instance, a set of existing systems can be identified as competitors to the one under design and used to perform a competitive assessment. The competitive assessment at the right of the house of quality is based on experts' assessment of competing systems relative to the customer requirements used for the current design project. A similar assessment can be done for competing systems relative to system attributes and documented at the bottom of the house of quality (Breyfogle 2003, p. 349–356; Dieter 1999, p. 69–75).

Overall, this qualitative technique helps the designer distinguish between strict requirements and customer preferences, while providing a prioritization of the entire requirement set that guides the allocation of design effort on specific elements. It also serves as a working document where the designer records basic assessments that can provide a basis for quantitative trade-off analyses in later phases of design, and where changes can be made as new information becomes available. These initial trades also serve to establish at a basic level the internal consistency of design trends within a single system. Most important, QFD provides traceability between customer requirements and system attributes, thus enabling the designer to address challenges as the relationship between requirements and system is understood and mapped in the house of quality. However, QFD is recognized as being a time-intensive technique, requiring an important effort commitment from the designer and facilitators. In the case of large projects, the resources needed to execute processes involved in the generation of the different parts of the house of quality can be significant, and should be managed carefully (Breyfogle 2003, p. 349–356; Dieter 1999, p. 69–75).

Deployment of the QFD House of Quality

When the house of quality is completed and the critical system attributes have been identified, a System Requirements Review (SRR) takes place to perform a final check on the results. Then, the QFD process calls for the deployment of system attributes as the inputs, namely the "What's" of a second house of quality. Similarly, the technical importance or the technical risk values of system requirements are passed as importance ratings in the second matrix. The "How's" in this second house of quality are the subsystems, which enable the system requirements. In this manner, the second house of quality is generated with the same procedures utilized in the first one, and analogous insight and results are produced, granted, though, at a lower level in the system decomposition. After critical subsystems are identified and checked in a Conceptual Design Review (CDR), the results of the second house of quality are deployed as inputs to a third one, where the subsystem attributes represent the "How's." The generation of matrix elements is repeated yet again, yielding results at still a lower level in the system decomposition. The deployment of house of quality results, as inputs to the next one, is repeated as many times as necessary within allowable expenditure of resources. The deployment process is illustrated in Figure 19.6, where key elements in each house of quality are shown (Dieter 1999, p. 69–75).

The parallelism between the deployment progression of QFD and the design process explained earlier in this chapter is evident. Both feature the recursive use of a basic process to progressively decompose a system and provide a traceable path between elements from different levels in the system hierarchy. This inherently lays

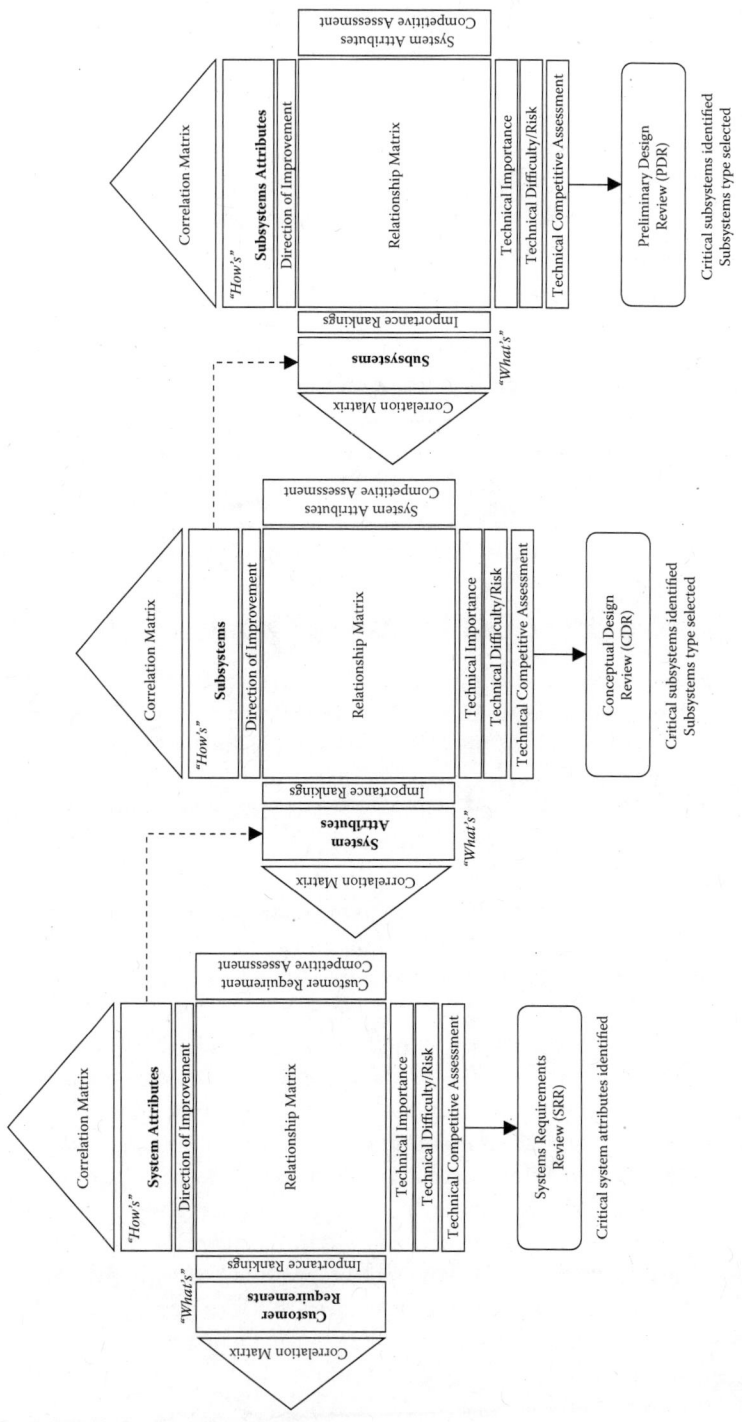

Figure 19.6 Deployment progression in QFD.

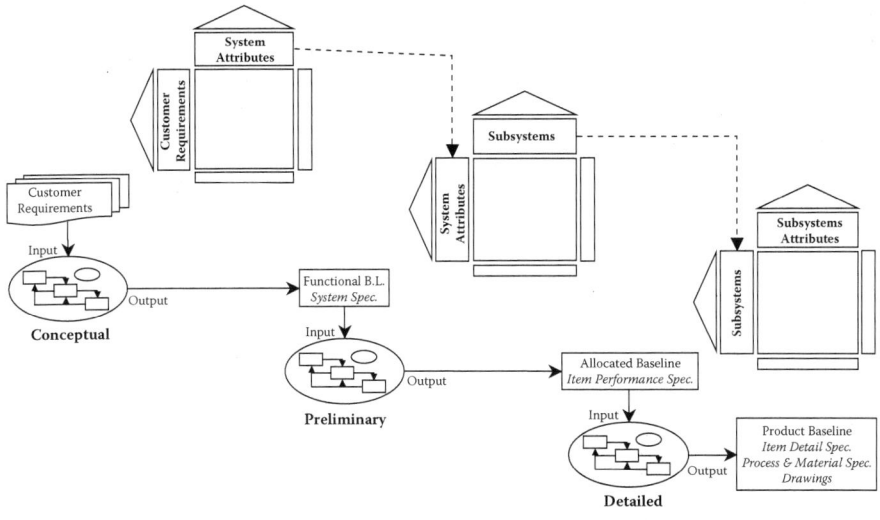

Figure 19.7 Comparison of DoD systems engineering design process sequence and QFD progression.

a trace between activities and decisions in different phases of the entire design process. Placing both process flows side by side reveals the correspondence between elements as shown in Figure 19.7, which uses the DoD systems engineering (design) process presented in Chapter 17 ("Fundamentals of Systems Engineering"). In this manner, QFD represents a valuable supporting technique for systems design, both in content and structure.

Summary

Design is generally characterized as being a cognitive process of information transformation, decision making, and problem resolution. As such, it significantly overlaps with accepted systems engineering processes and practices. The design process presented is therefore heavily reliant on systems engineering but considers relevant tools and techniques from a variety of sources. The process of design begins by establishing a relevant context that supports all future steps with relevant data and knowledge, and calls for the implementation of rigorous structured approaches for that purpose such as SWARM. The analysis of customer requirements allows the designer to transform this information into a set of system attributes that provides a system specification and enable the generation and selection of concepts through techniques such as analysis of alternatives. After generating a functional baseline, the basic design process is cycled to generate system definition and specifications at progressively greater levels of decomposition. With each iteration, a verification process is executed and reported in a design review before documenting the design in a new baseline and proceeding to the next design phase. Traceability through different system structure hierarchy levels is pointed to as a key feature in the process, and supported in part by QFD, which documents system mappings between hierarchy levels at each step and reveals the most important items, thus guiding the allocation of design resources. The importance of exploring relational structures and generating system knowledge through appropriate model transformations highlights the value of QFD, which effectively supports and documents them.

References

Beissner, F.L., Lovell Jr., W.A., Robins, A.W., and Swanson, E.E. 1984. Application of Near Term Technology to Mach 2.0 Variable-Sweep-Wing Supersonic-Cruise Executive Jet. NASA Contractor Report 172321, Langley Research Center, Hampton, VA.

Breyfogle III, F.W. 2003. *Implementing Six Sigma—Smarter Solutions Using Statistical Methods,* 2nd ed. Hoboken, NJ: John Wiley & Sons Inc.

DoD (Department of Defense), United States. 2001. *Systems Engineering Fundamentals.*

Dieter, G.E. 1999. *Engineering Design: A Materials and Processing Approach,* 3rd ed. New York: McGraw-Hill.

Haggerty, J.J. 1996. Spinoff 1996. National Aeronautics and Space Administration, Commercial Development and Technology Transfer Division.

Jimenez, H. and Mavris, D.N. 2005. Conceptual design of current technology and advanced concepts for an efficient multi-mach aircraft. *SAE Transactions* 114: part 1, 1343–1353.

Jimenez, H. and Mavris, D.N. 2007. A framework for collaborative design in engineering education. Paper presented at the *45th AIAA Aerospace Sciences Meeting and Exhibit,* January 8–11, in Reno, NV.

Kossiakoff, A. and Sweet, W.N. 2003. *Systems Engineering Principles and Practice.* Hoboken: John Wiley & Sons.

Nahm, Y.-E. and Ishikawa, H. 2004. Integrated product and process modeling for collaborative design environment. *Concurrent Engineering: Research and Applications* 12(1):5–23.

National Research Council, Committee on Aeronautical Technologies. 1992. *Aeronautical technologies for the Twenty-First Century.* Washington, D.C.: National Academy Press.

Raymer, D.P. 1999. *Aircraft Design: A Conceptual Approach,* 3rd ed. Reston: American Institute of Aeronautics and Astronautics, Inc.

Soban, D., Biltgen, P., and Mavris, D.N. 2005. A technology assessment methods survey: concepts and tools. Paper presented at the *1st International Conference on Innovation and Integration in Aerospace Sciences,* August 4–5, in Belfast, Northern Ireland.

Suh, N.P. 1990. *The Principles of Design.* New York: Oxford University Press.

Zwicky, F. 1969. *Discovery, Invention, Research through the Morphological Approach.* New York: The Macmillan Co.

Zwicky, F. 1967. New methods of thought and procedure. In *The Morphological Approach to Discovery, Invention Research and Construction,* Eds. Fritz Zwicky and Albert G. Wilson, 273–297, New York: Springer-Verlag.

Chapter 20

Advanced Design Methods

Dmitri Mavris and Hernando Jimenez

As the complexity of systems designed increases, and more stringent requirements are set forth, the methodologies used in design must evolve accordingly. Advanced design methods address this need by incorporating new and innovative tools within novel approaches, or applying established methods from other fields in new ways. As such, they represent the state of the art in systems design.

Key Concepts

Matrix of Alternatives
Multi-Criteria Decision Making
Concept Space Exploration
Sensitivity profiling
Surrogate models
Probabilistic design

Introduction

Previous chapters illustrated the foundations of systems architecture and systems engineering, key tools and enablers, and the general approaches encompassing the current paradigm. Though they represent a body of accepted practices, none of these approaches are flawless, nor do they provide a "silver bullet" solution. It is vital to the field of system design to recognize the limitations, shortfalls, and opportunities for improvement in the way systems are developed. This chapter presents a selection of advanced design methods that enhance the process of systems design across its multiple steps. The first two, Interactive Reconfigurable Matrix of Alternatives and Genetic Algorithm Concept Space Exploration, address the initial design phase by enhancing concept space definition, concept generation and selection, and supporting the analysis and trade-offs that take place in the early definition of system taxonomy. Use of Sensitivity Profiling is presented next as a method implemented primarily during preliminary design to capture and visualize requirement-attribute sensitivities and leverage trade-off analyses at any level of system decomposition.

Finally, a series of probabilistic design methods is introduced as a means to address design tasks involving different sources of uncertainty. These latter methods are built upon a common core of enablers and are invoked during preliminary and later phases of design. For each method, a description is provided in terms of the shortfalls it addresses and improvements it affords. The key enablers required are then identified and clearly explained, followed by description of the method's implementation illustrated through representative applications in actual design projects.

Interactive Reconfigurable Matrix of Alternatives

Formulation

Selection of component types and general system taxonomy in the conceptual design phase is of critical importance because it locks degrees of freedom and cuts off entire regions of the concept space early in the process when knowledge about the system is very limited. These early decisions are a major driver in the evolution of the design during the rest of the process. In Chapter 19 ("Systems Design"), the concept of *Analysis of Alternatives* (AoA) was introduced as an approach to structuring the tasks of defining the concept space, generating alternatives, and making a selection. The first two tasks are based on morphological analysis, an approach that facilitates the rigorous investigation of large multidimensional spaces by enumerating and studying the possible alternatives for system elements and the relationships among them. A *morphological matrix* documents the main components of the system and the possible alternatives for it, while the *cross-consistency matrix* captures the relationships among them (Zwicky 1967; 1969). In AoA, the relationship is usually limited to compatibility among alternatives, and is documented with binary entries for compatible or incompatible combinations (see Figures 19.1 and 19.2 in Chapter 19 for examples of these two matrices). The third task, that of selecting among the numerous alternatives, was left to subject matter experts whose experience and knowledge provide the analytical basis upon which choices are made.

There are a number of shortfalls in this approach. First, as was mentioned in Chapter 19, the number of possible configurations grows very quickly as the designer incorporates more system components and alternatives into the morphological matrix. Often, in an attempt to be rigorous in the definition of the concept space, the designer will include so many alternatives that the total number of configurations becomes unmanageable. Second, the selection of one alternative over all others for a system component implicitly carries a valuation scheme based on some criteria and their importance. Subject matter experts will only use one valuation scheme at a time; and if more than one expert is queried, it is likely that they will use different ones. Most important, the scheme is not explicitly stated or documented. Third, the selection process should incorporate contextual information beyond the technical realm, so that the availability of alternatives subject to previous selections can be driven by criteria other than their compatibility with other alternatives. In fact, it is quite possible that political pressures, programmatic risk considerations, or budgetary constraints at a high level can eliminate some of the more exotic alternatives populating the matrix of alternatives. Fourth, the matrix of alternatives is an inherently static document and does not lend itself very well in its basic formulation for electronic design reviews where decisions can be challenged and rectified *in situ*.

These needs and shortfalls drive the formulation of the Interactive Reconfigurable Matrix of Alternatives (IRMA). As suggested by its name, IRMA instantiates the matrix of alternatives in an interactive form so that it can be dynamically reconfigured during concept selection activities and aid in the down-selection of alternatives. By providing this type of tool, decision makers can engage in a collaborative fashion and play "what-if" games by recursively conducting down-selection exercises under a variety of contextual assumptions. IRMA also documents information to implement filters that explicitly eliminate alternatives by incorporating certain assumptions into the selection task. Finally, IRMA offers the option of using decision-making support techniques that quantify how alternatives rank relative to each other based on a quantitative valuation scheme (Engler et al. 2007). There are two major enablers for IRMA: decision-making techniques and interactive visualization.

Enabler: Multi-Attribute Decision Making

The importance and difficulty of configuration decisions early in the design process, as well as the limitations of subject matter experts in making these decisions, reveal the need for decision-making supporting techniques "able to handle multiple, potentially conflicting criteria" (Li and Mavris 2006). Fortunately, there is no shortage of Multi-Criteria Decision Making (MCDM) techniques, with over 70 in the published literature (Roman et al. 2004). Choosing an appropriate MCDM technique is a selection problem in itself that has been approached in the past (Li and Mavris 2006; Hwang and Yoon 1981; Ozernoy 1987; Poh 1998) but that lies beyond the scope of this chapter. A common classification of MCDM techniques features two main branches: multi-attribute and multi-objective. Multi-Attribute Decision Making (MADM) deals with the *selection* problem based on prioritized attributes of a finite number of alternatives (Sen and Yang 1998, p. 18), and thus is directly relevant to alternative selection for system components. Among MADM approaches, the Technique for Ordered Preference by Similarity to Ideal Solution (TOPSIS) (Hwang and Yoon 1981) is found to be best suited in the majority of component alternative selection applications (Engler et al. 2007). Additionally, it incorporates a relative weighting mechanism that allows decision makers to capture a preference profile across attributes, conveniently addressing one of the AoA shortfalls identified in the previous section. TOPSIS requires a matrix to document information about all alternatives, the data of each alternative coded in a vector of attributes. Using attribute values for each alternative, and the weighting profile selected, TOPSIS estimates the Euclidian distance between all alternatives and an ideal solution. The ideal solution is a composite of the best values in each attribute for the alternative set under consideration. As a final result, this technique provides a ranking of all the alternatives based on the proximity to the ideal solution, that is, the minimization of the Euclidian distance to the ideal (Li and Mavris 2006).

Enabler: Visual Analytics

When considering the need for visualization in an interactive and graphical method such as IRMA, some may judge it a somewhat trivial observation. After all, how can decision makers collaborate in the selection of system concepts and interact with the matrix of alternatives if they cannot see it? In the context of the advanced design methods presented in this chapter, *visualization* has far-reaching implications that extend beyond merely "seeing" documents or design artifacts. Analysis in design involves vast amounts of information, in which conflicting and sometimes counterintuitive trends are found. Available data can be overwhelming and at the same time be incomplete or inconsistent. Yet it requires human judgment to sift through this information, make the necessary associations and transformations, and discern sensible meaning from it. *Visual analytics* is an emerging field that helps decision makers address these challenges by synthesizing and visualizing data to gain understanding from it. It is "the science of analytical reasoning facilitated by interactive visual interfaces" (Thomas and Cook 2005, p. 4) that "combine[s] the art of human intuition and the science of mathematical deduction to directly perceive patterns and derive knowledge and insight from them" (Wong and Thomas 2004, p. 20–21). Analysts are therefore able to recognize information that confirms prior knowledge and that which represents discoveries and the creation of new knowledge. Visual analytics began as development from the combination of scientific and information visualization, but currently represents a multidisciplinary approach that includes techniques for analytical reasoning, visual representations and interaction, data representations and transformations, and production, presentation and dissemination of results (Thomas and Cook 2005, p. 4; Wong and Thomas 2004, p. 20–21).

Implementation

Before describing the implementation of IRMA, it is essential to note that it does not just refer to an electronic tool but rather to the method of creation of the tool, the transformation and manipulation of information necessary to populate it, and the specific process of using of the matrix of alternatives for concept down-selection. First, IRMA requires that designers collect three main sets of information and document them appropriately in the tool before being able to perform selection exercises. The first one is the compatibility matrix, which identifies three possible conditions: (1) compatible, (2) incompatible, and (3) required. "Required" refers to certain component alternatives that require the selection of a specific alternative on another component to ensure system feasibility. For example, consider how the selection of composite materials for the fuselage of an aircraft requires that certain manufacturing techniques be selected as well. The second piece of information refers to the attributes and values of all the alternatives for a given component, which will be used in TOPSIS to provide decision-making support if the designer requires it (Engler et al. 2007). For instance, the "platform" component on the matrix of alternatives for a long-range strike model may include alternatives such as a B-1B, a B-2, and a B-52. In building the IRMA, the designer must decide what are the criteria with which alternatives will be evaluated. In this case, criteria include range, speed, payload, radar cross section (RCS), and cost. Then, values for these criteria must be researched and documented. When the MADM option is available, the decision maker can adjust the importance of these criteria, and IRMA will make recommendations for the most appropriate platform, based on the resulting rankings. If range is heavily weighted, for example, the B-52 will rank higher; whereas if designers determine that RCS is far more important than cost or speed, then the B-2 will rank highest. Note that the B-2 has a low RCS with respect to the other alternatives considered, but is significantly more expensive and is limited in speed relative to alternatives.

The third and last piece of information regards any criteria that decision makers wish to have as a filter. A popular choice is Technology Readiness Level (TRL), a measurement for the assessment of technology maturity originally developed by NASA (Mankins 1995). This measurement is commonly used to capture programmatic risk, where immature technologies will require a longer time to become available and present greater uncertainty on the benefits they are expected to provide once implemented. If designers gather TRL information on all available component alternatives, applying the filter at some TRL value will eliminate all immature technologies under the prescribed value. This is referred to as a "coarse filter," whereas "fine filters" make use of more detailed metrics such as R&D funding and implement a specific value as the threshold. Additional features in IRMA include a calculation for the current number of possible system configurations based on available alternative choices, providing decision makers with a measure of the current size of the concept space (Engler et al. 2007).

After populating IRMA with all necessary pieces of information, the procedure for its use in concept down-selection calls for the use of coarse filters first to focus the problem. Next, use of MADM rankings are appropriate for further narrowing down the number of options for each component. This requires that decision makers dial in the criteria weightings for each component for which TOPSIS has been implemented. Determination of weightings can be performed ad hoc or be supported by the technical importance measurement of the Quality Function Deployment introduced in Chapter 19 ("Systems Design"). By progressively eliminating lowest-ranking alternatives in the TOPSIS rankings, decision makers can engage in more focused discussions about the most promising alternatives. During such discussions, alternatives that are deemed not to meet customer requirements are eliminated manually. This assessment can be performed by continuously combing the remaining alternatives, row by row. The process may be repeated by examining again the values of coarse filters and revisiting MADM rankings to eliminate more low-ranking alternatives. At the end of the process, summarized in Figure 20.1, a single family of concepts is revealed.

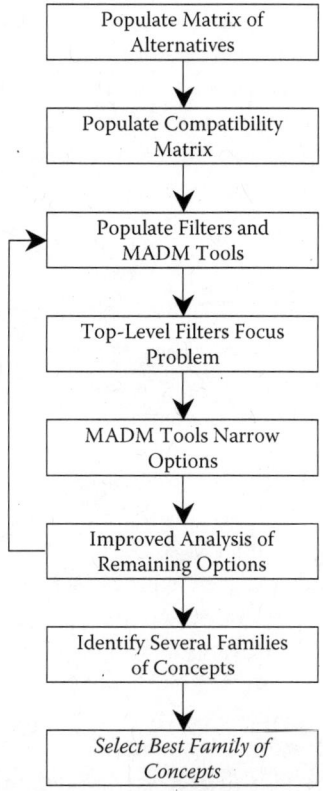

Figure 20.1 IRMA concept in down-selection process. (From Engler III, W.O., Biltgen, P., and Mavris, D.N. 2007. Paper presented at the *45th AIAA Aerospace Sciences Meeting and Exhibit*, January 8–11, in Reno, NV.)

Genetic Algorithm Concept Space Exploration
Formulation

The previous section stated that expert opinion had limitations in its supporting role within concept selection. Not only are experts limited by their empirical experience, but also by the natural inability of humans to efficiently deal with complex concept spaces ruled by numerous and nonintuitive relationships. This is particularly evident in systems where concept selection and subsystem definition are equally significant drivers for system performance. In such a case, the designer can choose several concepts, all of which may meet customer requirements, but only certain designs within each concept are sure to meet system requirements. Without fully developing multiple concepts in parallel, the designer cannot determine what concept represents a more promising option with lower risk.

A notable example of this type of design problem is that of a supersonic aircraft with requirements on sonic boom minimization and range targets. Sonic boom is known to depend on mission parameters such as Mach number and altitude, on aircraft weight, and on geometric parameters (Jimenez and Mavris 2005). In fact, boom minimization has been successfully demonstrated through geometric manipulation of the airframe (Northrop Grumman 2003). It follows that the intuitive way to reduce sonic boom is to reduce aircraft weight as much as possible, adjust the mission profile, and select favorable aircraft taxonomy. However, all these elements also have an important effect on range, revealing a fairly complex trade-off. Because so much information about the system needed to perform this trade analysis has not yet been defined in concept selection (i.e., aircraft geometry, sizing routine, mission optimization), the designer is unable to study sensitivities and justify the choice of some system components. Moreover, small variations in the specific values of required range and boom minimization stringency can quickly change what configuration is most appropriate, and, to make matters worse, different concepts can provide equally feasible designs.

To address these challenges in concept selection, a genetic algorithm concept space exploration method is used. This approach brings Modeling and Simulation (M&S), a method enabler, into the conceptual design phase to provide a quantitative assessment that adequately supports concept selection. The second enabler of this method is the genetic algorithm. Use of genetic algorithms for concept generation, evaluation, and selection of supersonic and multi-Mach aircraft has been successfully implemented in the past (Buonanno 2005; Buonanno and Mavris 2004; Jimenez and Mavris 2005). A third and last enabler is visualization of results to determine the performance frontier of the current space under consideration.

Enabler: Genetic Algorithm

The genetic algorithm is a search and optimization technique developed by Holland (1975) in the 1970s, whose formulation is based on Darwin's theory of evolution

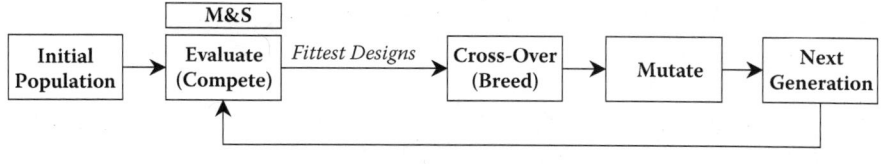

Figure 20.2   Summary of the genetic algorithm procedure.

and survival of the fittest. This approach enables the exploration of nonconvex, multimodal, and disjoint design spaces by searching the range through multiple points simultaneously. This approach contrasts with traditional gradient-based techniques that search the design space one point at a time, fail when encountering discontinuous or nonsmooth regions in the range, and cannot guarantee global optimality. However, the advantages of the genetic algorithm come at the expense of increased function calls, which may be prohibitive in some cases. The characteristic feature of the genetic algorithm is the emulation of genetic evolution of designs over multiple generations. The search begins with the definition of a population of designs, each defined by a vector of attribute values. These attributes are used as inputs in the objective function, and are coded in a binary string, which enables the design definition to include continuous and discrete variables. The output of the objective function is used directly or is manipulated to generate a fitness parameter, which indicates how fit a given design is or how well it performs. The dominance of stronger samples in a population is captured by allowing the fitter designs to "breed" in an operation referred to as cross-over. This operation uses the definition of two-parent designs, as coded in their respective binary strings, and generates two offspring designs based on some set of rules for combining strings segments. This step, effectively capturing the essence of parent genetic material combination in offspring, is followed by a potential genetic mutation implemented as a random alteration of a part of the binary string. Designs in the new generation are then used in a new iteration of the genetic algorithm, where the process of competition, breeding, and mutation is repeated. The cycle is repeated for a number of times until dominant designs stabilize in the population. These dominant designs are at the local or global minima of the design space (Hajela 1990; Lin and Hajela 1993). A summary of the genetic algorithm procedure is shown in Figure 20.2.

Implementation

Genetic algorithms, and the availability of computationally inexpensive M&S tools, offer a convenient means to address the key challenges of the concept selection problem at hand. First, genetic algorithms allow the designer to identify the dominant designs that may confirm subject matter expert intuition or reveal unexpected trends. Second, the technique allows for a combination of discrete and continuous variables to be handled simultaneously, thus enabling the selection of discrete alternatives within a component, and the selection of continuous parameters for that component alternative. Third, the criteria for truncating a genetic algorithm run can be made arbitrary, meaning that the designer can choose to stop the run at any point without losing any information or results developed until then. This is desirable because in this context only the dominant configurations and key design features within each configuration are sought, and are usually achieved within a few design generations. The genetic algorithm is being used as a coarse search method in the concept space and the design space within each configuration, not as a full optimization mechanism that requires a significantly greater number of design generations, and should probably be conducted in later design phases.

In this method the designer first generates the matrix of alternatives and the compatibility matrix for the relevant system components as would normally occur during AoA. A supersonic aircraft would require matrices such as the ones presented in Figures 19.1 and 19.2 of Chapter 19 ("Systems Design") and in Jimenez and Mavris (2005). Additionally, relevant parameters and value ranges for each alternative must be defined. Using the available alternatives and compatibility rules, the genetic algorithm can be set up by providing the necessary mapping between the attributes coded in the binary string of the genetic algorithm logic and the alternatives of the matrix of alternatives. A similar mapping between alternative's parameters and attributes coded in the binary string is completed. For instance, in the binary string of the current example, the first two bits are set to code four possible alternatives of aircraft wing planform, as shown in Figure 20.3. Also shown is the coding of four possible alternatives of empennage in the third and fourth bits, and continuous alternative attributes such as wing sweep angle and aspect ratio in other parts of the string. The setup is graphically represented in Figure 20.3.

As the genetic algorithm is executed, different concepts representing the plethora of morphological combinations are modeled, and their range and sonic boom performance are recorded. The population contains multiple samples corresponding to the same morphological combination, but differing in the value of alternative

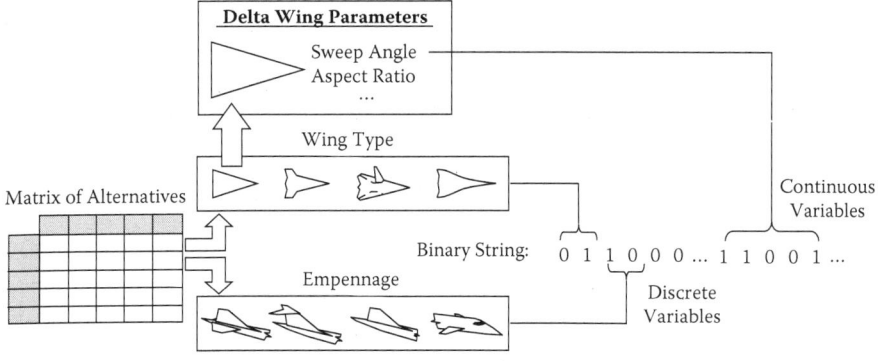

Figure 20.3   Genetic algorithm setup for concept space exploration.

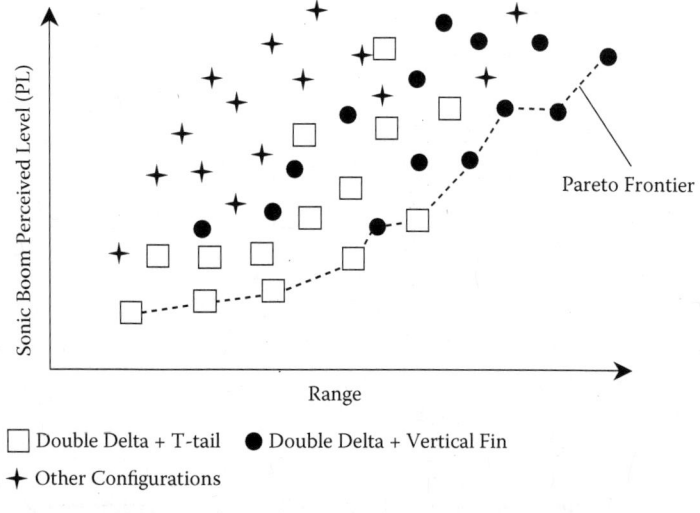

Figure 20.4 Notional visualization of genetic algorithm concept space exploration.

parameters. For instance, there will be several designs featuring a delta wing and a conventional empennage, but these designs will differ in the geometry of the delta wing and the geometry of empennage surfaces. Based on values of the fitness parameter, the designs featuring the best range and sonic boom performance are used for cross-over and generation of offspring designs. After some number of design generations, the dominant configurations start to become evident. For the present example, and based in part on the results from Jimenez and Mavris (2005), a combination of double delta wings and T-tail empennage arise as a dominant configuration for low sonic boom. The combination of a double delta wing and a single vertical fin surfaces as a dominant design for range. The designer can use the data points of the last generation(s) and visualize them in a plot of sonic boom versus range, as shown notionally in Figure 20.4. The dominant solutions reveal the Pareto frontier, which describes the design performance envelope for the current concept space. Most important, the trade-off between the two metrics can be quantified and understood across multiple concepts, and across design configurations within a given concept.

As shown, this method supports the exploration of the concept space in cases where expert opinion is particularly limited. By leveraging on computationally inexpensive M&S tools, and making use of the genetic algorithm, a large number of concepts can be generated and quantitatively evaluated with an acceptable level of detail. Dominant concepts are identified, helping designers confirm or discover trends while greatly reducing the risk of poor concept selection in what would have otherwise been a strongly speculative context. Finally, adequate visualization of results enables the designers to gain deeper insight into the trade-offs and sensitivities that support configuration decisions moving forward in the design process.

Sensitivity Profiling

Formulation

In any phase of the design process it is necessary to perform trade-offs and understand parameter sensitivities so that the system definition can be decomposed into lower-level functions, requirements, and physical elements. Designer experience is a valuable resource for estimating the general behavior of a system and predicting performance trends. However, quantitative assessments become a necessity, even in the early phases, and it is therefore necessary to resort to adequate modeling tools. Unfortunately, many of them are legacy tools that have been inherited over the years as part of organizational inertia or lack of resources in research and design entities to create new tools. Though they may represent verified models accepted by leaders and experts in a field, the inner workings of legacy tools are usually unknown. The inevitable result is a loss of transparency and the tool becomes what is commonly known as a "black box." In this context, designers face a formidable challenge of understanding and documenting sensitivities with which trade-off analyses can take place.

The creation and use of a sensitivity profiling environment addresses this gap by capturing the behavior of tools in surrogate models, explained next, and visualizing it in a graphical interactive display where the design space can be adequately explored. These two features—namely, the use of surrogate models and visualization techniques—are the two primary enablers of this method. The profilers graph the relationship between system attributes (model inputs) and metrics of interest (model outputs) for attribute-metric combinations as notionally shown in Figure 20.5. By including input controls, sensitivity profilers become an interactive, collaborative, visualization tool, bringing transparency to design space exploration and trade-off analyses, and leveraging informed discussions among designers. This type of environment was originally formulated as a Unified Trade-off Environment (UTE; Baker 2002) that enabled the analysis of mission requirements, system attributes, and technology improvements. However, the general sensitivity profiling method has independently evolved as an outgrowth of the UTE, and has been successfully applied in a variety of other applications and problem types.

Enabler: Surrogate Models

As their name suggests, surrogate models are proxies for other models and tools, such as the aforementioned legacy codes. They are "mathematical approximations that capture the behavior of the models they substitute, by fitting functions

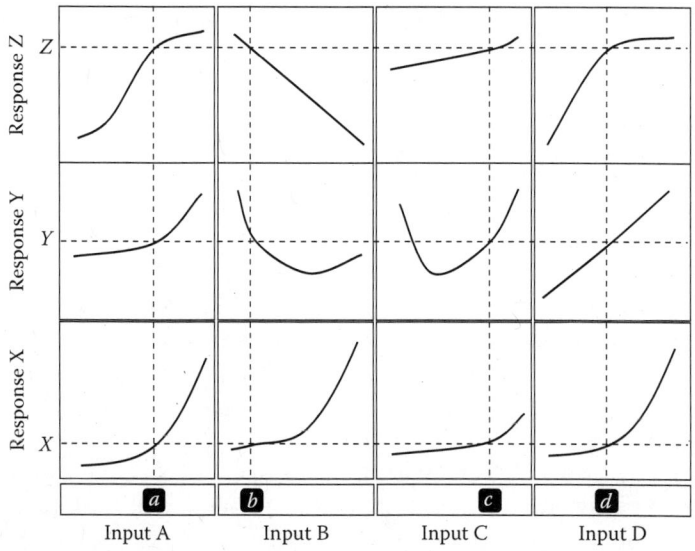

Figure 20.5 Notional display of sensitivity profilers.

of known form to empirical data samples through statistical methods" (Jimenez 2008). Many may wonder why some analyses are conducted with surrogates when the original tool is readily available. One of the reasons, as was just discussed, is the lack of transparency of legacy codes. When executed, "black box" codes provide a single output data point for a given input, and do not explicitly characterize the relationship between inputs and outputs. Because surrogate models have a fully known structure, they can be made to show this relationship explicitly. Another reason, which will become relevant in later sections, is the significant reduction in computational expense required by surrogate models relative to some legacy tools, whose extended runtimes make any consideration for large numbers of executions prohibitive. Improvements in execution time provided by surrogate models are rooted in their mathematical structure, usually much simpler and smaller than that of legacy codes.

However, the advantages afforded by surrogate models require that an upfront computational investment be made in their creation, and their accuracy in representing certain relationships is not universally guaranteed. To understand these considerations, it is first necessary to note that a surrogate model is empirical, meaning that it is an approximation based on observations of data (Myers and Montgomery 2002). There are therefore two fundamental components in the generation of all surrogates: the adequate selection of data samples, and the mechanism by which a function is made to fit these observations, thus capturing the behavior of the underlying phenomena. The selection of these two components depends heavily on the type of surrogate model being created. Popular forms include artificial neural networks (Bishop 1996; Haykin 1998), Gaussian processes (Rasmussen and Williams 2006), and Spline models (Wahba 1990). Given their long history and widespread use, this section focuses on a type of surrogate called responses surfaces to illustrate basic concepts in surrogate generation.

Response Surface Methodology (RSM) is the collection of statistical techniques related to the modeling and optimization of processes via response surface equations. Response surfaces are linear regression models of the form shown in Equation 20.1, where the "x" terms are the independent variables, also called predictor variables; "y" is the response, or dependent variable; and the "b" parameters are the unknown regression coefficients. The term "ε" captures the approximation error of the response surface. Values for regression coefficients are estimated using data samples, expressed as pairs of corresponding input–output (x, y) values, in model-fitting techniques. Note that even though the structure of Equation 20.1 is an nth order polynomial function for the independent variables "x," it is a linear relationship for the regression coefficients "b." Thus, the model fitting, or regression process, which attempts to calculate estimates for regression coefficients, is considered linear. The least-squares method is the most commonly used approach for coefficient regression in RSM (Myers and Montgomery 2002).

$$y = b_0 + \sum_{i=1}^{k} b_i x_i + \sum_{i=1}^{k}\sum_{j=1}^{k} b_{ij} x_i x_j + \cdots + \varepsilon \tag{20.1}$$

The selection of an appropriate set of data observations is crucial in estimating the value of regression coefficients with adequate accuracy. The careful design of experiments, from which data will be gathered for model fitting, implements RSM statistical techniques and is appropriately referred to as Design of Experiments (DoE). Designed experiments are usually documented as tables that describe the settings of independent variables for each experiment or data sample. It is common practice to use normalized values in designed experiment tables so that they can be universally applied regardless of the specific independent variables used or the range of values of interest. This concept is illustrated in Figure 20.6, where

2^2 Design of Experiments

Run	X_1	X_2
1	0	0
2	1	0
3	0	1
4	1	1

X_1: Baking Temperature
X_1: Range: 300°F – 400°F
X_2: Baking Time
X_2: Range: 30 min – 45 min

Run	Temp.	Time
1	300° F	30 min
2	400° F	30 min
3	300° F	45 min
4	400° F	45 min

Figure 20.6　Sample implementation of a designed experiment.

a 2^2 design is applied to experiments varying the baking temperature and time of a cake. A number of experimental designs exist, not only to fit response surfaces of varying characteristics, but also to extract valuable information about the process being approximated. A "main effects" design, for instance, can be used to conduct an ANalysis Of VAriance (ANOVA), which reveals the sensitivity of process outputs to different inputs and enables analysts to screen-out statistically unimportant parameters. Once the data is collected and the model fitting task is completed, surrogate models are tested for accuracy based on statistical estimates such as the coefficient of multiple determination—R^2, the model fit error distribution, and the model representation error distribution. Detailed discussions regarding these and other considerations about surrogate models are beyond the scope of this section. Complete theoretical and practical explanations are readily available in published literature (see, for instance, Box and Dapper 1987; Khuri 2006; Myers and Montgomery 2002).

Implementation

The implementation of a sensitivities profiler in the context of the design process is best explained through a representative example. Suppose that designers are facing decisions regarding certain wing parameters in the preliminary design phase of an aircraft. This corresponds to the definition of subsystem attributes that meet system-level requirements, which include takeoff and landing distance, emission of nitrogen oxides, acquisition cost, and direct operating costs. Designers first identify the necessary modeling capabilities to evaluate aircraft performance and cost estimates as a function of subsystem attributes. In this example, wing parameters such as area, sweep angle, aspect ratio, taper ratio, and thickness-to-chord ratio, illustrated in Figure 20.7, are relevant. Once the modeling environment is available, and assuming it is a black-box legacy code, a surrogate model type and a corresponding design of experiments are selected. This selection is based on the expected behavior of the system requirement metrics (output) as a function of the aforementioned subsystem attributes (inputs). For this design task, it is anticipated that the relationships in question can be adequately approximated with second-order response surfaces. Designers thus proceed to select a design of experiments and execute model runs following the experiments table as applied to the wing parameters and the value ranges defined for them. When the response surfaces are fitted and tested, the values of all regression coefficients are known.

Note that a sensitivity profile describes how a given response y changes with variations in a specific independent variable x_i. It follows that the curve in a sensitivity profile is simply the partial derivative of the response y with respect to the independent variable x_i for which the profile is generated. Each profile therefore plots a 2D slice of an n-dimensional design space. Because all regression coefficients are known, calculation of the partial derivatives is straightforward. This process

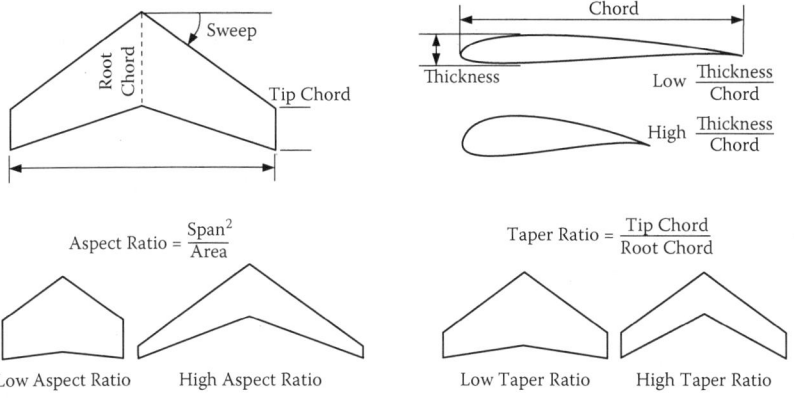

Figure 20.7   Wing attributes for sample sensitivity profiles.

can be done manually, though some statistical software packages such as JMP™ estimate these partial derivatives and plot the sensitivity profiles automatically. Response surfaces for the present example were created and used to generate the profiles shown in Figure 20.8. The dotted lines indicate the settings for attributes and resulting values for requirements metrics.

By observing these profiles, designers are able to determine what attributes have a greater impact on requirements metrics. For instance, profiles for wing area feature large positive and negative slopes, indicating significant changes in requirement metrics for variations in that wing attribute. In contrast, parameters with profiles whose slope is nearly zero, such as taper ratio, have little impact on requirements. The direction of the profiles also provides valuable information to designers regarding trade-offs between competing requirements. For example, trends suggest that a higher value of aspect ratio is preferable for reducing NO_x emissions, but also result in increases for acquisition and direct operating costs. Designers can combine information about the magnitude and direction of the slope to determine how much one requirement can be traded for another. For instance, reductions in aspect ratio below its current value provide negligible improvements in direct operating costs and acquisition costs, and result in a relatively bigger degradation of NO_x emissions.

Given that the profiles are constructed from partial derivatives, the trends depicted for a given attribute x_i will change as the value of another parameter x_j is modified. This property of the sensitivities profilers is of the greatest importance and value to the designer, given that the trends observed under a combination of attribute settings can vary in magnitude (and even direction) for other combinations of settings. For example, the trends shown in Figure 20.8 result from the combination of values for all the wing parameters under consideration. By varying wing sweep value from its lowest setting, as shown in Figure 20.8, to its highest value, the trends for all

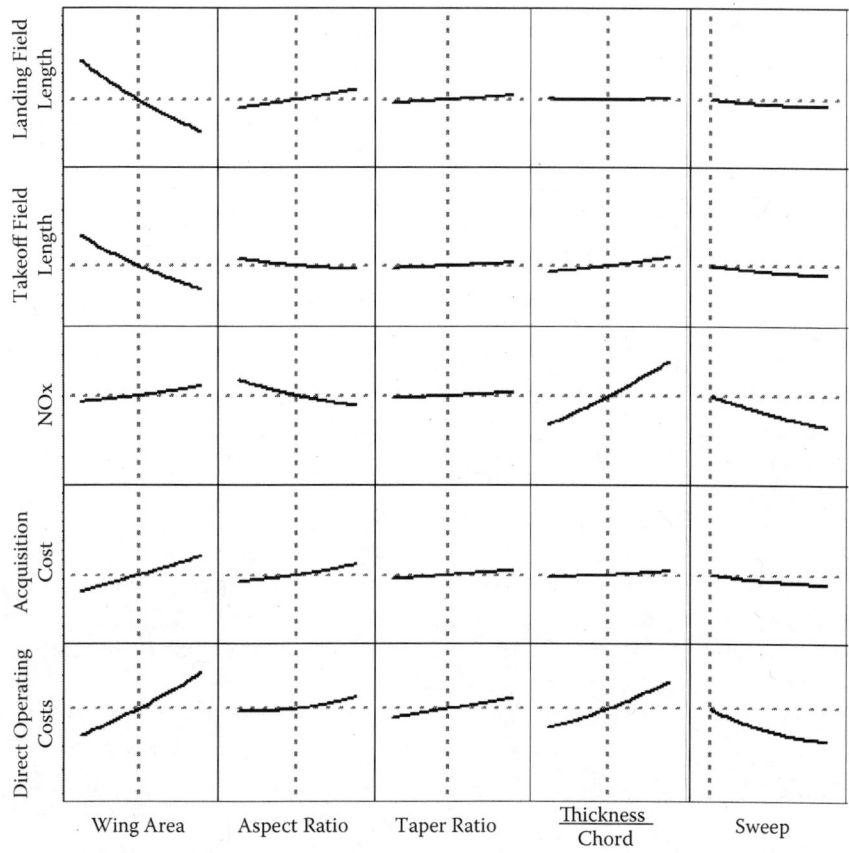

Figure 20.8 Sensitivity profiles for sample aircraft.

other attributes change as illustrated in Figure 20.9. Designers can now observe that the thickness-to-chord ratio increases direct operating costs for low sweep angles, but that at sufficiently high sweep angles, the trend flattens out almost entirely and even seems to reverse slightly. Similarly, the significant effect of the thickness-to-chord ratio on NOx emissions is vastly mitigated for high sweep angle values.

This type of observation, enabled by a dynamic and interactive sensitivities profile environment, provides critical support to designers during trade-off analyses by directly revealing trends and sensitivities in a multidimensional design space. Though the example shown was in the context of preliminary design, the method can be applied at any level of decomposition and any phase in the design process. Most importantly, designers gain more insight about the system at hand and are able to make informed decisions that leverage the success of the design program.

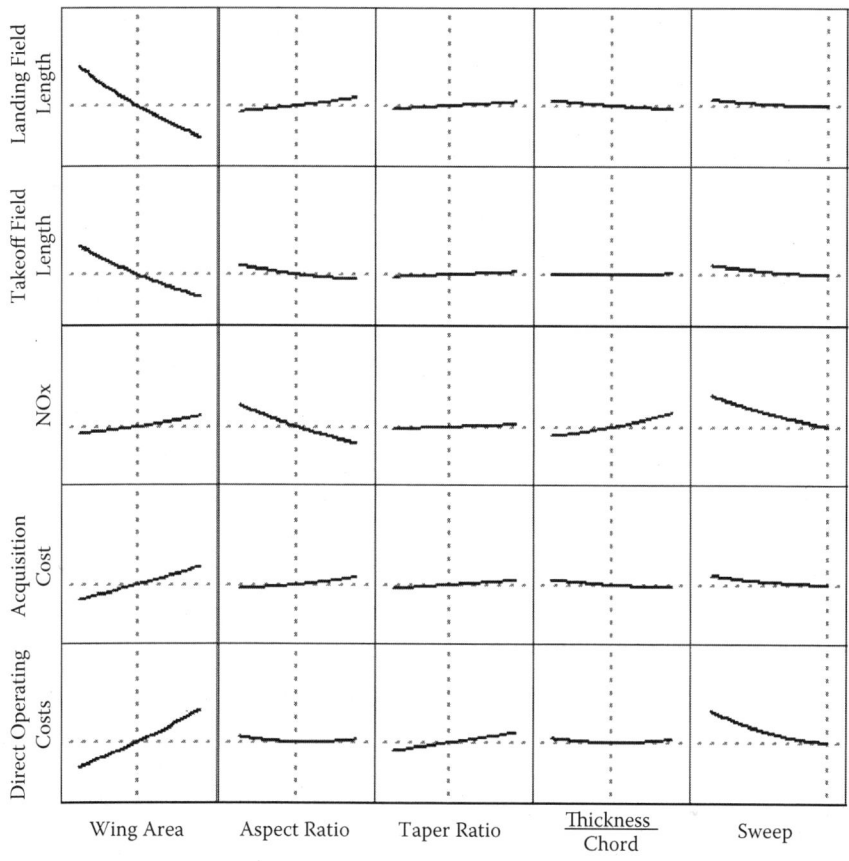

Figure 20.9 Changing sensitivity profiles with attribute settings.

Probabilistic Design Methods

Formulation

The presence of uncertainty in design is recognized as such a fundamental issue that it has arguably become a field of its own, consolidating and building upon established mathematical techniques and design practices. The methods presented in this section are but a select few representing the state of the art in research methodologies that attempt to manage and mitigate uncertainty in the design process. All of them share a common set of fundamental principles and core enablers, but offer a wide variety of applications and uses.

Designers face uncertainty every time that there is limited or no knowledge about an element in the design process. For example, there may be uncertainty about the combination of system attributes that meet customer requirements. Uncertainty may also be found in parameters that describe the context in which a design exists, and that have an impact on its feasibility. Examples include economic variables and operational conditions. Moreover, the interconnectivity of components inherent in the architectures of systems leads to the propagation of uncertainty from one element to another across levels in the hierarchy structure. If the weights of different system components are not fully known, for example, then the weight of the assemblies they comprise will also be uncertain. Uncertainty is also associated with technologies whose impacts on system performance are not fully known, particularly for more immature technologies. Other sources of uncertainty include airframe design evolution, potential changes in customer requirements, and regulatory licensing and certification (Nam et al. 2006). A fundamental concept in probabilistic design methods is that uncertainty can be characterized in the parameters where it originates, and can be propagated through the definition of the system using relationships, functions, and processes that are sufficiently understood.

Owing to its probabilistic nature, uncertainty is commonly characterized and handled through random variables. The study of random variables and their properties in *probability theory* assumes that density and mass functions are known. A fully known probability function implies that all possible or potential observations of the random variable, called the *population*, have been made and collected. However, when studying real processes and phenomena, analysts can only perform a finite *sample* of observations. Thus, they cannot completely describe the probability function but rather *infer* its general properties. The science that focuses on these studies is called *statistical inference*, another fundamental building block in probabilistic design methods (Hayter 1996).

The effectiveness of inference depends on the extent to which the sample is *representative* of the population. There are two characteristics that drive this feature in a sample: randomness in the selection of observations and the number of observations. By definition, samples whose observations follow a pattern have a lower degree of representation, and thus stronger patterns lead to less representation. The randomness of a sample is readily addressed by *probability theory* and the techniques that enable experimenters to sample from prescribed probability distributions. One such technique is *Monte Carlo simulation*, which represents the first major enabler of probabilistic methods. A larger number of observations also leads to increased representation, the limiting case being a sample containing all observations and thus completely representing the population. It follows that inference is enabled by the ability of experimenters to collect or produce large numbers of observations. In the current context, these observations often result from execution of modeling tools whose potentially high computational expense makes the task of gathering large numbers of observations prohibitive. Surrogate models were introduced in an earlier section and are said to have significantly shorter runtimes while capturing the underlying behavior of the modeling tools. It follows that *surrogate models* are the second major enabler of probabilistic design methods. Finally, *visualization* of probabilistic results involves large sets of data points within multidimensional plotting schemes, and thus represents the third and last enabler.

Probabilistic Design Space Exploration

The first application of probabilistic design methods concerns the exploration of the design space in search of feasible solutions. This step takes place after requirements have been fully defined, and requires that a modeling environment or appropriate surrogates be readily available. The design space is represented by the variation of design parameters, which can be system or subsystem attributes and are inputs in the modeling environment. Outputs of the models are metrics of interest relevant to system feasibility. Probability distributions spanning across adequate value ranges can therefore be assigned to the parameters that define the design space and used in a Monte Carlo simulation to recursively sample values for all parameters and perform a model run. The result is a probability density function or its corresponding cumulative density function for metrics of interest. Requirement threshold values can be plotted to assess the percentage of system designs that meet that requirement (Kirby 2001; Mavris and Kirby 1999). Moreover, the set of Monte Carlo observations can be recorded in a table and sorted to further study the values of a given parameter that meet the requirement. This process is illustrated in Figure 20.10 where uniform distributions have been assigned to all input variables.

It is important to recognize that system feasibility depends on its ability to meet all requirements concurrently. Consequently, a shortfall in this approach is that it shows the set of designs that meet a single given requirement at a time. To address this gap, the probabilistic approach to design space exploration can be extended to evaluate designs resulting from the Monte Carlo simulation in terms of two metrics of interest at the same time. Consequently, the requirements associated with these metrics can also be evaluated concurrently, and joint probability distributions can be obtained in place of single-variable probability functions. In this manner,

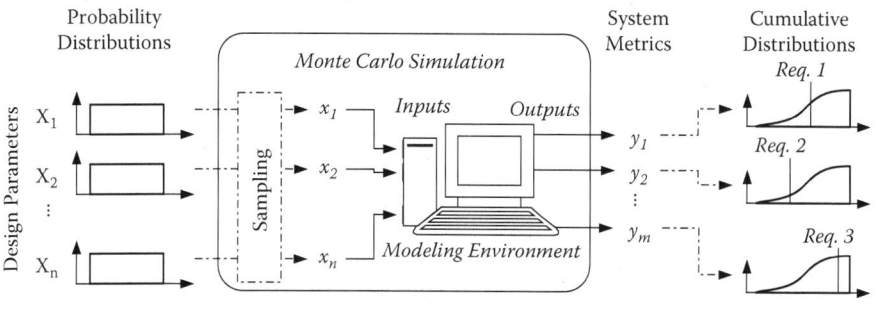

Figure 20.10   Probabilistic design space exploration process.

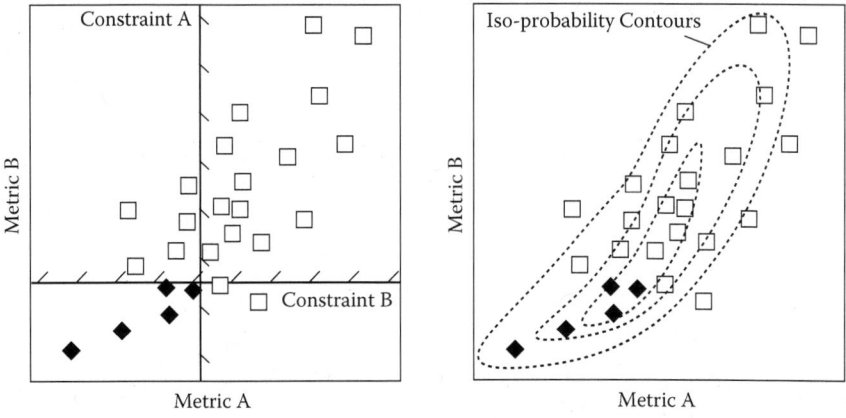

Figure 20.11 Notional design space exploration via joint probability distributions.

designs that meet both constraints concurrently are readily identified in one of these plots (Bandte et. al. 1999a, b). This approach is illustrated in Figure 20.11 using notional metrics. Note that requirements threshold values are shown as horizontal and vertical lines, and that hatch marks denote the unfeasible side of the constraint. Feasible design points are denoted with a different marker for clarity. Notional iso-probability contours are also shown to illustrate the joint probability distributions that result.

It is conceivable to use multiple such plots so as to identify designs meeting all constraints, not just two at a time. Current software packages such as JMP™ allow for multivariate scatter plots to be generated based on a single dataset, so that any design (or set of designs) can be identified in all plots concurrently, readily revealing how they fall on feasible or unfeasible regions of the design space. This technology enables the *filtered Monte Carlo* technique, which combines the Monte Carlo simulation input and output data points, and visualizes them concurrently in a series of plots spanning all input–input, input–output, and output–output combinations. Designers can then select a point in one of these plots, and observe where in the design space it lies as viewed from all other plots. The key capability afforded by this approach is that requirements can be implemented by means of filters on the plots, which effectively apply the requirements' threshold values and eliminate points not meeting said values. After applying various filters, the remaining designs are those that concurrently meet all requirements. Designer can then identify the combinations of input parameters that yield designs satisfying all constraints placed on output parameters. The reverse what-if game can be played out by applying filters on input parameters and observing what type of solutions result relative to constraint thresholds. These concepts are notionally illustrated in Figure 20.12, where data points are coded by color and marker type to denote different subsets of solutions.

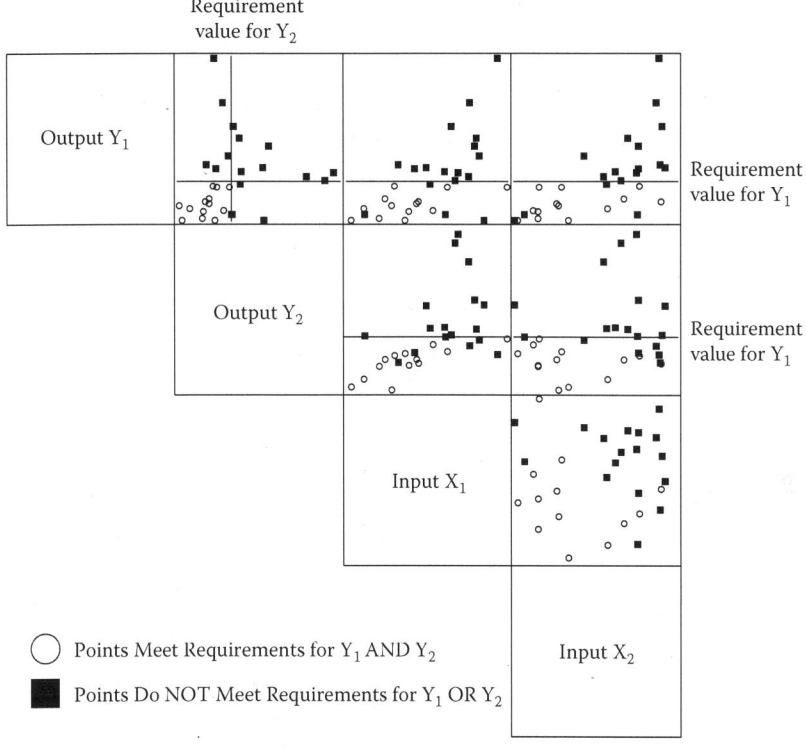

Figure 20.12   Notional display of a filtered Monte Carlo design space exploration.

Chapter 19 ("Systems Design") introduced Quality Function Deployment, a technique used for requirement prioritization and to relate system requirements across multiple levels in the design structure. It was also shown that QFD supports the exploration of relational structures and model transformations on a qualitative basis, and that it can be deployed to reflect the phased approach to the design process. The multivariate scatter plot used for the filtered Monte Carlo technique is particularly powerful because it represents a quantitative complement to QFD, also reflecting the hierarchical structure of the system's design, the phases of the design process, and most important, enabling the quantitative exploration of relational structures and model transformations.

First, consider the multivariate plot shown on Figure 20.12. The plot of X_1 versus X_2 provides information about one system attribute relative to another (i.e., it gives information about their relationship). Both of them are considered inputs from the point of view of a modeling tool or a surrogate model, but they are in essence two parameters of a model. This plot is therefore analogous to the relationship matrix $\underline{M}$ in the mathematical formulation provided in Chapter 4 ("Structure, Analysis, Design, and Models"). It is also analogous to the correlation matrix in the QFD greenhouse. Note however that the $\underline{M}$ matrix codes this information with a very low resolution, namely a binary system that asserts "There is a relationship" or "There is not a relationship." The correlation matrix in QFD increases this resolution by providing information about the quality of this relationship, using a scheme such as {strong negative, weak negative, zero, weak positive, strong positive}. The plot of X_1 versus X_2 provides yet another improvement in the quality and resolution of information about the relationship between the two parameters. For example, the statistical correlation between X_1 and X_2 can be calculated and expressed quantitatively.

Similarly, the plot of Y_1 versus Y_2 in Figure 20.12 provides information about the relationship between these two parameters, and thus speaks about their relational structure. These are outputs from the perspective of a modeling tool or surrogate; but as was the case with the X's, they are essentially two parameters of a model. It is worth clarifying that the model using X parameters is different and distinct from that which uses the Y parameters, even though both models are describing the same system. In this case the plot of Y_1 versus Y_2 is analogous to the $\underline{N}$ matrix introduced in Chapter 4, and analogous to the correlation matrix for the roof of the house in QFD. The previous observations regarding the resolution and quality of information between parameters afforded by qualitative and quantitative approaches is equally applicable.

The four plots of X_1 and X_2 versus Y_1 and Y_2 provide information about how parameters of one model relate to those of another model. It should be obvious that these plots collectively embody a model transform, represented by the matrix Q in the notation introduced in Chapter 4. They are also the quantitative complement of the relationship matrix in the QFD main room, which relates (qualitatively) customer requirements to system attributes. It is important to note that relational structures and model transforms in QFD follow a top-down flow, meaning that models whose parameters are at higher levels of abstraction are transformed into others at lower levels (greater levels of decomposition). This feature is evident in the transforms of each house of quality as well as in the overall deployment of QFD through the phased design approach. Conversely, modeling tools (or their surrogates) will often depict a bottom-up flow where inputs are parameters at a low level in the system hierarchy, and outputs are parameters at a higher level. For instance, a jet engine modeling tool will output performance parameters for the entire engine based on inputs describing geometric, mechanical, thermodynamic, and other properties of its parts and subsystems.

Two important observations can be made at this point. First, the general mathematical formulation for relational structures and model transforms (i.e., $\underline{M}$ transformed by Q into $\underline{N}$) is equally applicable to qualitative and quantitative forms. Moreover, it is equally applicable to a top-down and a bottom-up approach because it does not incorporate (and thus is not limited by) a specific flow for the definition of relational structures and the direction of the model transform. These concepts are illustrated in Figure 20.13.

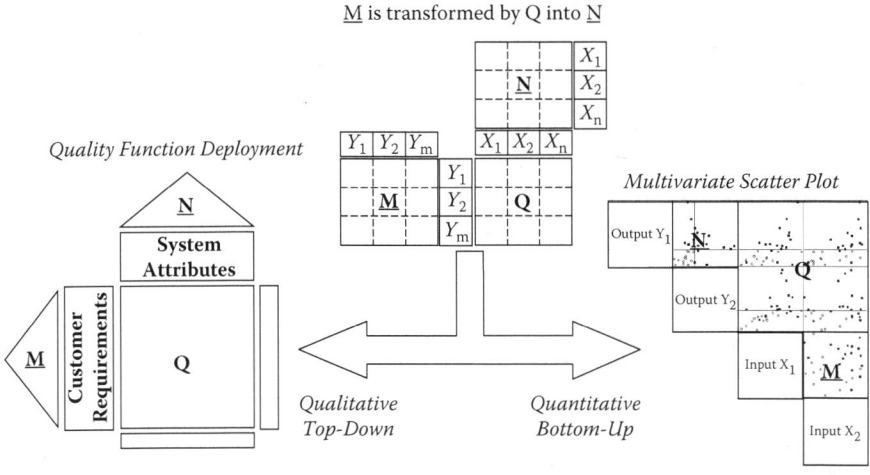

Figure 20.13 Relational structure and model transforms in matrix notation, QFD, and multivariate plots.

The quantitative study of relational structures and model transforms with multivariate scatter plots can be greatly enhanced if the mathematical form of the modeling tools is known. This is particularly the case when surrogate models have been created to replace black-box tools. In fact, the surrogate model explicitly states and describes the model transformation mechanism between dependent and independent variables, which embody parameters of different system models related by means of the surrogate's mathematical equation. A very simple example can be used to illustrate this, using only first-order response surface equations. Assume that a modeling tool has been used to create the three surrogates shown below. Note that only Y_3 includes an interaction term and that the regression coefficients are set to a value of 1 and can therefore be ignored.

$$Y_1 = X_1 + X_2$$

$$Y_2 = X_2 + X_3$$

$$Y_3 = X_1 + X_2 + (X_1 X_2)$$

In order to study the interaction term $(X_1 X_2)$, it can be equated to a parameter X_4 such that

$$X_4 = X_1 X_2 \text{, and } Y_3 = X_1 + X_2 + (X_4)$$

This effectively makes X_4 a dependent variable (function of X_1 and X_2) but it is still an input to Y_3. Variables that are the output of some modeling tool but serve as input for others are often referred to as intermediate variables.

A Monte Carlo simulation can be performed to independently vary the values of X_1, X_2, and X_3. The value of X_4 cannot be independently varied for obvious reasons. In this example, all independent variables were varied over the same range of positive values with a uniform distribution. The multivariate scatter plot and the corresponding statistical correlation matrix are shown in Figure 20.14a,b, respectively. The parallels between these quantitative artifacts and the matrix notation elements introduced in Chapter 4 have been included once again for clarity. First, note that the plots of X_1, X_2, and X_3 relative to each other appear to be fairly random and well distributed across the plot area, and that the corresponding values in the correlation matrix are very close to zero. This is expected as these three variables were independently varied. However, the relationship between the interaction term X_4 and its constituent variables, X_1 and X_2, is explicitly characterized both

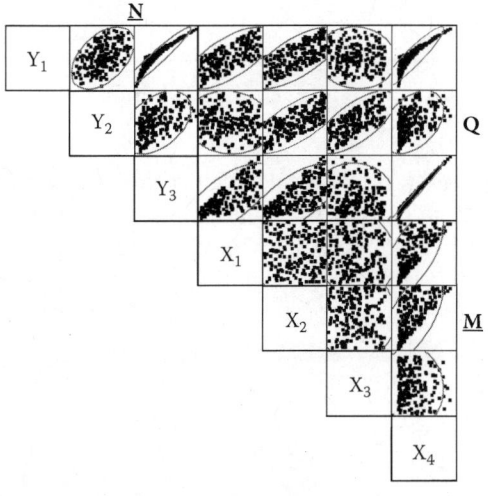

Figure 20.14a   Multivariate scatter plot for sample surrogate model.

	X1	X2	X3	X4	Y1	Y2	Y3	
X1	1.0000	0.0596	0.0200	0.6967	0.7089	0.0529	0.7102	
X2	0.0596	1.0000	0.1420	0.6563	0.7463	0.7602	0.6907	
X3	0.0200	0.1420	1.0000	0.1064	0.1137	0.7511	0.1100	Q
X4	0.6967	0.6563	0.1064	1.0000	0.9283	0.5076	0.9947	
Y1	0.7089	0.7463	0.1137	0.9283	1.0000	0.5723	0.9616	
Y2	0.0529	0.7602	0.7511	0.5076	0.5723	1.0000	0.5328	N
Y3	0.7102	0.6907	0.1100	0.9947	0.9616	0.5328	1.0000	

M (column header, centered above X2–X4)

Figure 20.14b   Statistical correlation matrix for sample surrogate model.

by the trends depicted in the plots and by the relatively high correlation values in the correlation matrix. Not surprisingly, the relationship between X_4 and X_3 is extremely weak, or for practical purposes nonexistent, because the interaction term does not include X_3.

In a similar fashion, the plots and correlation values relating X's and Y's can also be studied. Notice, for instance, that all equations are a function of X_2, and hence there is correlation between X_2 and all Y's. This is not the case with X_1, which is not in the equation for Y_2 and thus yields a weak correlation value and random scatter in its corresponding plot. The relationship between the interaction term X_4 and the dependent variables is somewhat more complex. There is evidence of correlation between X_4 and Y_1, even though X_4 is not in its equation. It should be evident, however, that because Y_1 and X_4 are both a function of X_1 and X_2, they should be correlated as well. In fact this situation illustrates the concept of relation preservation presented in Section 1 of the book. Also note that the relationship of X_4 is stronger with Y_3 where the interaction term does appear in the equation and whose influence enhances the relationship. Similar analyses can be performed for the outputs Y_1, Y_2, and Y_3, where commonality of input variables and the effect of the interaction term strengthen the relation between of Y_1 and Y_3 relative to other output relations.

Technology Evaluation

On some occasions, system performance assessments reveal that the current state of the art is insufficient to meet a requirements set, particularly in programs with very aggressive goals and little flexibility for requirements redefinition. Designers, therefore, turn their attention to promising technologies under development, whose expected entry into market date is somewhat compatible with the system's design and production schedule. Methodologies like Technology Identification, Evaluation, and Selection (TIES) lay out a structured process by which the performance gap is closed through technology infusion. Modeling technologies under development, and evaluating their impact on the system, is a main component of such an approach, but is one that introduces uncertainty into the design process because its exact impact on system performance is not fully known (Kirby 2001; Mavris and Kirby 1999).

The approach featured in TIES models technology impact through k-factors, which are basic multipliers and scaling factors applied to intermediate parameters in the modeling environment. A technology that reduces an automobile's engine block weight by 15 percent, for example, would be captured by the factor $k_{\text{Engine Weight}} = 0.85$, applied to the engine weight in the weight build-up calculation of an automobile modeling tool. Technologies can have more than one effect on a system and thus are modeled as vectors of k-factors, each capturing a different effect. However, the exact value of each k-factor is not known with complete certainty. Following the approach shown in the previous section, adequate distributions are assigned

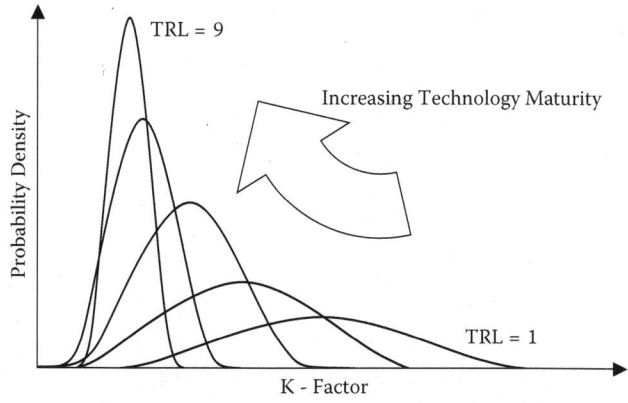

Figure 20.15 **K-factor distributions for different levels of technology maturity.**

to different k-factors and used in a Monte Carlo simulation that runs the system modeling tool for each set of k-factors sampled. The resulting dataset thus includes values of system metrics for each modeling tool execution (Kirby 2001; Mavris and Kirby 1999).

Unlike design space exploration where distributions for design parameters were defaulted to be uniform, k-factors distributions must be selected with care and be properly justified. In its basic formulation, TIES calls for designers to query technologists and determine technology maturity via the TRL scale, and a best, worst, and most-likely value for each k-factor. Together, these pieces of information allow the determination of adequate k-factor distributions as shown in Figure 20.15. Note that uncertainty is greater for more immature technologies, featuring lower TRL values, and thus the spread of the resulting k-factor distribution is larger (Kirby and Mavris 1999).

TIES is classified as an *exploratory* forecasting approach, which begins with the assessment of a current state and proceeds to extrapolate into the future to predict what will happen. TIES fits this scheme because it defines a baseline of the system and then predicts how the system will change by implementing technologies on it. In contrast, *normative* techniques "begin with future goals and work backward to identify the levels of performance needed to obtain the desired goals" (Kirby and Mavris 2002). Technology Impact Forecasting (TIF) is a normative approach that was originally developed as a stand-alone method but later became one of the evolutionary steps in the creation of TIES (Soban et al. 2005). In its general formulation, TIF performs a k-space exploration analogous to the design space exploration described earlier, but varying technology k-factors according to probability distributions while maintaining system baseline design parameters fixed. K-factor distributions are usually set to be uniform so as to evenly sample the k-space. Filtering and visualization techniques previously explained can then be applied to determine what improvements allow a baseline to meet prescribed performance goals. Though this method can be directly used to guide the identification of promising technologies by revealing crucial areas of improvement, it does not carry any explicit definition of specific technologies. It therefore provides the benefit of objectively quantifying performance deltas that meet prescribed goals, irrespective of any technologies that may be considered at a later time (Mavris et al. 1999a,b).

Summary

The methods presented in this chapter represent important advances over the basic set of design techniques and practices presented earlier in this book. In each case, a particular shortfall or persistent challenge in the design process provides the motivation for the development of such a methodology. By integrating key enablers into the design paradigm, designers can gain insight about the system and perform informed decisions throughout the evolution of system development phases. Advanced design methods in conceptual design were shown to support early decisions that lock degrees of freedom and commit development programs to particular concept family choices amid a notable lack of information. Trade-off analyses at any level of system definition are supported by the use of sensitivity profiles that explicitly revealed the relationships and trends of parameter across levels in the system hierarchy structure. Finally, the use of probabilistic design methods was shown to offer a core set of techniques that can be applied in a number of ways to address uncertainty in multiple elements of design. The advanced design methods presented do not represent an ultimate or exclusive form of engaging the challenges in systems development, but rather represent crucial contributions to the field upon which further improvements and future methodologies should be built.

References

Baker, A.P. 2002. The Role of Mission Requirements, Vehicle Attributes, Technologies and Uncertainty in Rotorcraft system design. Ph.D. diss. Georgia Institute of Technology.

Bandte, O., Mavris, D.N., and DeLaurentis, D.A. 1999a. Determination of system feasibility and viability employing a joint probabilistic formulation. Paper presented at the *37th AIAA Aerospace Sciences Meeting and Exhibit*, January 11–14, Reno, NV.

Bandte, O., Mavris, D.N., and DeLaurentis, D.A. 1999b. Viable designs through a joint probabilistic estimation technique. Paper presented at the *4th World Aviation Congress and Exposition*, October 19–21, San Francisco, CA.

Bishop, C.M. 1996. *Neural Networks for Pattern Recognition*. New York: Oxford University Press.

Box, G.E. and Draper, N.R. 1987. *Empirical Model-Building and Response Surfaces*. Hoboken: John Wiley & Sons.

Buonanno, M.A. 2005. A Method for Aircraft Concept Exploration using Multicriteria Interactive Genetic Algorithms. Ph.D. diss. Georgia Institute of Technology.

Buonanno, M.A. and Mavris, D.N. 2004. Small supersonic transport concept evaluation using interactive evolutionary algorithms. Paper presented at the *4th AIAA Aviation Technology, Integration and Operations (ATIO) Forum*, September 20–22, Chicago, IL.

Engler III, W.O., Biltgen, P., and Mavris, D.N. 2007. Concept selection using an interactive reconfigurable matrix of alternatives (IRMA). Paper presented at the *45th AIAA Aerospace Sciences Meeting and Exhibit*, January 8–11, Reno, NV.

Hajela, P. 1990. Genetic search—an approach to the nonconvex optimization problem. *AIAA Journal* 26(7):1205–1210.

Haykin, S. 1998. *Neural Networks: A Comprehensive Foundation*, 2nd ed. Upper Saddle River, NJ: Prentice Hall.

Hayter, A.J. 1996. *Probability and Statistics for Engineers and Scientists*. Boston: PWS Publishing Company.

Holland, J.H. 1975. *Adaptation in Natural and Artificial Systems*. Ann Arbor: University of Michigan Press.

Hwang, C.-L. and Yoon, K. 1981. *Multiple Attribute Decision Making, Methods and Applications, A State-of-the-Art Survey*. New York: Springer-Verlag.

Jimenez, H. 2008. Technology-enabled improvements for terminal area capacity and environmental impact. Paper presented at the *26th International Congress of the Aeronautical Sciences (ICAS)/8th AIAA Aviation Technology, Integration and Operations (ATIO) Forum*, September 14–19, in Anchorage, AK.

Jimenez, H. and Mavris, D.N. 2005. Conceptual design of current technology and advanced concepts for an efficient multi-mach aircraft. *SAE Transactions* 114(Pt. 1):1343–1353.

Khuri, A.I., Ed. 2006. *Response Surface Methodology and Related Topics*. Singapore: World Scientific Publishing Company.

Kirby, M.R. 2001. A Methodology for Technology Identification, Evaluation, and Selection in Conceptual and Preliminary Design, Ph.D. diss. Georgia Institute of Technology.

Kirby, M.R. and Mavris, D.N. 1999. Forecasting technology uncertainty in preliminary aircraft design. Paper presented at the *World Aviation Conference*, October 19–21, in San Francisco, CA.

Kirby, M.R. and Mavris, D.N. 2002. An approach for the intelligent assessment of future technology portfolios. Paper presented at the *40th AIAA Aerospace Sciences Meeting and Exhibit*, January 14–17, Reno, NV.

Li, Y. and Mavris, D.N. 2006. An intelligent, knowledge-based multi-criteria decision making advisor for aerospace systems design. Paper presented at the *25th International Congress of the Aeronautical Sciences*, September 3–8, Hamburg, Germany.

Lin, C.-Y. and Hajela, P. 1993. Genetic search strategies in large scale optimization. Paper presented at the *34th AIAA/ASME/ASCE/AHS Structures, Structural Dynamics and Materials Conference*, April 19–22, La Jolla, CA, USA.

Mankins, J.C. 1995. Technology Readiness Levels: A White Paper. National Aeronautics and Space Administration, Office of Space Access and Technology. April 6.

Mavris, D.N. and Kirby, M.R. 1999. Technology identification, evaluation, and selection for commercial transport aircraft. Paper presented at the *58th Annual Conference of the Society of Allied Weight Engineers, Inc.*, May 24–26, San Jose, CA.

Mavris, D.N., Baker, A.P., and Schrage, D.P. 1999a. Implementation of a technology impact forecast technique on a civil tilt rotor. Paper presented at the *55th National Forum of the American Helicopter Society*, May 25–27, Montreal, Canada.

Mavris, D.N., Soban, D.S., and Largent, M.C. 1999b. An application of a technology impact forecasting (TIF) method to an uninhabited combat aerial vehicle. Paper presented at the *World Aviation Conference*, October 19–21, in San Francisco, CA, USA.

Myers, R.H. and Montgomery, D.C. 2002. *Response Surface Methodology: Process and product Optimization Using Designed Experiments,* 2nd ed. Hoboken: John Wiley & Sons.

Nam, T., Soban, D.S., and Mavris, D.N. 2006. A non-deterministic aircraft sizing method under probabilistic design constraints. Paper presented a the *47th AIAA/ASME/ASCE/AHS/ASC Structures, Structural Dynamics, and Materials Conference,* May 1–4, Newport, RI.

Northrop Grumman. 2003. Aircraft-shaping theory proven sound in first flight demonstration. News release, 8 August, El Segundo, CA.

Ozernoy, V.M. 1987. A framework for choosing the most appropriate discrete alternative multiple criteria decision-making method in decision support systems and expert systems. In *Toward Interactive and Intelligent Decision Support Systems*, vol. 2, Ed. Y. Sawaragi, 56–64, Berlin: Springer-Verlag.

Poh, K.L. 1998. A knowledge-based guidance system for multi-attribute decision making. *Artificial Intelligence in Engineering* 12(3):315–326.

Rasmussen, C.E. and Williams, C.K. 2006. *Gaussian Processes for Machine Learning.* Cambridge: The MIT Press.

Roman, F., Rolander, N., Fernandez, M.G., and Bras, B. 2004. Selection without reflection is a risky business.... Paper presented at the *10th AIAA/ISSMO Multidisciplinary Analysis and Optimization Conference*, August 30–31, Albany, NY.

Sen, P. and Yang, J.-B. 1998. *Multiple Criteria Decision Support in Engineering Design.* London: Springer.

Soban, D., Biltgen, P., and Mavris, D.N. 2005. A technology assessment methods survey: concepts and tools. Paper presented at the *1st International Conference on Innovation and Integration in Aerospace Sciences*, August 4–5, Belfast, Northern Ireland.

Thomas, J.J. and Cook, K.A., Eds. 2005. *Illuminating the Path: The R&D Agenda for Visual Analytics.* National Visualization and Analytics Center, IEEE. http://nvac.pnl.gov/docs/RD_Agenda_VisualAnalytics.pdf

Wahba, G. 1990. *Spline Models for Observational Data.* Montpelier, VT: Capital City Press.

Wong, P.C. and Thomas, J. 2004. Visual Analytics. *IEEE Computer Graphics and Applications*, 24(5):20–21.

Zwicky, F. 1967. New methods of thought and procedure. In *The Morphological Approach to Discovery, Invention Research and Construction*, Ed. Fritz Zwicky and Albert G. Wilson, 273–297, New York: Springer-Verlag.

Zwicky, F. 1969. *Discovery, Invention, Research through the Morphological Approach.* New York: Macmillan.

Chapter 21

Wicked Problems

Andrew J. Daw

Defense acquisition has a turbulent past, and there are many instances where the processes, organizations, and activities of acquisition have come under the scrutiny and spotlight of the media, government, and National Audit Offices in the United Kingdom. Often, the response to this scrutiny has been expressed

Key Concepts

Capability perspective
Balanced multiperspective integration
Lack of definitive formulation
Ambiguity of closure criteria

in the form of reviews, changed programs, organizational initiatives, and new relationships within the community. Always the required end result has been an improved defense acquisition "something"—an environment, an organization, a product, a tangible artifact of some sort and physicality—and it is recognized that such "tangibility" must indeed be produced. Yet, despite the multiple attempts at solving this problem—be it Jordan, Lee, and Cawsey's (Jordan 1988), Downey's (Downey 1968), the Strategic Defence Reviews (Ministry of Defence 1998; 2002), Smart Procurement/Acquisition (McKinsey 1998), or most recently *Enabling Acquisition Change* (McKane 2006)—it appears that few changes (or the changes that are enabled and effected) do not have the desired results in either extent, speed of implementation, or completeness of change; they do not establish and consolidate a new state of the organism.

The analysis work of Warner Burke and George Litwin described a model of organizational performance, and identified three levels of response that characterize any organization: leadership, process, and systems. In change scenarios—as prompted by the reviews noted previously—the relative weightings of these levels on overall organizational performance can be seen. Too often, well-intentioned individuals seeking a quite obvious and necessary change can attempt to make up for poor processes or a lack of management attention. Regrettably, they are doomed to fail. This has led to the concept of an "immune system" (Barton and Daw 2005), in that the underpinning U.K. defense acquisition community responds to the change programs and initiatives as does the body in response to a virus: it takes it in, recognizes the existence of, and then analyzes the infection, applies antibodies to the wound, negates or nullifies the infection, and then returns to something that is very similar to the previous "healthy" state. As a set of loosely coupled "fiefdoms," well distributed and each operating to its own suboptimal targets, the defense acquisition communities have evolved to exemplify the perfect immune system.

It is suggested that a reason for these continuing failures—or at least the inability of these multiple initiatives to effect the desired organizational, process (and cultural) changes, and make them stick—is that the mindset considering the solution/end state is based on the concept of a fixed, rigid solution—that is, that it is possible to "solve" the defense acquisition problem as one would a complex differential equation or other mathematical construct: that there exists an answer. It is suggested that the evidence all points to a conclusion that this is not the case, and that the issues, constraints, and environment of defense acquisition cannot be solved in this way.

The Acquisition Context

Currently, modern defense acquisition tends to be expressed less in terms of items of equipment and more from a "capability" perspective. This move from the tangible to the abstract gives rise to several issues and conflicts—not least of which is one of definition (Daw 2004). Daw (2005) highlights issues arising from the differences between the customer and supplier definitions of capability in the U.K. defense domain. A précis of that discussion follows.

In the United Kingdom, the Joint Doctrine and Concepts Centre offers a framework (Figure 21.1) in which seven "capabilities" are defined as the primary goals of the military domain, and they are expressed as verbs: command, inform, prepare, project, protect, sustain, and operate (Ministry of Defence DCDC 2001). This so-called Defense Capability Framework represents a "customer" perspective of capability; to express these capability needs in terms of a satisfactory implementation whereby solution options may be discussed requires further analysis and breakdown.

The industrial perspective, however, is generally constructed around the five elements of people, process, product, technology, and facilities (as represented in Figure 21.2). The relationship between the five elements is also more clear-cut than those expressed in abstract terms, as each offers some contribution to the development and subsequent sale of product; the product is the focus of the industrial capability definition.

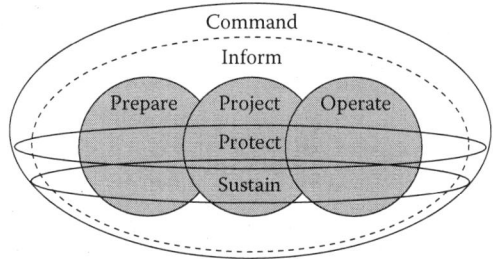

Figure 21.1 U.K. Joint Doctrine and Concepts Centre "Defence Capability Framework," expressing the seven "verbs" of operational capability.

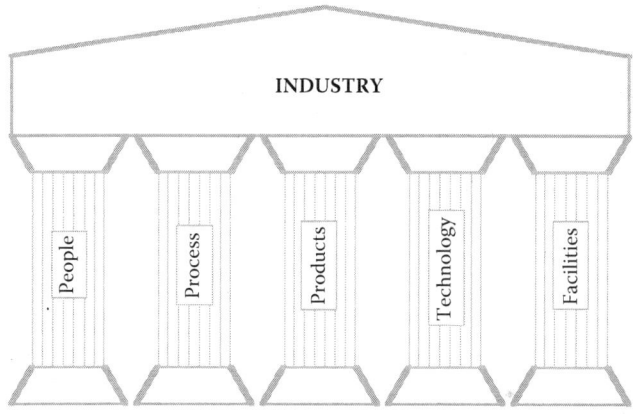

Figure 21.2 The five resource classes of the industrial definition of capability.

Clearly, without some degree of effort, these perspectives are not compatible (indeed, in some circumstances they are in conflict) and a middle ground must be sought if capability-based acquisition is to be a reality.

In the United Kingdom, an approach to this middle ground is found within the concept of the Defence Lines of Development (DLoD). Originally structured from military operations, the DLoD concept has offered a third interpretation of capability, and currently comprises eight elements: equipment, manning, training, logistics, tactics and doctrine, and force structure, infrastructure, and information. From these components capability is "derived" by having the

> … right equipment and information, operated to known rules of operation by trained personnel who are supported in theatre within a recognized and known organization and environment.

Each of these interpretations is appropriate for a particular audience and perspective. A primary issue is the reconciliation and combination of these perspectives so that an appropriate customer, in the supplier–user relationship, can be established for the mutual benefit of the contributing parties. Even here, though, it is essential that "mutual benefit" is expressed in the appropriate value terms (MoD wishes to achieve military effect ["win wars"], whereas industrial value is expressed in profit and shareholder value); the definition of common value statements and understanding is a start in the reconciliation of these disparate views, but the wide difference in understanding and language contributes to the underpinning "wickedness" of the problem. If the DLoD concept and structure does represent the middle ground, bridging the gap between customers and suppliers, they form the implementation mechanisms for the delivery of capability, rather than an explicit definition. If capability delivery is to be achieved from the design of a system solution that "integrates" the eight Lines of Development, a comprehensive overarching view of the system to ensure that the emergent properties and system trades are managed is essential.

However, the set of Defence Lines of Development is not static. By considering this as a core set of "buttons and levers" through which to deliver capability, it can be shown that additional components can and should be included on a context/application/domain basis. In particular, current thinking (Collins, Daw, and Westgarth 2003) defined a ninth Line of Development as industrial readiness, (but as this construct has not been recognized formally, it is not included in this analysis, although such a development line adds considerable weight to the proposed hypothesis. Further, detailed trade-space, option selection work in the left hand—National Trade space—blob, has highlighted significant opportunities for defining an extended set of DLoD, where the established core set of the formally published eight (TEPID OIL) was extended (to a total of 13) by the inclusion of appropriate, potentially domain-specific additions (Oliver 2007) to capture all pertinent and valued contributing elements to the capability required.

The operational environment in which these perspectives must be reconciled is also changing. Today, the concepts of Network-Enabled Capability or Network-Centric Operations and Effects-Based Operations are central to the defense domain and the seven U.K. capability verbs must apply within such an operational environment. Thus, a primary transformation is one between "verbs" and "nouns," and the overall integration and evaluation of the outputs to establish the resultant capability (as expressed through the aggregation of contributions from each of the Defence Lines of Development).

It is the overall recognition of the impact of all these constructs that clearly demonstrates the multitude of stakeholders, perspectives, trades, solution structures, issues, etc., that suggests that defense acquisition is not a mathematically tractable and solvable problem, but rather a "wicked problem." It is these issues that must be resolved and harmonized to achieve a managed successful outcome and an acceptable transition from the abstract to the tangible system solution.

"Wicked Problems"

In 1973, Rittel and Webber commented that "… designing systems today is difficult because there is no consensus on what the problems are, let alone how to resolve them."

This is also true of defining such issues as "capability" and other abstract notions of effect, and articulating and bounding the multidimensional space of interfaces and interactions that are present within the defense domain. Rittel and Webber (1973) suggested that wicked problems have ten characteristics:

1. They have no definitive formulation.
2. They have no stopping rule (i.e., they do not have unambiguous criteria for deciding if the problem is resolved, and getting all stakeholders to agree that a resolution is "good enough" can be a challenge).
3. They have solutions that are not true or false, but rather good or bad [or better or worse—current author addition].
4. They have no immediate and no ultimate test of a solution.
5. There is no opportunity to learn by trial and error, every attempt [at solution] counts significantly, and every solution is a "one-shot operation."
6. They neither have an enumerable (or exhaustively describable) set of potential solutions, nor is there a well-described set of permissible operations that may be incorporated into the plan.
7. Each wicked problem is essentially unique.
8. Each wicked problem can be considered to be a symptom of another problem.
9. They can be explained in numerous ways, though many stakeholders have various and changing ideas about what might be a problem, what might be causing it, and how it might be resolved.
10. The planner/designer has no right to be wrong.

In some sense, these characteristics can be considered as expressing an ill-posed question with an open-ended set of interrelationships and dependencies requiring a formulation of constraint mathematics that might preclude a solution space in the first place.

Thus, the classical systems approach, based on the four distinct phases of (1) understanding the problem, (2) gathering information, (3) analyzing and refining information, and (4) developing solutions, can be seen as being inappropriate at best and inoperable at worst. In drawing from the literature, John (2007) noted that

> One cannot understand the problem without knowing about its context, one cannot meaningfully search for information without knowing what to look for in the solution space; one cannot first understand, then solve …. With an ill-defined problem, from the beginning, it is

not clear what the problem is and, thus, what a solution is. Solving and specifying the problem develop in parallel and drive each other.

He concluded that "attempting to baseline requirements and then use an analytical approach to deal with wicked projects is a recipe for disaster" (Aaby 2002).

It is fair to say that the history of defense acquisition is littered with such instances where either the output did not function, could not be afforded, or achieved a delivery that was neither timely nor appropriate—all basic reasons behind the change initiatives and remedial activities highlighted above.

If one considers defense acquisition in view of these "wicked problem" characteristics, it becomes clear that many of these attributes have a clear correspondence (as presented in Table 21.1) with several observed instances in the acquisition space.

Thus, the premise is that defense acquisition in the round is a wicked problem unsuited to traditional methods and techniques, and whose essence is that it cannot be expressed as an analytical function to be evaluated and (re)solved. Notwithstanding this, however, the underpinning issue remains that there is a need to acquire the artifacts of defense in a timely, effective manner to a standard that achieves the desired effects. In the United Kingdom, Guidance to Defense acquisition is encapsulated in Ministry of Defence (2007), the new Acquisition Operating Framework (AOF) that supersedes the SMART Acquisition Handbook. This

Table 21.1 Defense Acquisition Relationships and Examples

Defence Acquisition Relationship & Example	Wicked Problem Characteristic	Mitigation mapping
Current acquisition initiatives are abstract in nature expressing needs in Capability terms	No definitive formulation	Visualization of the problem space and engineering techniques to enable parallel problem refinement and solution set option development
As capability is enduring, the over-arching requirement remains extant. There is no statement of 'how much (of something) is enough.'	No stopping rule	Continuous 'concept' definition, measurement and refinement actively grouping problem issues in an evidenced manner to spawn acquisition activities
The context of the environment is ever-changing and the need is for a balance of solutions across multiple contexts where one 'solution' set is good in one context and bad in another	Solutions that are not true or false but rather good or bad, (or better or worse (current author addition))	Visualization of system balances in context across all DLsoD with contextual decision making information flowing throughout the acquisition processes allowing for a coherent understanding of the impact and implications of decisions upon system balance in response to change
The ultimate provision of capability is achieved through end user activities and integration across a range of products and services which cannot be demonstrated in the real world – e.g. key user requirements such as 'Survivability'	No immediate and no ultimate test of a solution	Issues of assurance rather than acceptance at the higher levels of abstraction. The extensive use of the virtual world, simulation and visualization techniques to provide confidence throughout the decision making
Acquisition timescales are such that the technology element of any solution has advanced by many generations in the course of the development – e.g. UK has developed 1 example of each generation of fighter aircraft rather than an incrementally developed family	No opportunity to learn by trial and error, every attempt (at solution) counts significantly and every solution is a 'one-shot operation'.	Coherent management of the information associated with the problem and solution space to enable appropriate development techniques across all DLsoD with ongoing integration across the lifecycle, reflected in improved organizational and contractual environments.
The set of buttons and levers available for the solution of Defence Acquisition problems and thence the provision of capability which is then applied to a wide range of contexts enables an unending suite of trade-off and system balancing opportunities	No enumerable (or exhaustively describable) set of potential solutions, nor a well-described set of permissible operations that may incorporated into the plan	Coherent management of the information associated with the problem and solution space, seeking to represent solutions through defined system characteristics rather than requirements. Trade off processes, parallel problem / solution set development and ongoing decision making to achieve system balance
As noted wrt the 'one shot operation' current acquisition timescales are such that underpinning advances make solutions obsolete / unnecessary before acceptance / deployment – e.g. even ships of the same class tend to be different between and within production batches.	Each is essentially unique	Appropriate commercial environments that encourage movement through the framework with structures in place to support the production and informed use of information, pertinent facilities and maintenance of skills.
Defence Acquisition is the result of the political climate in which it is initiated. The services / capabilities required reflect the political and national posture, stance and aspirations which is itself reflective of and changes with a wide range of other wicked problems.	Each can be considered to be a symptom of another problem	The mapping of capability need to solution set opportunity defines a many to many relationship, where significant inter-dependencies exist. The use of non-linear, recursive processes mitigates some of these linkages and inter-dependencies, but the concept of potential residual wickedness at the point of implementation recognizes that not all risks, difficulties and unknowns can be removed from the process. Coherent informed decision making and information management (including a detailed understanding of the contributions made of existing programmes, products and services) is required to assist in the overall management of the feedback opportunities within the Value Chain.
There are no single, static views of defence or Defence Acquisition. The historical context of how a capability might have been delivered (e.g. in a Cold War environment), or a like-for-like replacement mentality of current acquisition organizations, enables a multitude of solution explanations to be developed. These issues exist within a rapidly changing world where there is less clarity of objective and issues such as inter-service rivalries contribute to the overall ambiguity of explanation and compound the internal and external views of the problem.	Can be explained in numerous ways, through many stakeholders who have various and changing ideas about what might be a problem, what might be causing it and how it might be resolved	The use of multiple system representations to enable the various stakeholders to be engaged and their decision making reflected in the impact and implication statements upon the other stakeholders.
Solution providers are not integrated within the problem – concept development 'team' and hence are offered perspectives which limit solution opportunities (particularly at the abstract level) and which may dictate solutions that lie outside the constraint boundaries.	The planner / designer has no right to be wrong	This will always be the case in a creative process. The planner / designer is in a no-win situation, the best that can be achieved is a break even case.

document discusses issues associated with requirements, specifications, through life support, etc., and offers a range of guidance to contractual and commercial issues, and processes that must be resolved along the way to enable delivery of the necessary products (and services) to achieve the military need. At the early stages of these acquisition guidance processes is the notion of "trading," whereby the competing elements of the problem and the solution are compared and decisions made upon a "best fit" process to the constraints applied—be they cost, timescale, support, performance, political, etc. A diagrammatic representation is provided within the "What Is the Scope of Capability Management?" Section of the AOF (Ministry of Defence 2007), the colloquially titled "Four-Blob Diagram" but more formally called the "Capability Value Chain" (Figure 21.3).

Developed originally as part of a combined industry–the MoD working group (Equipment Capability Group, ECG) in 2003, Figure 21.3 expresses a series of relationships and interfaces that provide the context for defense acquisition and the trading that should occur across all components of the environment and community in establishing what (product and or service) should be procured. Each "blob" provides a context for the subsequent "blob," and the feedback loops represent a formal set of communication mechanisms that enables challenges and corrections from one "blob" to another in a series of informed decision-making activities. Reading from

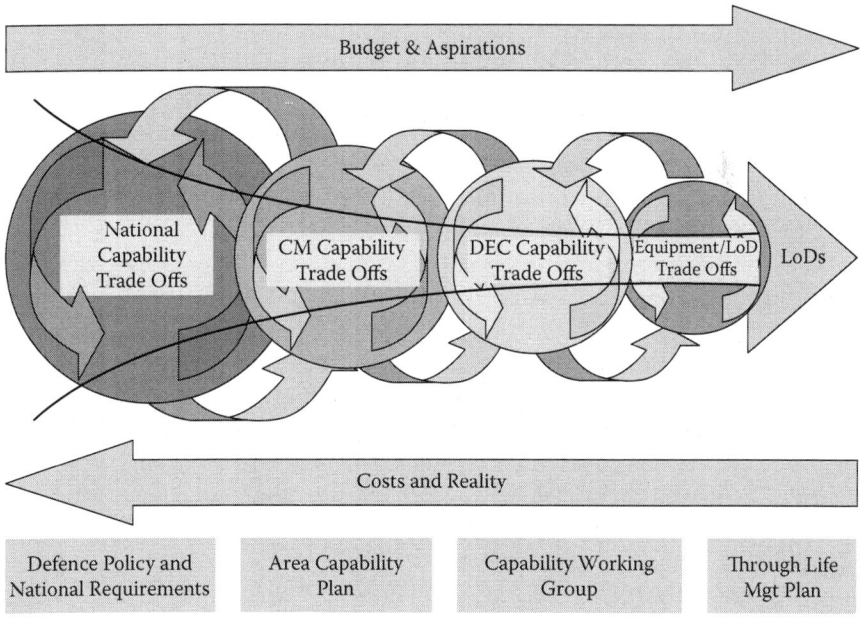

Figure 21.3 Capability Value Chain. (MOD AOF [U.K. Ministry of Defence 2007].)

left to right, therefore, develops the national (political) aspiration into the military capability contribution to the specific solution concepts, and then to the implementation and delivery mechanisms of the Integrated Project Teams within the new Defence Equipment and Support (DE&S) organization. The embracing context of this "left to right" view of the acquisition space is the underpinning budget and aspiration. Reading in the reverse direction, however, come the mitigating issues and boundaries (expressed as reality and costs), and these act as the constraints on the system context, defining what is and is not achievable. Thus, the system can be closed-loop and stable, with informed decision making and a full understanding of the impact and implications of decisions. From this, it is suggested there exists a well-established belief that an analytic, solvable mathematical representation for defense acquisition can be established that supports trading, multiple perspectives, and balanced solution development.

However, if defense acquisition is a wicked problem (and therefore cannot be expressed in solvable terms) but the Four-Blob Diagram represents a framework for its expression, are there methods by which the transition from the wicked capability expression of defense acquisition (the left-hand blob) can be managed to enable the development and acquisition of pertinent products and services within the constraints of the problem (as required at the right-hand blob), which may be considered a tame(r) problem? Functionally, the activities of each of the four blobs of the acquisition model are essentially the same—that of management and decision making in context. As the decision-making process transitions from left to right, the underpinning aspects of trading between options, constraints, and value continue to expose awkward questions, incorrect assumptions, incompleteness of data and information, and uncertainty in concept development—all within a potentially rapidly changing context and internal competitive rivalries. The interactions within and between these functional "blobs" therefore all contribute to the presence and style of "wickedness" in the domain.

Wickedness within the problem domain is not restricted to the left-hand side of this acquisition model, and the "type" and example of wickedness changes within the process. At the left-hand end, the wickedness is manifested primarily as diversity and uncertainty; in the second blob, wickedness is more characterized by choice, technology options, and trading; while at the right-hand end, it remains within the system through the issues of competition, changing context, and system solution maturity. The concept of Potential Residual Wickedness is introduced to recognize that the issues of acquisition in the defense environment (and potentially with relevance to other domains as well) are complicated and complex, and that even straightforward projects (at the right-hand end of the acquisition model) have problems that reflect to some degree or scale the attributes of wickedness. As will be discussed later, the problem at the right-hand blob—the implementation and delivery end—is not necessarily tame, in that it is well bounded and understood, but that it is potentially tamer (John and Daw 2007).

Within defense acquisition, wickedness may also be represented as "internal" and "external," which impact throughout the overall process to varying degrees. It may be considered that at the left-hand end, wickedness is essentially external, driven by the external environment and its relation to the capability desire and national aspiration, while at the right-hand end, it is "internal" (or residual), driven through the issues and relationships of solution development, delivery, and integration into a changing environment. Vested interest is one particular common example of internal and external wickedness, and this can be manifest at all levels and in many ways; further, as a cultural (intrinsically human) issue, it impacts the full range of competencies and behaviors, contract and project styles noted below.

Current defense acquisition practice may also introduce an eleventh attribute of "wickedness"—that of time, which appears in two forms: tempo and duration. Each of the "blobs" and their associated activities has a time constant associated with it—an associated tempo of activity. At the national—left-hand—end of the acquisition process, the problem definition is a "sample and hold" function and provides a sampling rate against which decisions are considered and objectives defined. The activities of the other blobs then act as a set of filters on the process that both provide the response to the decision stimulus and negate any benefit that might be derived from increasing the national sampling rate. This reflects the second aspect of time through the process—the duration of acquisition-decision making. The duration of the process makes decision making on solution attributes—such as technology for example—impossible to achieve in any meaningful way, and so limits the potential for flexibility and responsiveness, and change in the face of varying threats and issues. Such issues limit, therefore, both the customer and supplier options and—in the case of technology, particularly—almost ensure the delivery of products and services that are either at best out of touch with the changed problem space or at worst obsolete or wholly inappropriate.

The Four-Blob Capability Value Chain may also be considered useful as an OODA Loop (Edison 2002) in which

- The National Capability Trade Offs reflect the observation of the world and environment in which we operate, and initiates the decision-making process associated with a nation's position in that environment.
- The Capability Management trade-offs are then associated with orienting those aspects of society (in this case a military perspective) to that national view and those observations of the world / environment.
- By defining the appropriate portfolio of national capabilities, etc., the next set of trades (at the DEC level) reflect the decision to procure appropriate products and services to achieve the national aim.
- While the final set of trade-off processes occurring at the far right of the diagram express the actions taken to implement the procurement.

An analogy of this type offers a further mechanism for expressing the difficulties and subjectiveness of the acquisition environment and hence directing the mindset away from the view that the problem is solvable.

In defense acquisition, wickedness remains in the problem throughout the process, changes its style and characteristics throughout the process, and the underpinning complexity/complicated nature of the systems to be delivered ensures that some aspects remain uncertain even in the final delivery stages.

However, the present reality suggests that there are no current effective implementations of this model and that the underpinning perspective is based on equipment provision/acquisition at the right-hand end in response to almost any question and abstract capability requirement at the left-hand end. A fuller set of available capability components (expressed formally within the Defence Lines of Development) are not rigorously analyzed in the solution concepts, and opportunities for overall system balance through these other solution domains are rarely investigated.

Concepts for Managing the Wicked Problem

If it is accepted that defense acquisition can be appropriately characterized as a wicked problem, then it must be recognized that it cannot be "solved." It must, however, be managed. The current U.K. defense environment has a framework in which management could take place, which if implemented correctly would afford every opportunity to develop "Evidenced Information for Informed Decision Making," enabling a transition from the abstract to the tangible in a controlled fashion—the Four-Blob model. Therefore, if the wicked problem lies in the left-hand blobs and tame(r) project delivery problems exist in the right-hand blob—notwithstanding potential residual wickedness—what are the techniques that would enable an organization and community to move from one to another?

The following discussion addresses a set of perspectives for the management of the defense acquisition problem, offering this framework for identifying and addressing key elements of the wicked problem management space:

- Project style
- Systems engineering (including supporting technical mechanisms such as synthetic environments, and information management, and life-cycle changes)
- Competencies and skills of engaged personnel
- Organizational issues and constructs
- Commercial model and frameworks

The Characterization of the Four-Blob Model: Project Style

In addressing the first stage of management of the wicked problem, let us consider how the Four-Blob model might be characterized from a project style perspective.

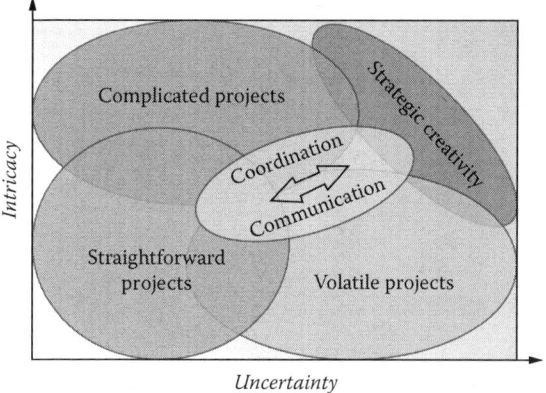

Figure 21.4 Considering a complexity perspective for project style/type. (From PA Consulting Group. 2006. Early lessons for establishing through life capability based programmes. Paper presented at the *RUSI Defence Project Management Conference*, October 10–11, London. Copyright PA Knowledge Ltd 2006. All rights reserved.)

What set of project types would one reasonably expect to see exist as a decision or capability need was articulated and refined from left to right in the model?

Figure 21.4 presents a view of this project style perspective and domain, and is drawn from a range of discussions and activities with PA Consulting (PA Consulting Group 2006). It is presented in a traditional four-box model, whereby the top right represents the unknown, difficult, wicked area and the bottom left the tamer, more predictable/solveable problem space. (Copyright PA Knowledge Ltd 2006. All rights reserved. www.paconsulting.com)

This model views the environment in an Intricacy–Uncertainty domain, where maximums of each reflect the wicked problem of acquisition, whereby the requirements and scope are uncertain, and the solution components and their interactions and interfaces have considerable intricacy. This is the domain of "Strategic Creativity" or the "Edge of Chaos."

The mapping through these perspectives can be considered as follows:

- Blob 1 National Trades   **Strategic creativity**
- Blob 2 Capability Decision Making (JCB)   **Volatile projects**
- Blob 3 Solution Option choices and balance   **Complicated projects**
- Blob 4 Solution implementation projects   **Straightforward projects** (notwithstanding PRW)

and so the issues of connectivity can be identified and addressed. What links the chaotic edge with the volatile and complicated projects to enable the management of the move/transition to more straightforward projects? This model suggests the importance of coordination and communication, to which should also be added the construct of context.

The ability to communicate and place in context the issues of the strategic domain is critical to the development of coherent solutions and programs, to the understanding of the impact and implications of decisions, and the feedback from the implementation domains. These feedbacks will be both positive and negative; in some circumstances, the positive feedback will allow for increased capability, cost effectiveness, etc., and a rebalancing of the solution space in favor of things that are going well to balance those that are falling short. In other circumstances, that feedback will be negative, reflecting an inherent inability to achieve within the constraints and parameters of the problem, and hence the feedback will highlight changes and adjustments to the national capability desired—either by recognizing and understanding the constraint and thence relaxing the issue or by appreciating that a solution is not yet viable and thence address issues of research and development (or whatever is possible) to improve the (technology, social, resource *inter alia*) base from which solutions can be developed.

In managing the wicked problem therefore, those at and in the left-hand domains must recognize that the environment is high in uncertainty and intricacy, and not attempt to apply reductionist theories to simplify the problem. There must be an embracing of the issues, not a shirking from them; an identification of all contributing environments to the solution; and a communication and coordination of activities across the framework to ensure consistency and coherence of the endeavors within the defined context.

The Characterization of the Four-Blob Model: Engineering Style

Consider now how these individual styles of project might be supported in engineering/process terms. What sort of project activity set would have to be conducted, and how, in order to pass through a decision or capability need that was articulated and refined from left to right in the model, particularly if the primary theme and underpinning requirement was to provide "Evidenced Information for Informed Decision Making"?

Figure 21.5 presents an overlay of possible systems engineering styles and process life cycles that might enable such a journey. Again drawn from PA Consulting Group (2006), these models are considered as follows:

- Blob 1   Strategic Creativity   **Emergence model**
- Blob 2   Volatile Projects   **Option model**
- Blob 3   Complicated Projects   **"V" model**
- Blob 4   Straightforward Projects   **Waterfall model**
             (notwithstanding PRW)

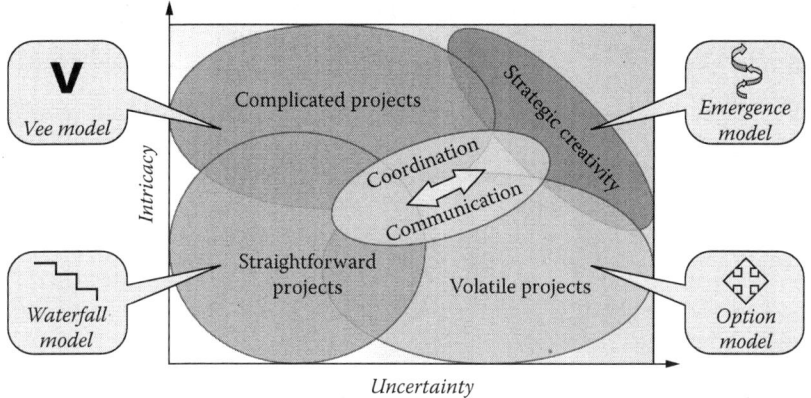

Figure 21.5   A (systems) engineering overlay of process styles. (From PA Consulting Group. 2006. Paper presented at the *RUSI Defence Project Management Conference*, Oct. 10–11, London. Copyright PA Knowledge Ltd 2006. All rights reserved.)

These various models do not necessarily have completely independent definitions and can, in general, be grouped into two pairs: the "higher-level" Option and Emergence Model, and the V and Waterfall models. (Copyright PA Knowledge Ltd 2006. All rights reserved.)

The traditional industry process for systems engineering and the development of a comprehensive *product* portfolio, within its definition of "capability," is the "V" model as illustrated in Figure 21.6, and the associated waterfall model whereby each activity is a natural sequential progression from the last. This has served well in various implementations for some time—particularly through the formal processes of Requirements Engineering, System Analysis, and Systems Design. The

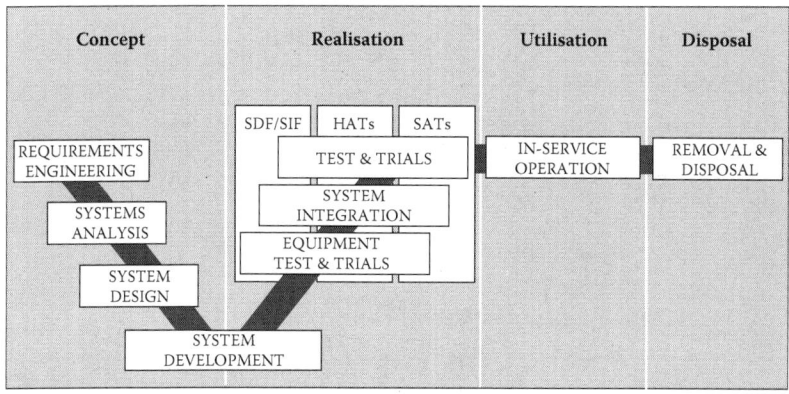

Figure 21.6   Traditional "V" diagram systems engineering process life cycle.

right-hand side of the "V" offers Test Acceptance and Integration—each of which enables validation and verification of the design and product.

However, it is product based and neither of these models inherently includes and reflects the issues associated with a nonlinear problem-solving requirement, or with the recursive nature of iterations, incremental integration across numerous stakeholders, or DLoD. How do the complex intersystem reactions take place and get valued within this structure? It is postulated that this type of "sequential" process is inadequate and ill-posed for the much broader canvas that is required when considering capability or the underlying issues associated with the Emergence or Options models.

A detailed comparison of the traditional "V" model with a new "Reaction Chamber" Systems Engineering Process Life Cycle is presented in Price and John (2002), which also describes a number of extensions to that original model that offer the required articulation of capability development than the "V" model suggests or permits. Figure 21.7 presents this new perspective.

Together with a formal recognition that certain aspects of those activities must occur in parallel, the new model places systems analysis processes in a much more dominant and important role: the continuous, through-life tail of the process lifecycle. This tail now enables a more defined and coherent through-life management policy to be adopted, as each of the systems analysis activities offers the opportunity for a measurable system delivery. It is the use of techniques in this area that enable the derivation and management of information, the identification of emergent opportunities, and option analysis necessary for the two higher-level models identified earlier.

This model also offers a direct linkage with the concepts of experimentation. If, in the context of a wicked problem, experimentation is considered in the broadest definition, opportunities to experiment with concepts, ideas, system balances, etc.

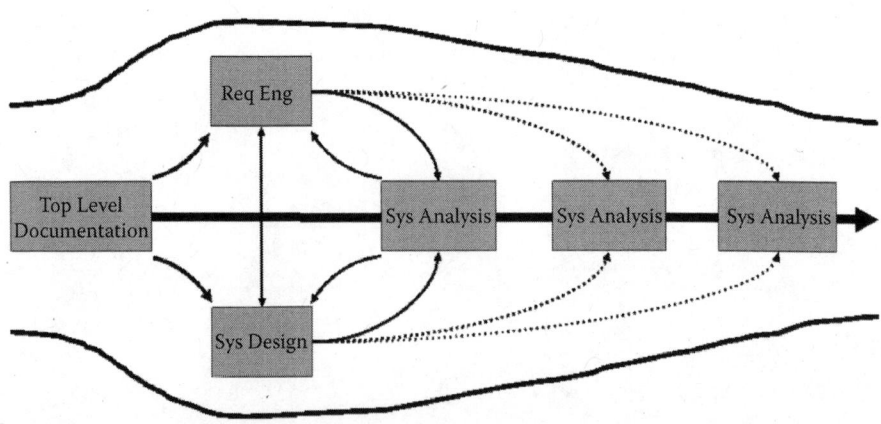

Figure 21.7   Reaction Chamber Systems Engineering Process Lifecycle. (Adapted from Price S.N. and John, P. 2002. Paper presented at *IFORS 2002*, Royal Military College of Science, Shrivenham, U.K.)

occur at every stage of the process life cycle. In the very early stages of the problem—solution development—experimentation can be applied across the first four boxes of Figure 21.7, embracing the Top Level Documentation development and ideas, the analysis of the requirements, the scope of the real-world solution opportunity, and then the traditional measurement of systems analysis. This construct of representation and iteration can and should be applied throughout the program life cycle and act as a key part of the overall through-life management issues.

System Analysis Support

The concepts suggested for the higher-level models—and their subsequent interfaces with the lower-level "V" and Waterfall representations of the tame(r) problems—places a new emphasis on coherent and continued analysis. This analysis is a broader construct than that traditionally associated with systems analysis in the systems engineering process life cycle, and reflects the ability to seamlessly drive the Reaction Chamber model from the Top Level Documentation to the Effectiveness and Performance representations of the model tail. It reflects the need to evaluate and analyze options across multiple disciplines and techniques, building a coherent picture that combines the measurable objective technical parameters with the less tangible, more subjective—so-called soft—parameters of the system. Further, the analysis must encompass a much more significant temporal representation; time is the primary parameter along the axis of the model. Thus, in the environment of managing wicked problems, multiple system representations, their relationships, and analysis are fundamental tools in its achievement.

In this context, support to decision making is vital, and analysis provides a major contribution to the evidenced information of informed decision making, supporting the ongoing development processes and system balances that provide cost-effective long-term solutions to the capability needs. The traceability of decision making—based on a broad range of analysis techniques executed across the life cycle—also provides a major component of the flexibility necessary to accommodate change. Figure 21.8 presents the established Nine-Layer Model of synthetic environments and its relationship to the Four-Blob diagram, and the transition from the abstract to the tangible is obvious. Overall, modeling and simulation of this type, style, and content enable improved understanding of the system in both the problem and the solution domains, through providing

- Structured focusing on system constraints
- Structured environments for trade space analysis, balancing the system solution options against the constraints
- Reduced reliance on physical world acceptance, doctrine, training activities, supporting the assurance, experimentation, and development processes
- An understanding of the problem and solution space, identifying both strengths and weaknesses/impact and implications of the decision making

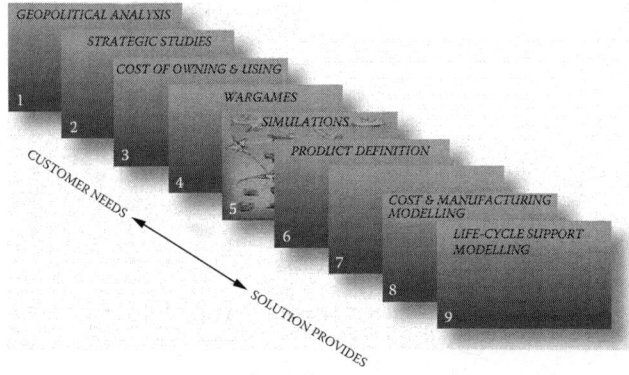

Figure 21.8   Established Nine-Layer Model of synthetic environments.

The applicability of this model to the Four-Blob Capability Value Chain and hence to the management of wicked problems is clear. Levels 1, 2, and 3 represent the national interests (the geopolitical and strategic views and aspirations; levels 3 to 5 offer support to the detailed characterization of the required national capabilities and inform the decision making in the second blob; the third blob uses levels 5 and 6 to define the products and services to be acquired (the overlap forming a crucial element of the interface); while the final (lower levels) support the detailed acquisition components of the fourth blob. Thus, this construct forms a cornerstone of the management of wicked problems; it also indicates the range of investment necessary to establish the appropriate system/environment representations in which all have confidence, so that such information and decision making can be supported.

As in all areas of management of the delivery of capability, the required outputs are achieved not only internally within the product or service components, but also across the DLoD, and hence system measurement and evaluation must be extended across the DLoD; the value of their contribution and the relative importance and weighting of their effects must also be assessed and incorporated into the overall capability calculation. The range of scenarios, operational contexts, etc. must also be weighed and balanced—and balance here is the key. Given the current operational, financial, political, and economic conditions, it is pointless optimizing system solutions; system balance is crucial (expressed in some sense by the MoD as flexibility and agility). The balance of the system is achieved and must be understood through the overall integration of the Defense Lines of Development, the policy, and constraints placed upon the system—all of which are represented in this analysis construct.

Acquisition Life-Cycle Considerations

Current acquisition life-cycle processes are linear and sequential. The CADMID cycle—of Concept, Assessment, Development, Manufacture, In-Service, and

Disposal (McKinsey 1998)—does not reflect the complex issues associated with the acquisition of abstract constructs such as defense and capability. As a life cycle for tightly specified and described products and services, it may well have considerable utility, but the issues of defense acquisition, particularly when viewed as a "wicked problem," are not so defined.

Therefore, there is a need to consider more radical life-cycle styles that enable improved consideration and understanding of nonlinear, recursive, incremental, and parallel activity methods. Work within both the ECG community and the BAE SYSTEMS research initiative at Loughborough (the Systems Engineering Innovation Centre (SEIC)) have highlighted this anomaly and proposed a more iterative and continuous initial acquisition phase. This is illustrated in Figure 21.9 (Westgarth 2006).

This model represents a method of maintaining the coherency and consistency required to manage the abstract concepts for defense acquisition. In this representation, the underpinning goals and values of the community may be expressed in some form of enduring (or at least slow changing) plan. Inputs to this plan encompass national strategies such as the Defence Industrial Strategy (Ministry of Defence 2005), defense doctrine, forward planning assumptions including financial components, environmental and intelligence reviews, research and technology strategies, *inter alia*. This forms the bedrock of aspiration and budget. This plan is then supported by a continuous suite of processes that measure, review, and balance the objectives (as per the plan) with the current reality. The cycle of baseline, audit, and analysis is brought together in the Capability Trade space where all the pertinent buttons and levers of contributing components are reviewed for

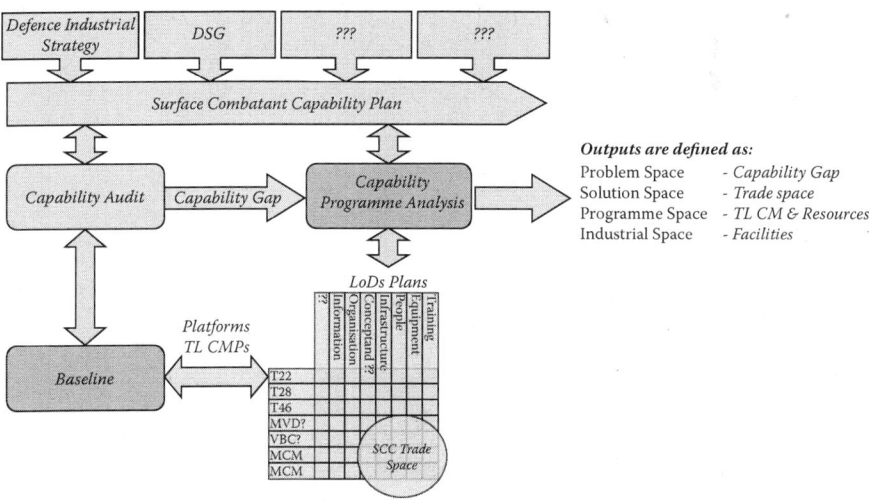

Figure 21.9 Exemplar new life-cycle approach to early defense acquisition activities.

balance, shortfalls, improvement options, and opportunities that, through the program analysis activity, spawn acquisition programs. The outputs of this continuous review and measurement process are defined by

- An agreed and formalized statement of the Problem Space, the Capability Gap assessed from both the aspiration and the current perspectives
- An initial statement of the Solution Space, expressed as opportunities and system characteristics across all the DLoD and other necessary elements (including the impact and implication of the current extant solution set and current plans)
- The Program Space, which expresses the development needs, timescales, constraints, initial through life capability management plans (built including the through life management plans of the current product and services contributing to the capability in question)
- The Industrial Space, which articulates particularly the facilities necessary to develop and then accept/assure the specific solution elements (facilities that may incorporate Synthetic Environments rather than physical components).

In such a life cycle, it would be these four elements and the necessary supporting information and context that were presented for review (at, e.g., a so-called Initial Gate).

The spawned project could then enter whatever was the most appropriate assessment/development life cycle, recognizing that within that process an underpinning activity was the constant reappraisal of the system characteristics, the integration issues, and the management of change, risk, and opportunity within the capability and the national context. These constructs and ideas would also support the current initiatives of Through Life Capability Management.

If this continuous refinement and measurement of capability and acquisition requirements is a characteristic of the front of any radical life cycle addressing the wicked problem, significantly more detail and emphasis must also be placed on the In-Service Lifecycle. A major element of McKane (2006) was the emphasis placed upon Through-Life Capability Management. This places additional "wickedness" within the defense acquisition environment and contributes greatly to the potential residual wickedness at the implementation and delivery end of the program, particularly when there are numerous definitions of capability (as noted above). Indeed, even the title of the plan is "wicked," in that capability is enduring (hence "through-life" is tautological), and it is the products and services that contribute to the capability that are "lifed" and hence to which a "through-life" perspective can be applied. To manage the "wickedness" here, therefore, new artifacts of planning in a definition phase are required and the Reaction Chamber style of systems engineering process life cycle offers considerable support to this issue.

Daw (2004) noted that one of the primary outputs from the Reaction Chamber Model is the development of integrated, Through-Life Management planning artifacts. At the left-hand side of the Four-Blob Model, these represent the Through-Life Capability Plans of McKane, while at the right hand (implementation and delivery side) they represent the through-life management plans for the products and services, across all the DLoD that contribute to that capability. Hence, the plans can be developed to constitute a coherent and consistent set of representations and visualizations of the problem and solution space. Currently these plans—where they exist—are equipment/system based with a primary view of cost. This is not sufficient to provide appropriate detail for planning purposes in capability terms when considering the timescales and timeframes of the component parts (e.g., recruitment, training).

Consider a simple construct for the definition of the plan; the customer has an important System Measure of Interest, which traditionally is required to improve over time to cope with increasing threats, improved performance, etc. Consider also (as in Smart Acquisition) the opportunity to deliver 80 to 85 percent of the required capability quickly supported by a series of increments to the solution over time. The resulting staircase could look as shown in Figure 21.10a.

The tread of the staircase is time; the pitch, the expected capability increase with respect to the Measure of Interest. To achieve this, a program must be established that can be costed. Hence, the Cost Time Performance trio of the U.K. Procurement Agency can be determined. The increase in capability is achieved across the DLoD through the standards tenets of Smart Acquisition (such as Incremental Acquisition and Technology Insertion), etc., but also through pull-through of research, development, focused investment, etc.

There are many additional elements that can be derived from such a representation, but a key benefit is one of flexibility, particularly of industry to respond to radical changes in the measure of interest. The development of a "staircase" requires

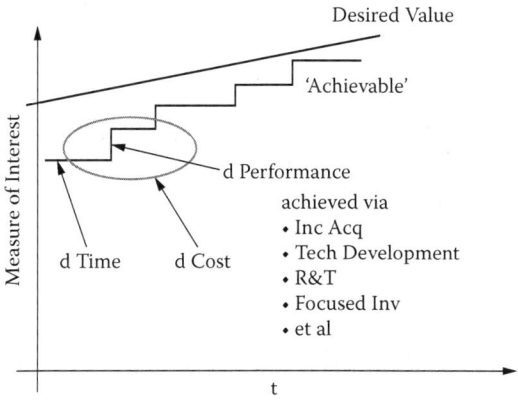

Figure 21.10a   Staircase model of through life acquisition.

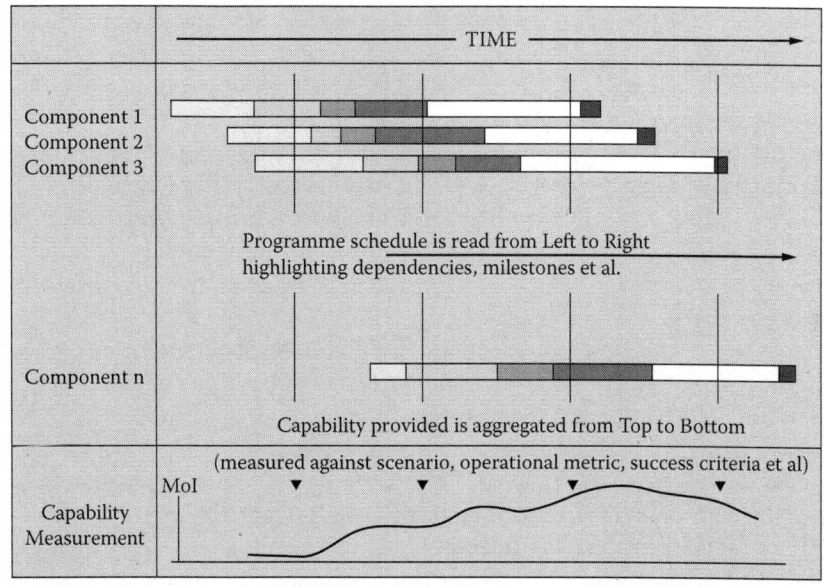

Figure 21.10b   Combined Gantt and capability aggregation representation.

a different set of risk methodologies and perspectives. Each element of a step—the tread and the pitch—is subject to uncertainty; there is uncertainty in the time of the delivery and again in the actual of capability increment against the measure of interest. The issue for the new management processes should therefore be to ensure that the uncertainty at all times is bounded, and recognize that the "opportunity" side to risk is one offering nimbleness of response to changing Measures of Interest, of flexibility in delivery, and direction and focus in investment.

A properly managed and contracted staircase offers considerable openness and visibility to the Ministry for planning purposes (both long term and short term), and to industry who—in some cases—could be competed and judged not just upon the initial product but also on the quality and confidence of support through life planning.

Schematically, the overall capability derived from these contributing components can be represented as a typical Gantt chart (Figure 21.10b). At the highest level, the rows of the Gantt chart represent not program activities, but the DLoD and their component elements that comprise the capability solution system.

Reading from left to right, therefore, provides a clear view of the schedule and interactions/dependencies between the component life cycles—offering views of proposed in-service, out-of-service dates, spend peaks in demonstration/manufacture, even start dates for future program to support an enduring capability need. This view is simply an extension of standard project management. Reading from top to bottom, however, affords the opportunity to "sum"/aggregate the contributions to capability provided by the individual components at any one time and thence through life, and the simple graphic at the bottom of the chart represents some presentation method against a particular measure of interest from the customer. This can be established with scenarios, operational metrics (e.g., casualties), success criteria, etc.

It should also be noted that this is not about forecasting the future—"As for the future, your task is not to foresee but to enable it" (Antoine de Saint-Exupery [1900–1944] French writer). Rather, it is about enabling a consolidated view of the management of the capability in and across all domains and contributing elements, not just the cost function of a particular piece of equipment or system, as is currently the case. Such a plan is predicated upon the ability to measure and evaluate the system and to understand the value or worth of an activity. Again, it is in this context that system analysis plays such a major role in supporting management opportunities.

Information Management

To support such a project life cycle, the management of information through the Four-Blob model is essential and becomes an issue of scale. Building on the multiple system representations of the analysis, the flow of information in terms of context, decision-making details, and the underpinning evidence is crucial to ensuring the consistency and coherency of the environment. Similarly, issues of configuration management become of great importance as this forms the basis of the management of change processes. In ensuring the maintenance of pertinent defense acquisition, particularly in capability terms, continued understanding of the external environment and the impact and implications of change in that environment must be built upon sound information and knowledge management techniques.

In the steady-state situation, much of the management of defense acquisition becomes an issue of the management of change. These changes occur in each of the decision environments encapsulated by the Four-Blob model, and a detailed understanding of the component contributors to the required capabilities, the product and service life cycles, the balances achieved, and the reasons behind decisions must be managed.

There are many information sources within this model. The information model must be initiated through an understanding of "value" in order that the context for the acquisition is well understood and that the decision making has sufficient context to be able to rank and prioritize alternatives at the highest level. This value statement—which becomes increasingly well defined in terms of measures of capability, effectiveness, and performance as progress through the process is made—is expressed in as many terms as is necessary; it is not necessarily a purely financial statement, and in some circumstances may not even contain a financial statement.

The uses to which this information is put changes throughout the model. Figure 21.11 (from BAE Systems and JA Consulting 2007) highlights a particular interpretation of the interfaces between the various areas of the model. The initial interface between national aspiration and the first military decision making represents **Inform**, whereby the flow of information is at the highest national level and

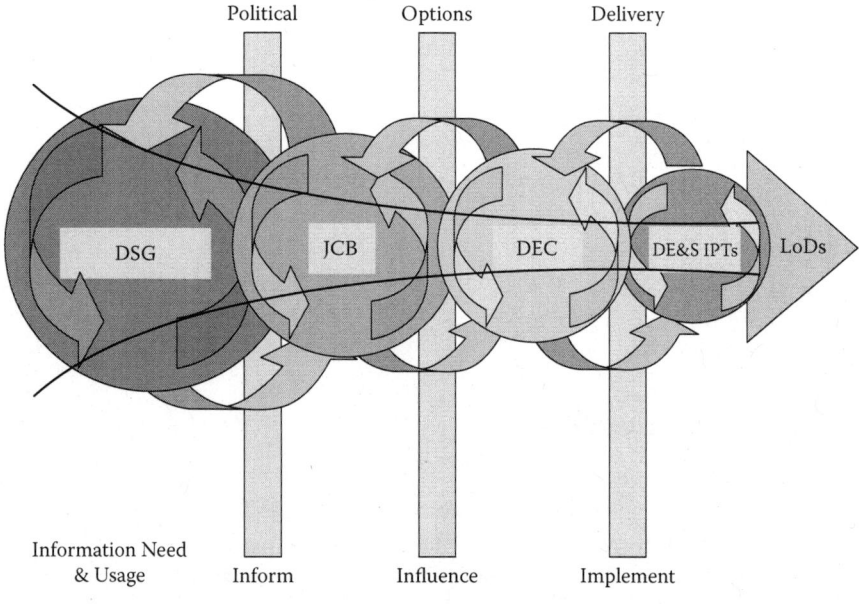

Figure 21.11 Information considerations of the interfaces.

informs the community—industry and national/international government—about aspirations and national intent. This represents information at the highest level of the geopolitical arena and provides the underpinning context of acquisition—the fundamental national requirement. Industry is informed by this as well as informing on the basic risks and opportunities of those aspirations, in terms of development, regeneration, setting to work, providing a considered opinion on viability, etc.

The **Influence** interface requires information flows and management that enable the impact and implications of decisions relating to the set of solution options to be debated and understood. This requires information that is broad in scope, comprehensive in context, amenable to study through sensitivity mechanisms, and available to all parties to contribute. Once again the impacts and implications of decisions must be well understood. This interface also reflects the first opportunity to bring together the problem and solution domains—not in definitive "this is the answer" terms but in broad inclusive conceptual terms, which also include the extant products and services that are contributing to the capability requirements. This use of the information relating to the present environment and its associated plans (acquisition, upgrade, disposal) forms the basis of the forward-looking program of endeavor and offers the initial view of the integration needs and mechanisms that bring existing in-development and proposed programs together.

The final interface discussed here is the **Implement** view—the need for information to flow to those charged with the physical act of implementation and delivery of the solution set. Again, this information contains significant contextual information, clear statements of the trade studies, the system balances defined, the scope, value and importance of the trade space constraints, and increasingly mature integration information to enable the final delivery to the user.

It is key to recognize that the objective of the information views at this stage of trade space identification and analysis is to determine the solution space in which a viable/feasible solution may sit. It is NOT about determining the exact solution. The trade space is defined in terms of the solution attributes and characteristics that any solution must exhibit in order to be compliant. This trade space will be bounded by a range of factors including policy, industrial constraints, performance issues, financial constraints, the scenarios of operation, the tactics doctrine, and policy of use—among others. Thus, the combination of information types and sources may be considered as providing a set of "system characteristics" against which the final acquisition is achieved/contracted, rather than formal, restricted, and rigid requirements—a specifications- based process where language and vested interest all too often color the solution environment. The system characteristics are defined by the constraints applied to the acquisition and in turn define the solution volume across all pertinent views, stakeholders, and disciplines in which it is reasonable to seek solutions. They also provide confidence that a solution from that space will have value to the community of users and stakeholders, and can be utilized within the operational construct required. Such a set of characteristics also feed into the relevant contractual model as a necessary part of the acceptance and assurance criteria.

Throughout this information model, the content and context of the interfaces is driven by the decision making of the blob to the left of the current position. Thus, as information flows through the model, this information set is available to and used by all stakeholders in the domain; there cannot be a "them and us," "information is power" situation. This requires (discussed in the following text) new commercial and financial models and openness between stakeholders that to date has not been observed in the defense acquisition community (but was alluded to in the Defence Industrial Strategy [Ministry of Defence 2005]).

The Characterization of the Four-Blob Model: Skillset and Competencies

If these are the project/program environments and techniques in which people and organizations are going to have to work and interact with and in, what sort of skills and competencies will they have to have? What sort of people are required in order to make these programs effective and the transitions from the abstract to the tangible viable?

In the emergence model, the skills reflect the abstract nature of the problem. The characteristics of these practitioners will lay primarily in their abilities to correlate abstract concepts, to engage with uncertainty and ambiguity and perhaps have

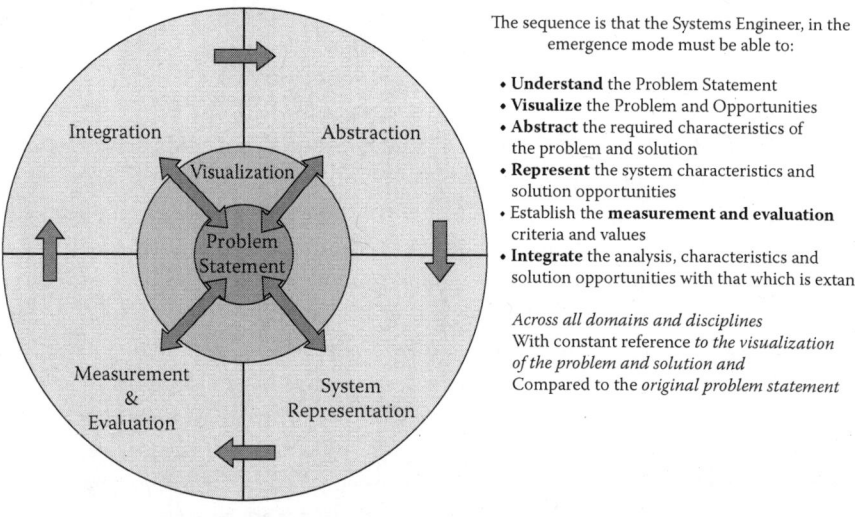

The sequence is that the Systems Engineer, in the emergence mode must be able to:

• **Understand** the Problem Statement
• **Visualize** the Problem and Opportunities
• **Abstract** the required characteristics of the problem and solution
• **Represent** the system characteristics and solution opportunities
• Establish the **measurement and evaluation** criteria and values
• **Integrate** the analysis, characteristics and solution opportunities with that which is extant

Across all domains and disciplines
With constant reference *to the visualization of the problem and solution and*
Compared to the *original problem statement*

Figure 21.12 Emergence model techniques and skills.

a more intuitive feel for the ways and directions in which to move. These are not typical characteristics of an engineer or an engineering community. There are no formal requirements or specifications; these are areas in which the wicked problem characteristic are predominant—multiple stakeholders, no right or wrong answer, multicomplex interrelating and interdependent issues, and confused value definitions and statements. Figure 21.12 provides a view into these characteristics by highlighting the required abilities as

- *Abstraction* of the key fundamentals of the required characteristics of the problem and solution space
- *Representation* from multiple perspectives of the potential solution sets and options
- Identification and implementation of the required *measurement techniques*, metrics, etc., that will represent the "how well" aspect of the solution options in response to the value required from the solution space
- The *integration* of the option sets into a preferred solution set, including the initial metrics, interfaces, etc., that develop a route through the overall problem space.
- Communication of the combined problem and solution space in a way that engages, informs, provides confidence and offers inclusive engagement of stakeholders. Given the multidiscipline and multidimensional nature of this problem, this is expressed as a *visualization* ability/skill that is linked directly to the original requirement/problem definition and then supports all the other skills as noted above.

Individual personal behaviors in this model are also likely to include:

- Refined interpersonal skills
- Strong intuitive and intuition based aspects
- Supportive team working traits

In this domain, the overlap between project management (PM) and systems engineering (Sys Eng) skills is likely to be significant. In the "V" and Waterfall models, however, the divergence of the PM and Sys Eng skills will be greater, as these models demand the harder-edged issues of "completer/finisher" to ensure delivery of the required artifacts.

In managing the varying issues as they move through the defense acquisition processes, the five skills noted above are constantly utilized but the degree to which one or other is the dominant skill varies with the level of decision making, problem management, and context. By the time the "V" and Waterfall models are realized at the right-hand side of the diagram, the delivery skills are of higher importance, the definitive actions and activities of the individual processes are well defined. and the issues of information configuration management and integration are the dominant needs, underpinned by visualization of the overall integration and development of the solution space.

The Characterization of the Four-Blob Model: Organizational Style

The management of wicked problems in the context of defense acquisition has so far been considered at a personal level, where the processes and activities to be conducted and the skills and competencies required have been discussed primarily from the individual's perspective. If this is the environment in which acquisition personnel operate, what sort of organizational issues should be considered and how should organizations—on both the customer and supplier side—be structured to best effect and achieve the end goals?

To manage these types of problems and their associated interactions and dependencies, organizations must demonstrate a range of styles. In all cases, the behaviors must be open and cooperative. and information must be available, managed and controlled, and understood. The communications between parties and thence the flow of information between parties must be based upon common language and understanding, and each must be cognizant of the others perspectives and values.

The relationships between organizations must be built upon trust. Given the scale of the problem to be managed and the multiplicity of perspectives that must be coordinated and balanced, it may be expected that the organizations will share resources. An example of this sharing can be found by considering the volume scale and scope of the system representations necessary to view the whole system problem and solution set. It is unreasonable to expect or assume that any one individual organization will have complete access to the representations required, the information content and context, and the facilities necessary to achieve an integrated picture/environment in which to conduct the necessary studies. In this case it is obvious that a group, consortium, or partnership of organizations will be necessary—and the customer organization must be a part of that consortium.

It is likely that there will be sharing of human resources as well as physical facility resources. Such arrangements offer considerable advantage to the organizations involved as they enhance the overall understanding and communication between the parties. This, in turn, enhances the overall mutual development of the problem and solution space, providing clear insight into the range of pertinent perspectives of the acquisition.

Issues of information exchange need to be resolved as security aspects will impinge upon distribution mechanisms and usage. Further, the commercial arrangements must allow for innovative IPR resolution to ensure the most appropriate exchange of information and thinking.

The Characterization of the Four-Blob Model: Commercial Style

If the preceding discussion reflects elements of the processes, project styles, organizational issues and skills and competencies of the people involved, the final element provides the commercial arrangement in which it must all fit together.

Current acquisition process at the far right-hand end of the Four-Blob Diagram is fundamentally rooted in competition. Competitive forces are seen as those elements of the commercial world that provide best value for money at the point of acquisition and delivery; often this is translated into cheapest bid regardless of understanding, context, background and experience. For those projects that are familiar in content or repetitive in nature, then clearly such an approach has much to support it.

However, as has been noted earlier, defense acquisition has many characteristics that take it way from this relatively simple model, and it requires the execution of processes in ways that do not fit into these established and simplistic frameworks.

Figure 21.13 (Barton and Daw 2006) illustrates a set of commercial models that can be thought of as a journey through the varying styles of project and issue that are prevalent within defense acquisition (and wicked problems). The axes of the model are output focused in that they address issues of delivery in terms of the "type" of output and the "numbers" of the output, where "type" relates to whether the output is of an intellectual or physical style, while "numbers" considers the repetitive (many similar) or non-repetitive (unique, singular nature) outputs from the contract.

The journey through this diagram, therefore, considers the implications of each of these output perspectives in the identification of an appropriate commercial

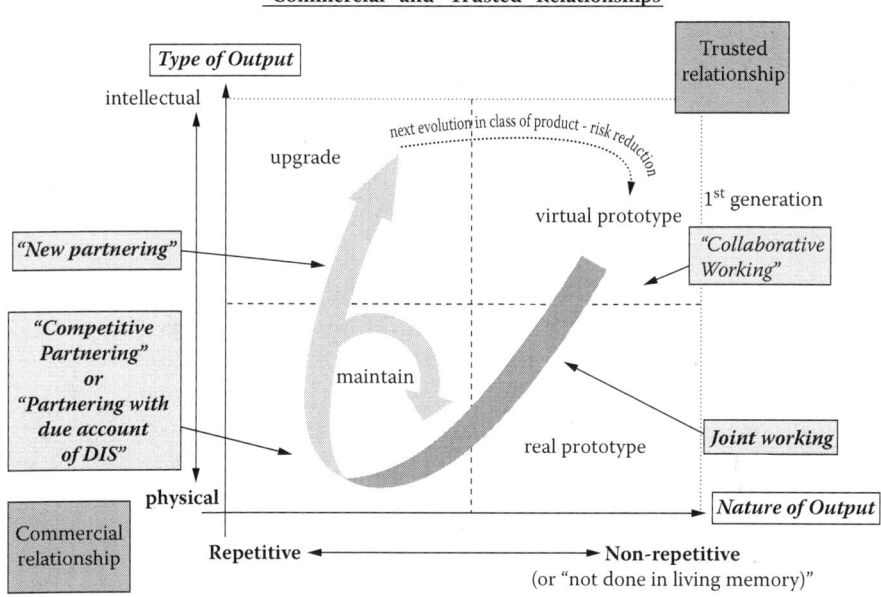

Figure 21.13   A structure of commercial models.

relationship (including the financial structures also). Top right, the non-repetitive, intellectual pairing, reflects the highly complex, edge-of-chaos model of the project style and the left-hand blobs of the basic model. Colloquially, the technical issues of these programs may be represented as "never having been done before" (or in living memory). In analyzing these types of problems, issues of industrial regeneration may come to the fore, implying that such projects have been completed before but are now beyond the industrial capability through skill fade, obsolescence of facilities, inappropriate understanding/integration of new technologies, etc. Programs such as the U.K. future aircraft carrier program (CVF) or any future manned Moon mission may be considered in this area of the graphic. To be effective and successful here requires considerable innovation in the commercial framework—recognizing, of course, the need for hard boundaries such as budget, etc., and the implications, therefore, that perhaps an answer may not be possible. Such commercial relationships require high levels of trust, joint governance, and flexibility of budgeting and work assignments. This is "collaborative working."

Success in this area may then move the program into a prototyping arrangement (non-repetitive, physical program) whereby the relationship now has significant delivery elements to it. Now, the commercial arrangement allows for greater control to ensure the development timescales are maintained project costs are controlled. The performance element of the conventional triad (performance cost time) remains more open, where the opportunity for shortfalls in the program remain, reflecting the innovation, the risk, etc. of moving in these uncertain areas. Flexing of the financial structure of the contract is essential (within the overall budgetary bounds) to ensure all appropriate avenues are explored (and are able/encouraged to be explored) to investigate the real opportunities for successfully developing the "prototype." This can be expressed as "joint" working where controls are more pertinent and prevalent.

With further success, the program—including the transition to full production in the context of the overall acquisition—now moves to the bottom left of the chart (repetitive, physical), although even here it must be recognized that the repetitive numbers do not imply large batch runs. Defense acquisition (in the United Kingdom, at least) tends to be seen as producing limited runs/batches where small numbers are commonplace. The land domain may be the exception to this, but even here, the thousands of vehicles produced over the course of a program is dwarfed by the equivalent civil motor manufacturing environment where hundreds of thousands of units are produced annually. Therefore, even here—because the military acquisition numbers are not those of commercial society—the commercial structure may not be best suited to the competition in all paradigms. Flexibility is still required—the potential residual wickedness of complex and complicated system development and procurement—and the balance between hard program management delivery and the flexibility of balancing budgets in accordance with technical and program issues must be available. This commercial relationship may be based upon competitive pressures and effects, but a measure of partnering and partnership between customer and the appropriate supply chain remains essential.

The final element involves the commercial arrangements necessary to achieve a flexible in-service, upgrade, and technology insertion/obsolescence management program. Now the project returns to a degree of intellectual output, which is associated in particular with the maintenance of fleets, configuration control in the face of change, and improved/modified build standards and functionality. Here, the commercial model must balance the need for flexibility to support the management of change and innovation, while at the same time being sufficiently structured and driven to maintain the delivery components of improvement and upgrades. This is a new partnering construct that has yet to be explored fully. However, it is clear that such an arrangement must be broad enough to accommodate the research and technology drivers of development, improvement (upgrade), and incremental acquisition; flexible enough to recognize the uncertainty associated with research pull through (reflecting the bounded uncertainty of the top right of the diagram); yet strict and controlled enough to ensure and enable the delivery of the required upgrades (the strong program management issues at the bottom left of the model).

Associated with each of these models will be artifacts of sharing, in terms of risk and reward, in terms of information and context, and in terms of revised trade spaces and system balances. It is to be expected that programs must be capable of adapting to changing requirements without the need for wholesale contract amendment or cancellation, and so both parties of the customer supplier model must adopt more flexible views of the contracting process. Issues of acceptance against contract will only have real meaning and value on the left-hand side of the model—the repetitive construct—while the right-hand side (nonrepetitive) will be more concerned with assurance.

The movement and transition between the various project styles and associated commercial/contractual issues brings into focus the need to ensure that the appropriate metrics are available for each style and relationship. Within the defense acquisition environment, the issues of value for money, affordability, shareholder value, etc., are primary financial requirements, constraints, and measures. It must be expected that any individual metric will have a different interpretation and potentially meaning at each level and, indeed, that the set of metrics used to monitor and manage a particular style will change between styles. Thus, the contractual/commercial relationships established to manage the wickedness must be flexible and agile enough to recognize these changes and to manage the transition between one type/style of program and contract, and the next. These perspectives will also contribute to and form part of the overarching context of the overall management activity and understanding. Although financial metrics as measures of a program have been highlighted here, the same principles of change, interpretation, and meaning apply to all metric types and classes used to monitor programs and contracts in this problem domain.

A Mapping of the Wicked Problem Characteristics and the Management Issues Discussed

Table 21.2 presents a putative mapping of these techniques against the attributes of a "wicked problem," demonstrating that the constructs developed here can be used to mitigate the issues associated with this class of problem in the defense acquisition domain. (Table 21.1 also presents a comparative view of the general set of attributes, exemplars, and relationships for defense acquisition.)

Conclusions

The history of defense acquisition initiatives and change programs has demonstrated that in all cases the required results have not been achieved to the extent expected or required. It is suggested that this is because an underpinning mindset exists which suggests that the problem of defense acquisition can be expressed and solved in a mathematical sense through a set of "equations." It is proposed that this is erroneous and that, in fact, defense acquisition should be considered as a "wicked problem," a perspective that is emphasized when comparing the wicked problem characteristics set and the observed nature of defense acquisition, particularly when expressed in abstract capability terms.

Table 21.2   Wicked Problem—Defense Acquisition Comparisons and Management Mappings

Wicked Problem Characteristic	Mitigation Mapping
No definitive formulation	Visualization of the problem space and engineering techniques to enable parallel problem refinement and solution set option development
No stopping rule	Continuous "concept" definition, measurement, and refinement actively grouping problem issues in an evidenced manner to spawn acquisition activities
Solutions that are not true or false, but rather good or bad (or better or worse [current author addition])	Visualization of system balances in context across all DLoD with contextual decision making information flowing throughout the acquisition processes allowing for a coherent understanding of the impact and implications of decisions upon system balance in response to change
No immediate and no ultimate test of a solution	Issues of assurance rather than acceptance at the higher levels of abstraction. The extensive use of the virtual world, simulation, and visualization techniques to provide confidence throughout the decision making
No opportunity to learn by trial and error, every attempt (at solution) counts significantly, and every solution is a "one-shot operation"	Coherent management of the information associated with the problem and solution space to enable appropriate development techniques across all DLoD with ongoing integration across the lifecycle, reflected in improved organizational and contractual environments
No enumerable (or exhaustively describable) set of potential solutions, nor a well-described set of permissible operations that may incorporated into the plan	Coherent management of the information associated with the problem and solution space, seeking to represent solutions through defined system characteristics rather than requirements. Trade off processes, parallel problem/solution set development and ongoing decision making to achieve system balance

Table 21.2 (continued) Wicked Problem—Defense Acquisition Comparisons and Management Mappings

Wicked Problem Characteristic	Mitigation Mapping
Each is essentially unique	Appropriate commercial environments that encourage movement through the framework with structures in place to support the production and informed use of information, pertinent facilities, and maintenance of skills.
Each can be considered to be a symptom of another problem	The mapping of capability need to solution set opportunity defines a many to many relationship, where significant interdependencies exist. The use of nonlinear, recursive processes mitigates some of these linkages and interdependencies, but the concept of potential residual wickedness at the point of implementation recognizes that not all risks, difficulties and unknowns can be removed from the process. Coherent informed decision making and information management (including a detailed understanding of the contributions made of existing programs, products, and services) is required to assist in the overall management of the feedback opportunities within the value chain.
Can be explained in numerous ways, through many stakeholders who have various and changing ideas about what might be a problem, what might be causing it, and how it might be resolved	The use of multiple system representations to enable the various stakeholders to be engaged and their decision making reflected in the impact and implication statements upon the other stakeholders.
The planner/designer has no right to be wrong	This will always be the case in a creative process. The planner/designer is in a no-win situation, the best that can be achieved is a break-even case.

While it is recognized that a wicked problem cannot be "solved," it has been shown that, using current MOD policy guidance for acquisition—a Capability Value Chain (Four-Blob) model—models, behaviors, processes, and techniques can be identified and applied to manage the issues of wickedness. Again, this is particularly appropriate when the issues of defense acquisition are expressed in abstract capability terms, and the solution set is not associated (solely) with simple equipment views. Defense acquisition (to enable the provision of military capability) requires a coherent and consistent environment in which all stakeholders can participate and contribute.

Five primary mechanisms are identified that can be applied to manage the wicked problem:

1. *Organizational structures*, behaviors and attitudes must be established that change throughout the value model and recognize a movement from "inform" through "influence" to "implement". It is essential that the inform and influence actions are understood to be both active and passive and apply to all stakeholders within the value chain

2. Within this organizational structure a range of multiple project styles exist and the issues associated with the transition between each reflect the various transitions through the value chain

3. These project styles require the development and application of *multiple engineering process* lifecycles and styles developing from an emergence model at the abstract level through to the traditional V and waterfall methods in delivery. This model set includes essential constructs such as Systems Analysis Support and Radical Acquisition Lifecycles (continuous solution concept measurement and development), underpinned by comprehensive Information management techniques

4. These multiple engineering process styles are executed through people, and the *skillsets and competencies* of those engaged in the overall process are different within the context of the value chain; that movement between the elements of the chain must be reflected in the changing skills of those involved. However, it should be recognized that the complementary skills of program/project management and systems engineering are key to the overall success of this model. The overlap between the two disciplines is significant at the left-hand side of the value chain, and this only reduces with movement toward the right-hand side to reflect the delivery requirements more formally. The five primary skills and competencies identified within this paper are key underpinning attributes at all stages of the model.

5. Finally, projects within the organization are conducted within a set of *commercial structures* and environments that reflect and are appropriate to navigation of the value chain model. The left-hand (abstract) side of the model reflects an environment of high intellectual requirements against a sparse knowledge background. The commercial framework must recognize as a transition from

this style, through an experimental prototyping phase (with its own commercial issues) to one of repetitive, physical artifact development (albeit in low volumes) where the current practices have more relevance, although they are not a universal panacea. Finally, the commercial constructs must transition and enable further iterations of the value chain through the upgrade, in-service support environment which in part returns to a more innovative, less certain development environment.

Each of these mechanisms is required to enable the management of wicked problems; on their own, each is necessary but is not sufficient to achieve the required result. Further, it is the development of all these constructs that will support the successful implementation of Through-Life Capability Management. A schematic representation of this integration of processes and styles is offered in Figure 21.14 where, borrowing from the MOD Defence Capability Framework icon, a structure for implementing organizational, program, competency, engineering and commercial activities is offered.

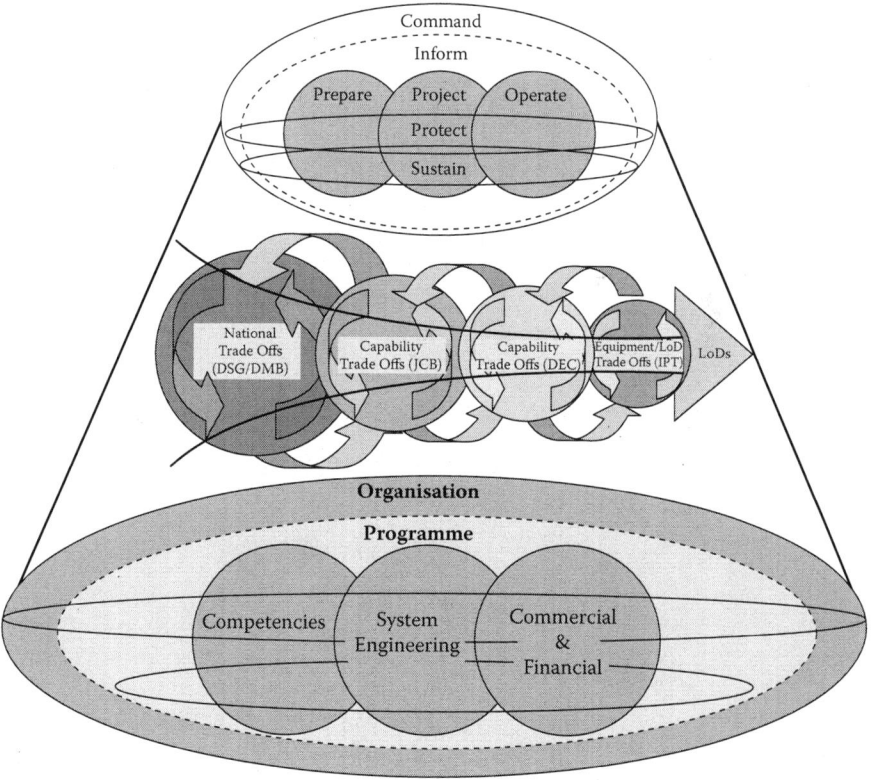

Figure 21.14 A transition from the Capability Value Chain to a defense acquisition capability model.

Overall, therefore, a preliminary mapping of these techniques and the underpinning activities has shown that the attributes of the wicked problem can be addressed and to some extent mitigated/managed. However, even when they are fully utilized and executed, there remains a potential residual wickedness at the point of implementation/delivery. This residual wickedness reflects the complexity and high levels of complication required in today's solution set for capability, whether that be driven by the number and type of interface, interaction, and interdependency between systems or simply by the nature of human action, understanding, and perception. The key issue is that it exists because the environment is uncertain; these management issues are not established to remove that uncertainty, rather to minimize its effect and enable informed decisions to be made about its impact and implications.

The combination of these techniques, models, and methods, underpinned by a comprehensive information management processes, will enable

- Improved management of the overall defense acquisition process and environment, through coherent and consistent (evidence) information for informed decision making
- Inclusivity of stakeholder, system perspective, and development opportunities
- Integration, capability acceptance/assurance
- Support to flexibility and the management of change
- A focused investment and identified opportunities for research pull-through

In wicked problem terms, while the planner/designer may still have no right to be right, he/she will at least be able to break even, so that the customer knows what he is getting when and to what level of quality and confidence, while the supplier knows that his delivery is managed and has confidence in its applicability and acceptance, and both have a strategy to manage the inevitable set of changes that will occur through life.

Acknowledgments

The author gratefully recognizes and acknowledges the support and contribution made to this work by many friends and colleagues. Particular thanks to BAE Systems for permission to publish, to Bob Barton, Richard Westgarth, and the other members of the Equipment Capability Group, to Professor Charles Dickerson and other academic reviewers (Professors Alex Duffy and Srinivasan Raghunathan) for their critiques and suggestions, and to Professor Phil John for his original inspiration and the discussions about the concept of Potential Residual Wickedness. All figures are used with permission, and I gratefully acknowledge that support. I hope I have done them all justice.

References

Aaby, A. 2002. Computational Complexity and Problem Hierarchy. http://cs.wwc. edu/~aabyan/Theory/complexity.html

BAE SYSTEMS and JA Consulting. 2007. Internal BAE Systems Project Development work with Strategic Business Development and JA Consulting, April 2007.

Barton, R. and Daw, A.J. 2005. Private conversations and adapted from Burke, W. and Litwin, G.H. 1994. Diagnostic models for organizational development. In *Diagnosis for Organizational Change*, Ed. A Howard and Associates. London and New York: Guilford Press.

Barton, R. and Daw, A.J. 2006. Private conversations Robert Barton (BAE Systems) and Andrew Daw.

Collins, M., Daw, A.J., and Westgarth, R. 2003. Unpublished research and study activity for the UK MoD Equipment Capability Group.

Daw, A.J. 2004. New process and structure thinking for capability development. Paper presented at the *9th International CCRP Symposium*, September 14–16, Copenhagen, Denmark.

Daw, A.J. 2005. On the use of synthetic environments for the through life delivery of capability. Paper presented at the *NATO RTO SAS Conference*, April 26-28, Norfolk, VA.

Downey, W.G. 1968. *Report of the Steering Group on Development Cost Estimating*. London: Ministry of Technology.

Edison, Toby. 2002. Rugby and the OODA loop. *Rugby Magazine*, February/March.

John, P. 2007. Contracting against requirements documents or shared models? Paper presented in the *BAE SYSTEMS R&T Conference*, February 12–13, in Loughborough, U.K.

John, P. and Daw, A.J. 2007. Private conversations Professor Phil John (Cranfield University) and Andrew Daw, February 2007.

Jordan, G., Lee, I., and Cawsey, G. 1988. *A Report on the Arrangements for Managing Major Projects in the Procurement Executive*. London: Ministry of Defence.

McKane, T. 2006. *Enabling Acquisition Change*. London: Ministry of Defence.

McKinsey and Company. 1998. *Transforming the UK's Procurement System*. London: Ministry of Defence.

Ministry of Defence. 1998. *The Strategic Defence Review*. London: Ministry of Defence.

Ministry of Defence; Development, Concept and Doctrine Centre (DCDC). 2001. Joint Warfare Publication 0-01 *British Defence Doctrine*, 2nd edition. London: Ministry of Defence.

Ministry of Defence. 2002. *Strategic Defence Review: A New Chapter*. London: Ministry of Defence.

Ministry of Defence. 2005. *Defence Industrial Strategy: Defence White Paper*. London: Ministry of Defence.

Ministry of Defence. 2007. What is the scope of capability management? In *Acquisition Operating Framework*. London: Ministry of Defence. http://www.ams.mod.uk/aofcontent/ operational/business/capabilitymanagement/capabilitymanagement_scope.htm [NB: Superseding the "Acquisition Management System Handbook" Vol. 6, HMG MoD]

Oliver, R. 2007. Trade space method for capability and lines of development. Paper presented at the *INCOSE UK Spring Conference*, April 16–18, Swindon, U.K.

PA Consulting Group. 2006. Early lessons for establishing through life capability programs. Paper presented at the *RUSI Defence Project Management Conference*, Oct 10–11, London, U.K.

Price S.N and John P. 2002. The status of models in systems engineering. Paper presented at *IFORS 2002*, Royal Military College of Science, Shrivenham, U.K.

Rittel, H.W.J. and Webber M.M. 1973. Dilemmas in a general theory of planning. *Policy Sciences* 4:155–169.

Westgarth, R. 2006. Internal presentation by Westgarth, R. (Equipment Capability Group) for the Surface Combatant DIS Pathfinder 2006.

CASE STUDIES

5

Chapter 22

Fleet Battle Experiment-India (FBE-I) Concept Model

This case study is the first of four. It will challenge class to apply what they have learned in the first two sections of the book. In this case study, the focus is on the logical modeling of a single sentence from a narrative of an operational concept for the STOM enabled by the NCO capabilities used in FBE-I.

Key Concepts

Logical modeling of a sentence
Operational concept
Network-Centric Operations (NCO)
Ship-to objective maneuver (STOM-

Background

The background on FBE-I was covered in Chapter 13 ("Modeling FBE-I with DoDAF"). The U.S. Chief of Naval Operations (CNO) series of experiments called Fleet Battle Experiments (FBEs) were designed to investigate Network-Centric Operations (NCO) as a framework for the development of new doctrine, organizations, technologies, processes, and systems for future war fighting. FBE-I (spoken as FBE-India) was the ninth in the series.

A more detailed discussion of Network-Centric Warfare (NCW) and its relation to NCO and Command, Control, Communications, Computing (C4), and Intelligence, Surveillance, and Reconnaissance (C4ISR) was presented in Chapter 15 ("Toward Systems of Systems and Network-Enabled Capabilities").

The vision of FBE-I was to "operationalize" NCW by building and maintaining a C4ISR architecture that provided Joint Forces with wide-area connectivity, enhanced bandwidth, and reach-back capability.

Class Exercise

The following exercise and instructor response can be used as reference material for an instructor to create similar exercises that are of interest to a class that is using this book for academic instruction or professional training. The exercise in this case study was performed by the graduate-level systems architecture class of 11 students at Loughborough University in the spring semester 2008. Of the four case studies in this textbook (which the same class was assessed by), this was the simplest but it was also the most challenging because the concepts of logical modeling were completely new to the class. This was the first assessment of the class. In what follows are the exact instructions, materials, and instructor's response to the exercise. The instructions include the responsibility of the class to organize itself to work in small groups. The practice of architecture and systems engineering is as much about personal and professional relationships as is it about technical details.

Instructions to the Class

The purpose of this exercise is to assess both individually and collectively the ability of the class to apply the Principle of Definition.

- The objective of the logical modeling of a sentence is to
  - Extract the relationships comprising the sentence
- You will need to
  - Use a formal language to derive a minimal model
  - Ensure that the model is complete
  - And captures the intended meaning
- The class will practice modeling a sentence using a UML modeling tool

The class will use the following procedure to respond to the exercise:

- Organize yourselves into groups of three members.
  - A fourth member is permitted only if the number of students in the class is not divisible by three.
  - Designate one member as the chief architect; the others should serve as assistants or domain experts.
- By agreement with the class, each group will model a sentence based on FBE-I.
- Each chief architect will present the group's UML logical model to the class:
  - 10 minutes for presentation plus 5 minutes for review

Class Assessment

The class will be assessed as follows:

- Each group will be assessed according to whether their logical model
  - Extracted *all* the relations comprising the sentence
- Assessment will also include:
  - Quality of the derivation of the model
    - The traceability, for instance, of the model to the sentence
  - Minimalism of the model
  - Completeness
  - Successful capture of the intended meaning
  - The quality of the UML diagram
- The assessment will be provided privately to each group after the morning break.

FBE-I Concept Statement

Figure 22.1 is the DoDAF OV-1 graphic of FBE-I. The two sentences below discuss the ship-to-objective maneuver (STOM), first as an enabler of a task force and then in terms of what enables STOM. Please derive a logical model of the second sentence describing STOM.

- The FBE-I ship-to-objective maneuver (STOM) employed classic fire and maneuver tactics to enable the Littoral Penetration Task Force (LPTF) to penetrate past the littoral region more deeply and swiftly than ever before.
- STOM is enabled by Force Elements afloat, Special Operations Forces (SOF) ashore, and Intelligence, Surveillance, and Reconnaissance (ISR) operating under a network-centric command and control (C2) structure that is able to target* the mission objective responsively as an integrated fighting unit.

Instructor's Response

Figure 22.2 displays the logical diagram of the instructor's response to the class exercise.

> **STOM** is *enabled* by *Force Elements* afloat, *SOF* ashore, and *ISR* operating under a *network-centric C2 structure* that is able to *target* the *mission objective(s)* responsively as an *integrated fighting unit*.

---

* Targeting is a process of six sequential activities (F2T2EA) that are used target (i.e., engage) the objective: find, fix, track, target, engage, and assess.

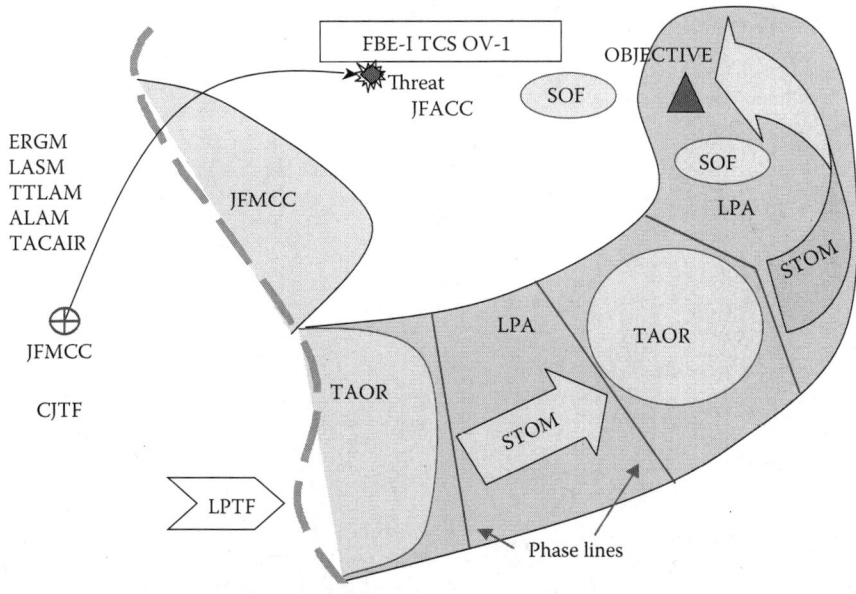

Figure 22.1 FBE-I operational concept.

<u>STOM</u> is *enabled* by *Force Elements* afloat, *SOF* ashore, and *ISR* operating under a *network centric C2 structure* that is able to *target* the *mission objective (s)* responsively as an *integrated fighting unit.*

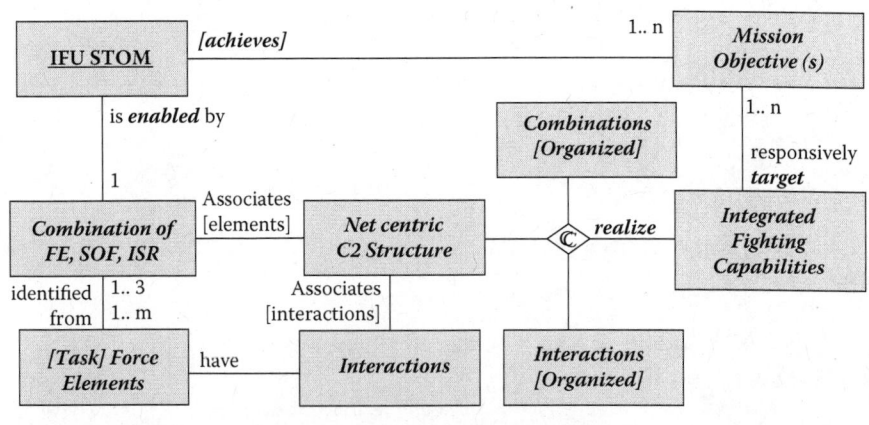

Figure 22.2 Instructor's model of FBE-I STOM concept.

The selected sentence describing STOM has strong similarities to the concept and terminology of the definition of *system* that derived in Chapter 3 ("Concepts, Standards, and Terminology") and the logical model presented Figure 3.3: Adaptation of the INCOSE and Hitchins definition to the Systems Engineering "V". For this reason, the instructor chose to interpret the definition of system from Chapter 3 and its logical model instead of developing a new logical model for STOM. In this approach, *system* in the logical model of Chapter 3, Figure 3.3 is interpreted as an Integrated Fighting Unit (IFU) STOM. The underlying idea here is to simultaneously focus on the IFU as system and the STOM as something the system does. The system elements in this case are chosen from the (Task) Force for the mission. The three elements identified by the sentence for STOM are: Force Elements afloat, SOF ashore, and ISR. The names of these elements have been shortened to FE, SOF, and ISR for simplicity of graphical presentation in the instructor's logical model (Figure 22.2). In keeping with the principles of network-enabled capabilities, the instructor interpreted the "integrating function" in the logical model for *system* as the "network-centric C2 structure." The "properties, behaviors, and capabilities" of *system* were interpreted as a new term, the "integrated fighting capabilities" of the STOM IFU, that were able to target the mission objective, responsively.

In Figure 22.2, "purpose" in the logical model of *system* has been interpreted as "mission objectives" and the terms "responsively target" have been placed on the relationship line between "mission objectives" and "integrated fighting capabilities." This terminology and construct was used to capture the concept "… that is able to target the mission objective(s) responsively as an integrated fighting unit." The terms and logical relationships of "interactions," "interactions [organized]," and "combinations [organized]" remain unchanged in the instructor's interpretation of the logical model of *system*, based on the information contained in the single sentence that was modeled. Other information could be used to interpret these terms to meet other objectives of the logical model. For example, if the model were used to explain the relationship of STOM to the other details of the operational concept (as depicted in Figure 22.1), then the SOF would be seen as having interactions with the adversaries defending the mission objective and the Force Elements afloat (that would be supporting the SOF operations).

Chapter 23

Air Warfare Model

Philip Johnson

This case study will challenge class members to apply what they have learned from Section 3 (Using Architecture for Systems Analysis and Design"). The focus is on what executable architectures are and how they can be implemented with executable UML and used with Model Driven Architecture (MDA™).

Key Concepts

Practical application of MDA™
Executable architecture
Conceptual integrity
Executable UML

Background

This chapter presents a case study that utilizes Model Driven Architecture (MDA) and executable architectures to quickly develop a suitable architecture to meet the case study's requirements. It will later be shown how this approach helps to ensure the conceptual integrity of the proposed architecture through simulation. The simulation will use Kennedy Carter's iUMLite software tool, which "allows production of executable UML® models and has full simulation capabilities" (Raistrick et al. 2004). The software was freely available at the Kennedy Carter Web site (http://www.kc.com) and was also provided with the book *Model Driven Architecture with Executable UML* by Raistrick et al. (2004). One of the key benefits of MDA is model reuse, which saves time and development effort. In MDA, reuse is enabled through the use of domains and can be expanded to entire architectural patterns, as this case study will illustrate. This case study will ultimately follow the development of an initial architecture to fulfill the requirements of an air warfare system. For this case study, the class exercise will consider the creation of UML use case and sequence diagrams based on the system concept description.

Class Exercise

The following exercise and instructor response can be used as reference material for an instructor to create similar exercises that are of interest to a class that is using this book for academic instruction or professional training. The exercise in this case study was performed by a graduate-level systems architecture class of 11 students at Loughborough University in the spring semester 2008. Of the four case studies in this textbook (which the same class was assessed by), this case study starts to bring together the concepts introduced to date to move toward the creation of system architectures, which will culminate in the final case study in Chapter 24 ("Long-Range Strike System Design Case Study"). This was the third assessment of the class. In what follows are the exact instructions, materials, and instructor's response to the exercise. The instructions include the responsibility of the class to organize itself to work in small groups. The practice of architecture and systems engineering is as much about personal and professional relationships as it is about technical details.

Instructions to the Class

The purpose of this exercise is to assess both individually and collectively the ability of the class to create a use case and sequence diagram using a UML modeling tool.

- A *use case* for a system is
  - A type of UML diagram that describes a specific type of behavior required of a system
  - Typically described as an interaction or transaction between the system and one or more actors
    - Sequences of interactions can also be displayed
- A *sequence diagram* for a system is
  - A type of UML diagram that describes the order of interactions for a system and the participants
    - The diagram is organized around the participants and their life lines, showing each interaction in order
- The class will practice creating a use case and sequence diagram using a UML modeling tool

The class will use the following procedure to respond to the exercise:

- Organize yourselves into groups of three members:
  - Designate one member as the (chief) architect
- Use the requirements narrative for the air warfare commander (AW) to:
  - Define use cases for AW
  - Identify the interactions and the associated participants

- Determine the order of the interactions
- Create a sequence diagram for AW
■ Tomorrow morning the chief architect will present the group's sequence diagram for AW to the class

Class Assessment

The class will be assessed as follows:

■ Each group will be assessed according to whether
  - The use cases exhibit *all* the key behaviors required of AW
  - The sequence diagram captures the key participants, their interactions, and the order of interactions
■ Assessment will also include
  - The quality of the UML diagram
■ The assessment will be provided privately to each group after the morning break

Air Warfare System Concept Statement

The air warfare control system manages the release of weapons, threat D3 (degrade, disrupt, deny) assessment, and weapon store levels.

Before an engagement, the engaging platform must be enabled by the air warfare officer (AW). When a platform is enabled, the weapon system is started, if it is not already on, with the weapon launcher disengaged. When the trigger on the platform is depressed, closing a micro switch, the weapon launcher is engaged and weapons are released. When the trigger is released, the weapon launcher is disengaged. There is an interlock that prevents weapons from being released unless a threat has been detected by the system. Once no threats are detected, the engagement is deemed completed and the platform is deauthorized from further weapon release. Further depressions of the trigger on the platform cannot release more weapons. After a short standby period, the platform's weapon system will be turned off unless the platform is reauthorized.

A metering device in the weapon system sends a pulse to the system for each weapon unit released. The cost of the weapon unit is calculated using the amount released and unit cost, which is displayed to the AW.

D3 assessments are stored until the status of the engaged threat is known. Threats sometimes escape without damage and the AW must annotate the D3 assessment with any available information—the threat's electromagnetic signature, for example. At the end of each day, D3 assessments are archived and may be used for ad hoc enquiries into engagements.

At present, two types of weapon are released from five platforms in the battlespace. Each platform takes its supply from one of two weapon stores, one weapon store for each weapon type. The weapon store level must not drop below 4 percent of the weapon store's capacity. If this happens, the platforms serviced by that weapon store cannot be enabled to release weapons.

Instructor's Response

The context chosen for this case study is an air warfare system consisting of a battlegroup operating within a battlespace, where the battlespace is considered the zone of influence of the battlegroup. The purpose of this exercise is to establish an initial architecture for the air warfare system using MDA techniques and to ensure that it is capable of providing an air warfare capability. The following two statements and Figure 23.1 further define the air warfare system:

- A battlegroup engages targets within a battlespace (or zone of influence) under the control of an air warfare officer (AW).
- The battlegroup consists of a number of platforms, each with sensors (S), processors (P), and weapons (W).

Note: This response significantly extends on the class exercise and builds on the UML diagrams created to introduce executable architectures.

The air warfare system described earlier looks quite simplistic at first glance, but considering it in more depth reveals greater levels of complexity, which will need to be resolved to provide the functional solution previously described. Rather than developing the air warfare system from the available requirements through a traditional engineering development path, a quicker approach would be to utilize an existing architectural pattern that is conceptually similar to the air warfare system, and can be reused through interpretation for this application. Providing the architecture chosen for use is of a suitable quality, this will provide a functional

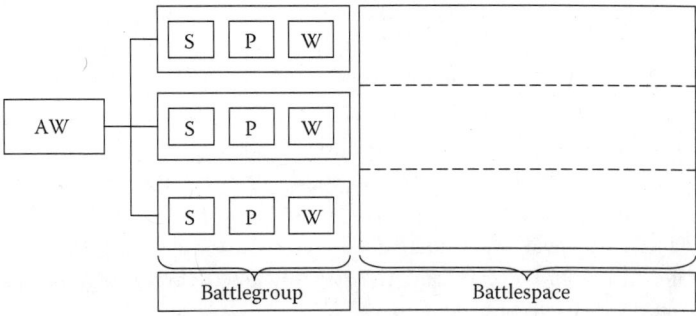

Figure 23.1   The air warfare system.

and logically consistent starting point for further investigation and development of the air warfare system architecture. If the architectural pattern chosen for reuse is executable, and the structure and underlying mechanics are conceptually similar enough to allow direct translation from the original application to air warfare, then an executable architecture can be derived from this process. This will significantly benefit the development process as, through simulation, the executable architecture can be tested against the original requirements and used as a communication facilitator with the customer. This process of reusing conceptually similar architectural patterns that can be executed is not meant to give an immediate solution to the problem the system is specified to meet, but rather to rapidly establish a suitable architecture that can be tested and developed. The use of such executable architectures allows the conceptual integrity of the proposed system to be tested against the requirements and for conceptual flaws to be identified by stakeholders through observation of the architecture's simulation.

The chosen architecture for interpretation into an air warfare system is the example architecture provided with the Kennedy Carter iUMLite software tool. This architecture is for a gas station system and describes the delivery of fuel to customers in a self-service garage forecourt. The following two statements and Figure 23.2 further define the gas station system:

■ An attendant controls the hardware operating within a gas station.
■ The gas station consists of a number of platforms, each with sensors (S), processors (P), and actuators (W).

While totally different subject matters, a gas station and an air warfare system, are conceptually similar; both involve the delivery of a consumable (fuel and weapons, respectively) to a client (the customer and threat, respectively). Working through

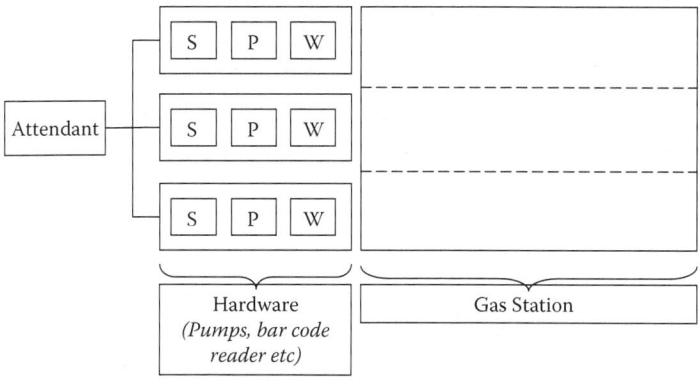

Figure 23.2 The gas station system.

Table 23.1 Interpretation of Gas Station Variables to Air Warfare Variables

Gas Station Variables	Interpreted Air Warfare System Variables
Customers	Threats
Attendant	Air warfare officer (AW)
Forecourt	Battlespace
Gas station	"Area of influence" of battlegroup (area within sensor and weapon range)
Fuel	Weapons
Gas pumps	Platforms (of battlegroup)
Pump fuel	Release weapons (the act of firing weapons from the weapon launcher)
Fuel delivery	Engagement (an event where a platform prepares to engage a threat)
Holster	Sensors (on platform, e.g., radar)
Motor	Weapon system (the overall weapon system, include tracking, etc.)
Clutch	Weapon launcher (the actual launch mechanism for the weapon)
Transaction	D3 Assessment (analysis of an engagement)
Gun	Trigger (fires weapon launcher)
Fuel grade	Weapon type
Fuel grade (four-star)	Weapon type (Aster)—Aster is a surface-to-air missile
Fuel grade (unleaded)	Weapon type (Rapier)—Rapier is a surface-to-air missile

the provided variables within the gas station architecture, a clear translation, or mapping, can be seen from one context to the other, as shown in Table 23.1.

The ability to interpret across subject matters gives reuse potential to successfully implemented architectural patterns, allowing them to be pulled across domains to significantly speed up the initial architectural design process on conceptually similar projects.

With a translation established, as presented in Table 23.1, the fully working executable architecture for the gas station can be converted into a fully working executable architecture for the air warfare system. This straight translation is unlikely to be the final air warfare system solution but it does rapidly provide us with a working architecture. As an executable architecture, it can be used to simulate the system in order to understand how it responds and performs in an air warfare environment, thereby helping to inform the observers of better modes of operation. Executable architectures are key to enabling the architectures to be simulated, and this topic is expanded upon next.

Executable Architectures

To explain what an executable systems architecture is, it is worth reiterating what a systems architecture is:

> The *architecture** of a system is a specification of the parts and connectors of the system and the rules for the interactions of the parts using the connectors.

This specification of the parts and connectors of the system and the rules for the interactions of the parts using the connectors is realized through the use of models. In an executable architecture these models have been created utilizing a standardized specification language that can be run as a program through the use of specialist simulator software. Although UML and MDA have been specified by OMG*, there is no current specification for executable architectures and the specification language underlying them. OMG has released a Request For Proposal entitled "Semantics of a Foundational Subset for Executable UML Models" (OMG Document: ad/2005-04-02), which aims to establish a specification, but this has yet to be approved and implemented. Due to the unavailability of a standard for executable architectures this case study uses a custom solution developed by Kennedy Carter. Kennedy Carter's iUMLite software tool uses a custom style of the UML notation, which they call Executable UML (or xUML), to represent executable models. xUML is an executable version of the UML, and xUML models are developed using the MDA process. Underlying xUML models are state charts where each state entry action is described in Action Specification Language (ASL). ASL is an implementation-independent language for specifying processing within the context of an executable UML (xUML) model. The aim of ASL is to provide an unambiguous, concise and readable definition of the processing to be carried out by an object-oriented system. It is the state charts and the ASL that the state entry actions are written in that is executed when the architecture is simulated. The exact relationship between the UML, xUML, and ASL will be covered later. For now the key to allowing xUML and ASL to be used effectively and efficiently is understanding how Kennedy Carter uses MDA to partition systems.

---

* The definition in the MDA Guide V1.0.1 is based on Shaw, M. and Garlan, D. 1993. *Software Architecture*. Singapore: World Scientific Publishing.

Functional partitioning	**Domain** partitioning
• Instability of system to changing technology	• Stability of system to changing technology
• Instability of system to changing requirements over time	• Stability of system to changing requirements over time
• High risk when integrating the functional areas to form the final system	• Low risk when integrating the domains to form the final system
– *Holistic system interfaces are not well defined*	– *Holistic system interfaces are well defined and loosely coupled*
• No generic reuse of system elements	• Reuse of system elements

Figure 23.3 Functional versus domain partitioning. (From Raistrick, C. et al. 2004. *Model Driven Architecture with Executable UML*. New York: Cambridge University Press.)

Partitioning Systems

Owing to the complexity, scope, and scale of systems, some sort of partitioning strategy is required to help break down the system into more manageable pieces. There are two main strategies for partitioning systems. The first is to partition systems by functionality; the second is by subject matter or *domain*.

Partitioning systems by functionality makes sense as requirements are often expressed as functions. The difficulty with functional partitioning is that functions tend to cross different areas of expertise, which results in a number of unfortunate consequences as outlined in Figure 23.3. Domain partitioning, also outlined in Figure 23.3, splits up the system by areas of expertise or subject matter, which helps alleviate the issues caused by functional partitioning.

Domains are the distinct subject matters present in any system that represent large reusable components and are depicted in a domain chart using an organization of UML packages and their dependencies. These dependencies show service provision between domains as client to server dependencies, but do not show flow (control or data). Client to server dependencies are seen as service provision to anonymous clients that allows domains to be loosely coupled by their dependencies. Note that this loose coupling of dependencies is a design issue, not an automatic result of domain partitioning.

Domain

Kennedy Carter uses four domain types to categorize domains and provide structure through layers in the domain model. The four domain types are:

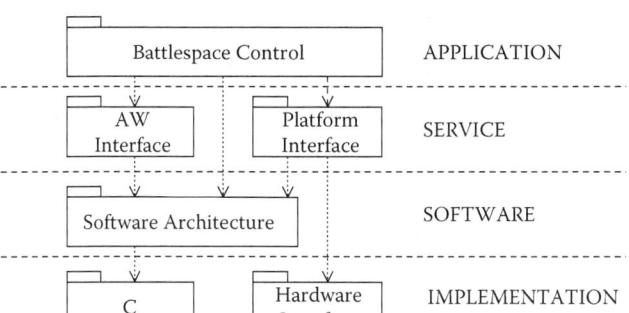

Figure 23.4   Air warfare domain model.

1. Application—purpose of the system from the users point of view
2. Service—basic services, independent of application (e.g., interfaces, logging, alarms, etc.)
3. Software*—provides the execution environment for all the application and service domains that have been formalized using xUML
4. Implementation—represents preexisting technologies and components

Domain models are organized through the use of domain types as layers. Kennedy Carter's convention is to show the application domains at the top of the model, followed by service domains, software domains, and finally, at the bottom of the model, implementation domains. A domain model for the air warfare system, complete with dependencies and domain type layers, is shown in Figure 23.4.

This layering provides structure to the domain model and provides a view from the more abstract user's perspective at the application domain layer down to the implementation-specific perspective at the implementation domain layer. This has significant commonality with the MDA approach. MDA prescribes certain kinds of models to be used, how these models may be prepared, and the relationships of the different kinds of models. MDA specifies three viewpoints on a system: the Computationally Independent Model, or CIM; the Platform Independent Model, or PIM; and the Platform Specific Model, or PSM. These viewpoints are analogous with the domain-type layers, each moving from a level of abstraction down to an implementation level. Figure 23.5 helps illustrate the relationship between the three viewpoints of MDA and the four Kennedy Carter domain types.

---

* Note that Kennedy Carter uses software architecture as the type. For system as opposed to software development, the architecture is the whole domain model and hence it is more appropriate to just call this type "software."

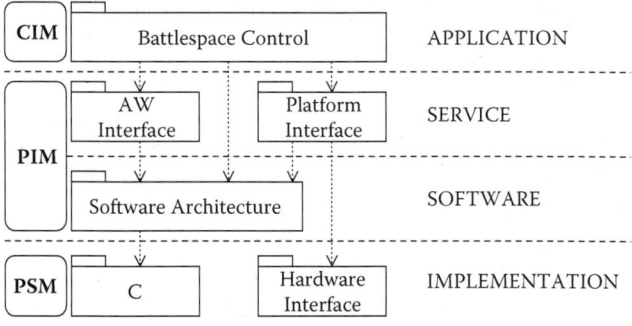

Figure 23.5   Air warfare domain model with viewpoints.

A crucial difference between the domain type layers and the viewpoints of MDA is that MDA prescribes methods for moving from one model to another model of the same system through model transformation. An MDA mapping provides specifications for transformation of a PIM into a PSM for a particular platform. MDA and model transformation is described further in Chapter 16 ("Model Driven Architecture").

The Implementation of xUML within Kennedy Carter's iUMLite

A model expressed in xUML has two layers below domains: classes, and states and operations. These layers form a hierarchy that helps further partition the system. Domains, as previously discussed, represent the subject matter or area of expertise. Classes are an abstraction of the set of things in the domain, and states represent the conditions of these classes. It is this hierarchy that enables the models to be executed within Kennedy Carter's iUMLite software tool. Figure 23.6 illustrates the xUML model layers for the air warfare system with examples of each layer.

The states and operations are expressed as a set of ASL statements that are executed when the simulation is run. It is the inclusion of ASL, which aims to provide an unambiguous, concise, and readable definition of the processing to be carried out by the system, that allows the architecture to be simulated within the iUMLite simulator. xUML is a carefully selected subset of the UML notation that allows models to be built that can be executed. UML specifies a number of diagram types but is quite informal as to how the different diagrams are to be used. In xUML, notations are always used for a specific purpose, which allows precise semantics to be defined for each xUML model element. For example, in xUML, a statechart is always associated with a class for a state model, which always describes the behavior of an object. The relationship between the UML®, xUML, and ASL is illustrated in Figure 23.7.

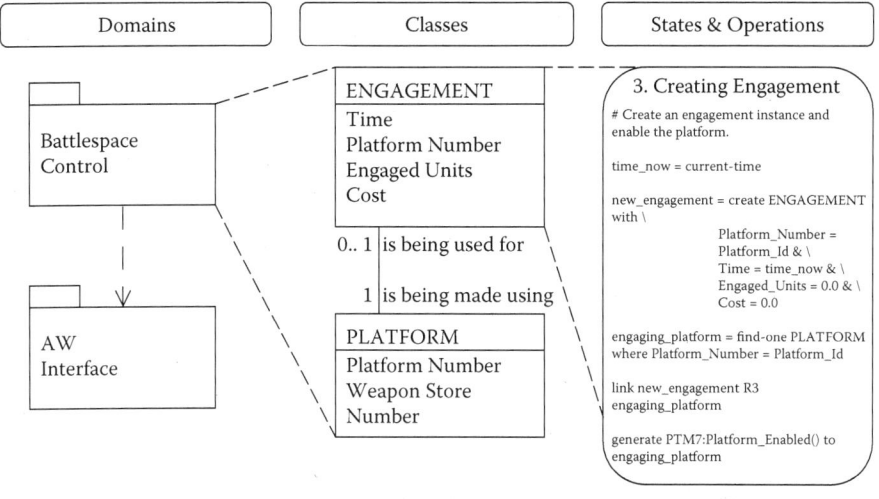

Figure 23.6   xUML model layers.

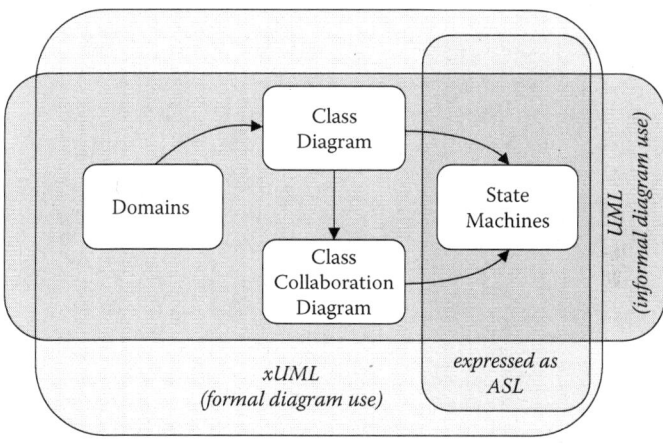

Figure 23.7   Executable UML (xUML).

A Practical Example of Reusing Executable Architectures with iUMLite

This section discusses the approach taken by the author to reuse the executable gas station architecture as an executable air warfare architecture. As previously stated, the software tool chosen for this exercise was Kennedy Carter's iUMLite software tool.

To interpret an executable architecture for reuse, a clear mapping must be made between the existing variables and the new application, as shown previously in

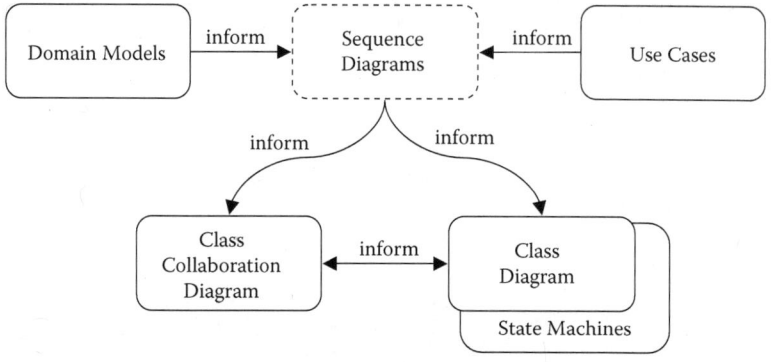

Figure 23.8   The role of sequence diagrams.

Table 23.1. The general method for developing executable architectures is shown in Figure 23.8, and it can been seen that the sequence diagram is the pivotal element that links the domain models and use cases down to the class and class collaboration diagrams. The sequence diagram establishes the sequence of events and the flows between the key domains of the air warfare system. This can be interpreted to create the class and class collaboration diagrams.

For the translation to be successful, the entire model needs to be converted in accordance with the interpretations shown in Table 23.1. This process should be fairly simple; after all, to achieve any level of logical integrity, there must be self-referencing within the model, and hence any change to a variable's name should be propagated throughout the model. Unfortunately, due to the multilayered nature of xUML, this conversion is not simple. While changes to variables in the domain model are generally propagated down through the model to the class level, the main stumbling block for the conversion process is that the state machines for the classes are written in ASL as text code. In iUMLite this code is not interlinked in any way, and hence there is no mechanism to change variables in and across state machines. To translate the state machines written in ASL requires every variable to be interpreted and translated by hand—a very slow, time-consuming, and error-prone exercise. This was found to be a significant barrier to effective architecture reuse, which was compounded by the time required to debug the ASL code during compilation due to errors introduced during the manual code conversion exercise.

For the interpretation of the gas station system into an air warfare system, the architecture was interpreted in the following order:

- Domain model
- Use cases
- Sequence diagrams
- Class diagram

- Class collaboration diagram
- State machines (expressed as ASL code)

The domain model has already been discussed and an example shown in Figure 23.5. The use case for the air warfare system engaging a threat and weapon delivery is shown in Figure 23.9 and depicts the main actors: the AW, or air warfare officer; the threat that the system must engage, and the weapons logistics officer who arranges the resupply of weapons to the air warfare system's weapon stores.

A use case for a system is a type of UML diagram that describes a specific type of behavior required of a system. The use case typically depicts an interaction or transaction between the system and one or more actors, as shown in Figure 23.9. A sequence diagram for a system is a type of UML diagram that describes the order of interactions for a system and the participants. The diagram is organized around the participants and their life lines, showing each interaction in order. In iUMLite the sequence diagram's lifelines are domains, and the interactions are between the system (shown as domains) and the environment. The sequence diagram for the air warfare system engaging a threat is shown in Figure 23.10.

The sequence diagram helps inform the class and class collaboration diagrams, as shown previously in Figure 23.8. The class diagram establishes the classes and the associations between them, and the class diagram for air warfare control is shown in Figure 23.11.

The class collaboration diagram establishes the signals between classes and terminators, and the class collaboration diagram for air warfare control is shown in Figure 23.12.

Underlying each class is a state machine. As the state machines had to be translated line by line by hand, the translations made in the preceding activities were documented exactly to provide clear and precise variable translations. For example, consider the state machine ASL code translation shown in Figure 23.13, which shows State 2 for the pump (gas station system) and platform (air warfare system), respectively.

Note in Figure 23.13 how the states are conceptually similar but the variables are specific to the application and hence have different names. Every one of these variables must be translated manually; and with numerous states across multiple state machines, it is a large undertaking. When this process is more automated, it will make the whole translation process much faster.

To allow the architecture to be executed, the model execution environment needs to be defined through the use of initialization segment methods and test methods. The initialization segment method defines the model execution environment through the specification of initial variable values (the variables as defined in the ASL code in the state machines) and the initial starting state of the state machines. The test methods provide stimuli to the initialized model through the assignment of variable values and the generation of signals. The test methods are defined and are manually initiated, allowing observation of the model's response to the stimuli.

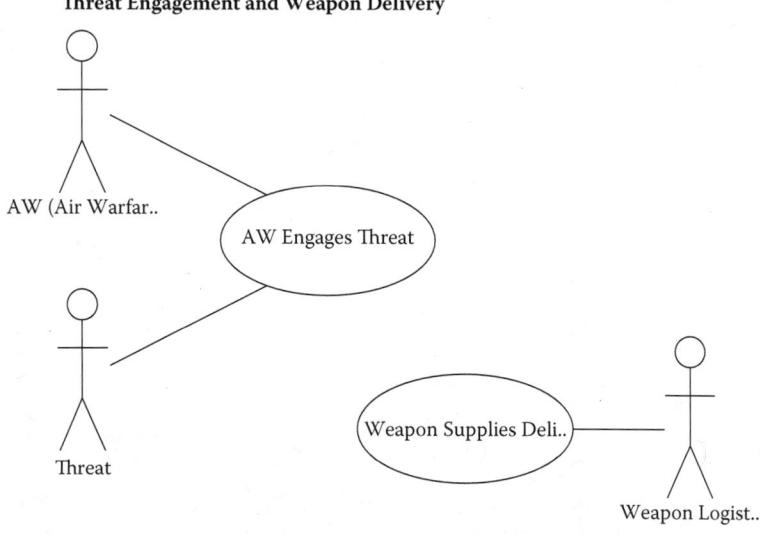

Figure 23.9 Use case: threat engagement and weapon delivery.

Again, both the initialization segment method and the test methods are written in ASL code and must be translated by hand from the gas station system to the air warfare system. The original test methods for a customer filling up and paying for a tank of fuel at the gas station were translated into test methods for engaging a threat with the air warfare system.

Simulating the Air Warfare Architecture

The main advantage of using ASL within the state machines is that it allows the architecture to be executed in a simulation environment and the system's response to predetermined stimuli (expressed as test methods) to be observed. Once the ASL has been debugged and the domain version has been built (a form of compilation), the architecture can be simulated in the iUML Simulator.

The simulation allows the observers to determine whether the system behaves as specified. This opportunity to assess the logical integrity of the proposed solution against the requirements at such an early stage in the project life cycle allows further clarification to be sought with the customer, and the simulation itself acts as a clear communication medium for the exploration of other possible solutions. At this stage in the overall systems development process, all that has happened is the requirements have been elicited from the customer, utilizing techniques such as logical modeling; and from these the development team has identified an existing *conceptually similar* executable architecture that can be interpreted to fulfill the

AW Engages Threat

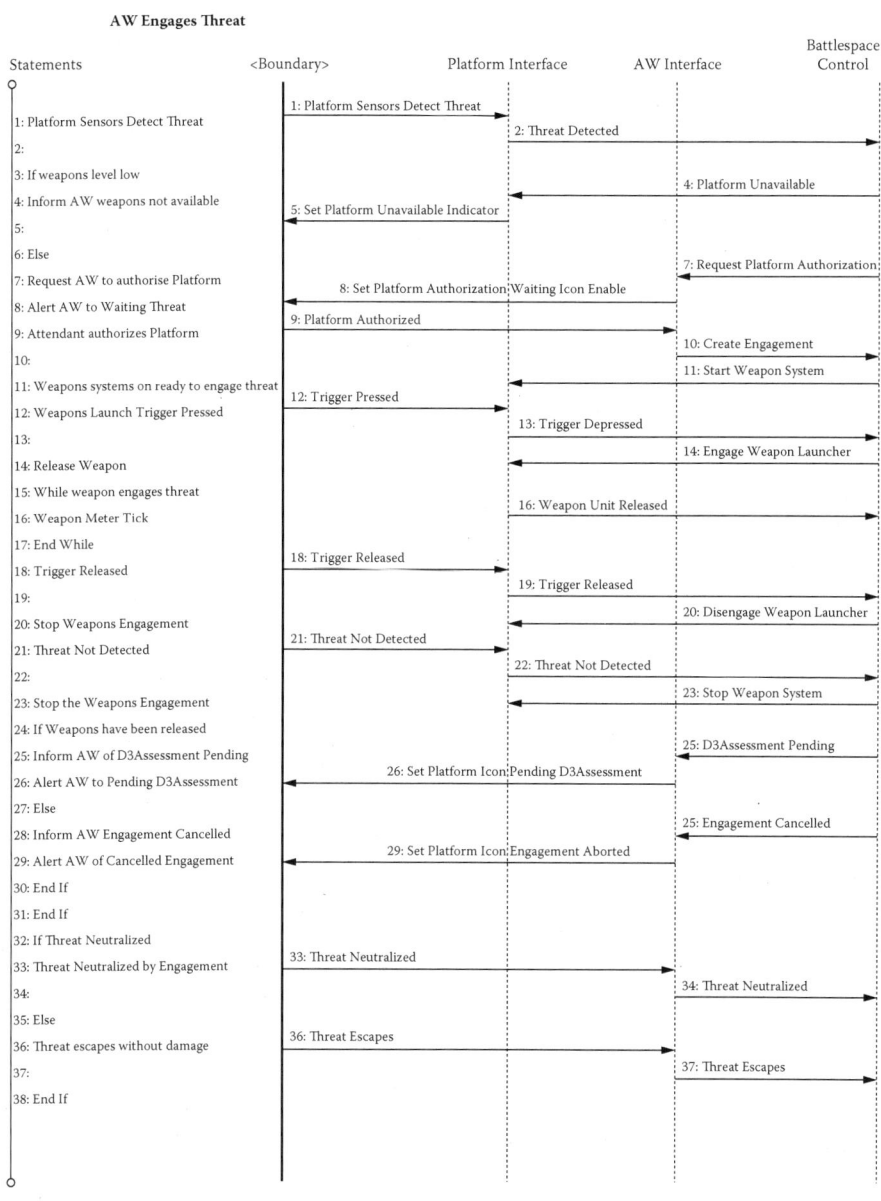

Statements	<Boundary>	Platform Interface	AW Interface	Battlespace Control

1: Platform Sensors Detect Threat — 1: Platform Sensors Detect Threat

2:

3: If weapons level low — 2: Threat Detected

4: Inform AW weapons not available — 4: Platform Unavailable

5: — 5: Set Platform Unavailable Indicator

6: Else

7: Request AW to authorise Platform — 7: Request Platform Authorization

8: Alert AW to Waiting Threat — 8: Set Platform Authorization Waiting Icon Enable

9: Attendant authorizes Platform — 9: Platform Authorized

10: — 10: Create Engagement

11: Weapons systems on ready to engage threat — 11: Start Weapon System

12: Weapons Launch Trigger Pressed — 12: Trigger Pressed

13: — 13: Trigger Depressed

14: Release Weapon — 14: Engage Weapon Launcher

15: While weapon engages threat — 16: Weapon Unit Released

16: Weapon Meter Tick

17: End While

18: Trigger Released — 18: Trigger Released

19: — 19: Trigger Released

20: Stop Weapons Engagement — 20: Disengage Weapon Launcher

21: Threat Not Detected — 21: Threat Not Detected

22: — 22: Threat Not Detected

23: Stop the Weapons Engagement — 23: Stop Weapon System

24: If Weapons have been released

25: Inform AW of D3Assessment Pending — 25: D3Assessment Pending

26: Alert AW to Pending D3Assessment — 26: Set Platform Icon Pending D3Assessment

27: Else

28: Inform AW Engagement Cancelled — 25: Engagement Cancelled

29: Alert AW of Cancelled Engagement — 29: Set Platform Icon Engagement Aborted

30: End If

31: End If

32: If Threat Neutralized

33: Threat Neutralized by Engagement — 33: Threat Neutralized

34: — 34: Threat Neutralized

35: Else

36: Threat escapes without damage — 36: Threat Escapes

37: — 37: Threat Escapes

38: End If

Figure 23.10   Sequence diagram: air warfare system engages threat.

Air Warfare Control v1

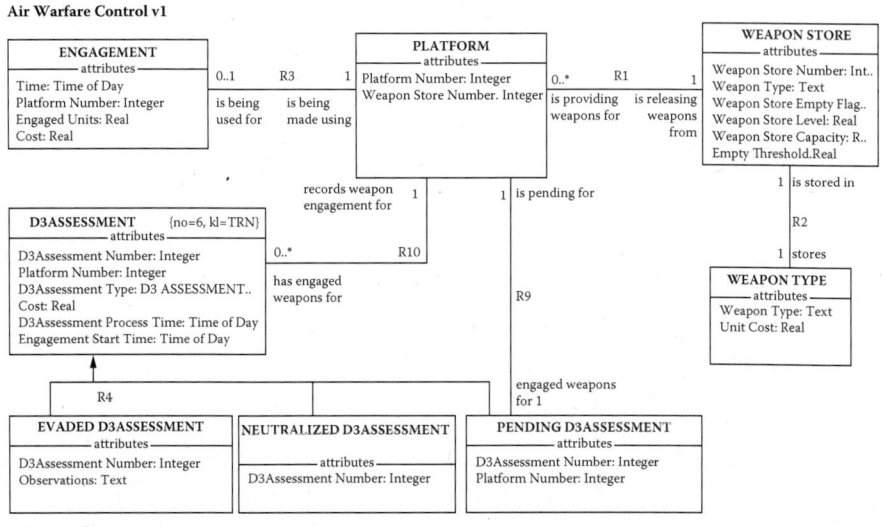

Figure 23.11   Class diagram: air warfare control.

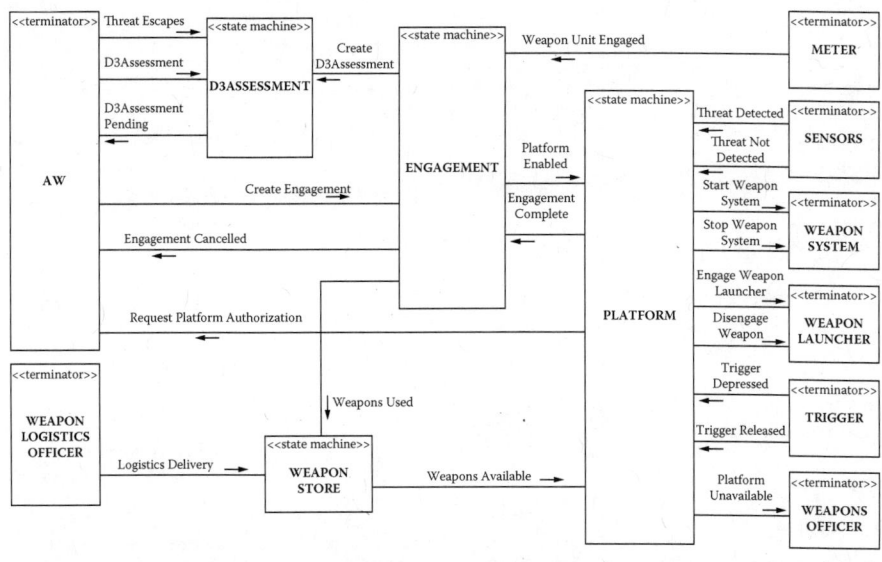

Figure 23.12   Class collaboration diagram: air warfare control.

project's application. The project team can now engage the customer through the architecture simulation to elicit any further implicit requirements and reach agreement that the simulated system behaves in the desired way. The simulation can be further utilized by the development team as a medium for exploration of other possible solutions as this simulation walkthrough will now illustrate.

2. Waiting Pump Enable	2. Waiting Platform Enable

# Determine whether the connected tank contains 	
# more than 4% of its capacity. 	
 	
supplying_tank = this -> R1
 	
if supplying_tank. Tank_Empty_Flag = TRUE then
    generate PMP4: Fuel_Level_Low () to this
else
    [ ] = AT1 : Request_Pump_Enable[ ]
endif
	# Determine whether the connected weapons store contains
	# more than 4% of its capacity.
	
	supplying_weapon_store = this -> R1
	
	if
	supplying_weapon_store. Weapon_Store_Empty_Flag = TRUE then
	    generate PTM4 : Weapon_Level_Low ( ) to this
	else
	    [ ] =
	AW1 : Request_Platform_Authorization [ ]
	endif

Figure 23.13   Example State Machine ASL code translation.

The simulation is initialized through the execution of a sequence segment. The original sequence segment method can be directly interpreted from the gas station system to the air warfare system. For example, now instead of units of four-star petrol, the system will now use units of Aster surface-to-ground missiles. Note that the simulation is purely to check the logical integrity of the proposed solution and does not include any detailed simulation of the system components. Hence, the use of Aster missiles is merely to give some level of familiarity to the solution and aid discussion, rather than being an implementation choice. Once the conceptual integrity of the system had been established, the choice of missile would, of course, be determined by the specialist working in that domain, but with the added benefit that they would know the intended use and interactions of the missile within the system.

The simulation of the architecture within Kennedy Carter's iUML Simulator allows the user to control the simulation, view the ASL being executed, and view and navigate the runtime object data. Viewing the runtime object data helps the observer decide whether the system is behaving as planned and to detect anomalies. A screenshot of the simulator with various classes displayed can be seen in Figure 23.14.

The simulation of the air warfare system quickly showed a potential conceptual flaw with the proposed solution. The gas station system utilizes a pump to deliver fuel to the customer; similarly, the air warfare system utilizes a platform to deliver weapons to the threat. In the gas station system, the pump counts the quantity of fuel released during the delivery and reports this quantity once the delivery is complete and it becomes a transaction. With the direct interpretation from the

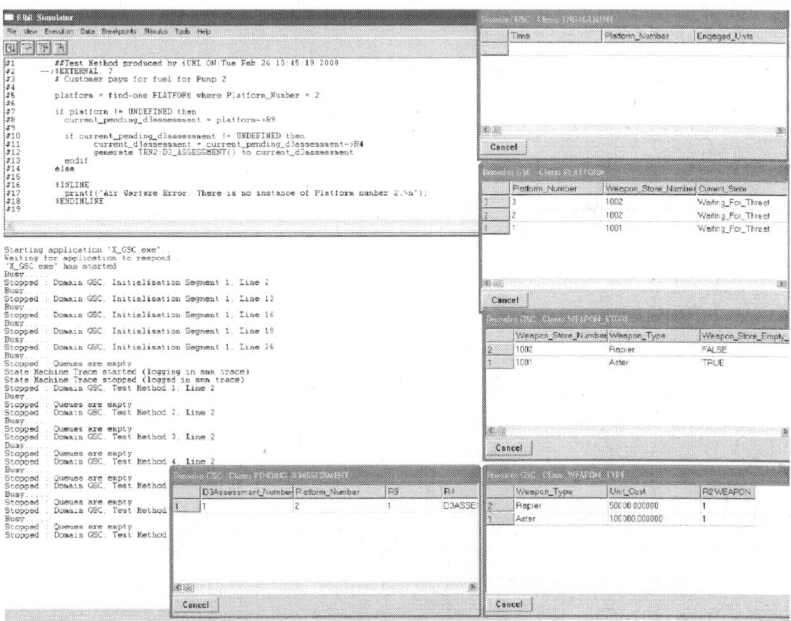

Figure 23.14   iUMLite Simulator screenshot.

gas station to air warfare application, the platform counts the number of weapons released at a target and reports this quantity once the engagement becomes a Degrade, Disrupt, Deny (D3) assessment. The problem with this process is that the weapon count used is only reported at the *end* of the engagement. If the weapon store amount is low to begin with and multiple weapons are used to engage a target, then there is a danger that the engagement could exhaust the weapon supply or that other weapon platforms could engage targets with a significantly depleted weapon store. The differentiator between the two applications here is scale. While the gas station pump can only deliver below the reserve fuel quantity during a delivery (due to the volume of the tank in proportion to the volume of fuel dispensed), the quantity of missiles in the air warfare system is much smaller and hence the risk of running out of weapons is high, either through a single prolonged engagement or multiple simultaneous engagements when the weapon store level is low. This can be ameliorated by moving the counter from the platform to the weapon store so weapons are counted out as they're released, rather than just at the end of an engagement. Through the executable architecture, a simple problem has been identified by observing a simulation of the system in use, and the architecture can be quickly corrected.

Toward Future Executable Architectures

The simulation has shown that a gas station architecture can be interpreted into a conceptually suitable architecture for an air warfare system, allowing rapid development of a prototype architecture for further testing, development, and communication with the customer. As with software development, the reuse of architectures allows projects to quickly identify conceptually similar existing architectures and quickly interpret them to the chosen application. The reuse of architectures does not determine the exact solution; in the MDA process, the solution will be derived from the individual domains, capturing the high-level, abstract form of the solution. Just as MDA has an implementation-independent PSM, so the architectural output is also implementation independent. The use of domain as opposed to functional partitioning helps maintain the integrity of the architecture during technology/implementation shifts and helps prevent obsolescence of the architecture due to technology. As with domain-based partitioning, the technology is encapsulated in one domain as opposed to across functions with functional partitioning. Together, the benefits of executable architecture reuse provide a significant competitive advantage over slower, more traditional development approaches.

Reference

Raistrick, C., Francis, P., Wright, J., Carter, C., and Wilkie, I. 2004. *Model Driven Architecture with Executable UML*. New York: Cambridge University Press.

Chapter 24

Long-Range Strike System Design Case Study

William Engler

This case study will challenge the class to apply what they have learned in Seciton 4 ("Aerospace and Defense Systems Engineering") with a focus on system design. In order to give a better idea of how the various methods and tools work together, as well as to give the reader an opportunity to practice the various methods, a notional air system is provided as an example. The system should satisfy the notional objective of a long-range strike aircraft carrying a dropped payload based on open sources and example data. The reader will use the requirements analysis and House of Quality techniques to better define and understand the problem before using a matrix of alternatives to develop an initial concept. Some of the results of further concept development using surrogate modeling will round out the process. It is important to note that, just as the solution to the customer requirements is not unique, the solutions to the individual steps in the system design process are not unique. Example responses are provided here, based in part on research done for the U.S. Air Force Research Laboratory (Mavris 2004), but there is no single correct answer to the exercises.

Key Concepts

Requirements analysis
House of Quality
Concept development
Surrogate modeling

Class Exercise

The following exercise and instructor response can be used as reference material for an instructor to create similar exercises that are of interest to a class that is using this book for academic instruction or professional training. The exercise in this case study was performed by a graduate-level systems architecture class of 11 students at Loughborough University in the spring semester 2008. Of the four case studies in this textbook (which the same class was assessed by), this was the only one structured as a series of class exercises. Rather than one large exercise, the case study was presented through multiple short exercises. In what follows are the exact instructions, materials, and instructor's response to the exercise.

Instructions to the Class

The purpose of this exercise is to assess both individually and collectively the ability of the class to create a Quality Function Deployment (QFD) House of Quality (HOQ) using standard QFD management and planning tools.

- ▪ Quality Function Deployment (QFD) is
  - A systematic approach to support needs prioritization that relates stakeholder needs to technical requirements.
- ▪ The QFD House of Quality (HOQ) is an integration of the outputs of QFD management and planning tools:
  - Each part of the house is a row, column, or matrix of data from the tools that supports the quality deployment of customer requirements to product engineering.
  - The class will practice creating a QFD HOQ using standard QFD management and planning tools, each of which will be presented through a sequence of exercises.

The class will use the following procedure to respond to the exercise:

- ▪ This class will organize itself into groups of three members.
  - Designate one member as the (chief) architect.
- ▪ For each of the nine QFD management and planning tools:
  - The class will have 10 minutes to create *sample data* outputs.
  - Each chief architect will be given 1 minutes for presentation.
  - It is recommended that you use the flipcharts provided.
- ▪ The class will nominate the best presentation.
  - The instructor will then provide a more complete dataset.
- ▪ No presentations will be required the following morning.

Class Assessment

The class will be assessed as follows:

- ▪ The class will be assessed as a whole based on
  - The sample data products nominated, and
  - Their understanding as a whole of each tool.

- Assessment will also include the class understanding and participation in the actual building of the QFD HOQ.
- The assessment will be provided to the class as a whole at the start of the next day.

Long-Range Strike System Concept Statement

As with all design problems, the process starts with the customer's requirements, both explicit and implicit. For this case study, the customer has defined the following explicit requirements statement.

> The long-range strike (LRS) system shall be capable of delivering a minimum of 20,000 pounds of payload a distance of 3000 nm and return while achieving a minimum mission reliability of 0.98. The system shall be capable of surviving in a threat environment as identified in Annex A and be capable of operating at night and in weather states as identified in Annex B. The system design shall achieve a balance of effectiveness, affordability, responsiveness, and regeneration in support of a nine-day campaign. The system shall be capable of operating from a 9000-foot military runway on sea-level standard day. The overall system Life Cycle Cost (LCC) shall be bounded by $XX B in Fiscal Year 2009 dollars. The system Initial Operating Capability (IOC) as defined by 12 mission-capable aircraft, crew, and supporting personnel shall be 2023.

The specific threat environment, operating state, and life-cycle costs required are not important for the examples shown here, but are included to ensure that the aspects of the design that they control are not ignored.

SWARMing

Based on the customer's explicit requirements, it is up to the engineer to gain an understanding of the customer's implicit requirements. For this portion, an extensive information search, sometimes known as SWARMing, is performed. As discussed in Chapter 20 ("Advanced Design Methods"), the engineer should use both written and nonwritten sources to gather as much insight into the customer's problem as well as the problems of the customer's customer. An individual with experience building, operating, or maintaining a similar system as to what is being developed can provide insight not otherwise available. As will be observed in the response, additional information often raises more questions that should be presented to the customer or used to drive the concept selection process.

Exercise 1

Either on paper or electronically, use the ideas of SWARMing to better understand the LRS system and the requirements for it.

- What requirements are missing or need more direction?
- What requirements are in conflict or agreement?
- What sources of information are at your disposal or are available to you?
- What are the general categories of information that are missing and must be derived?
- What are the hidden implications within the given problem statement?

Exercise 1: Instructor's Response

The quantity of helpful information that can be obtained as part of SWARMing would fill many pages. As a demonstration, only a few of the areas of interest are shown here. One of the first requirements that is notably absent is the required speed of the aircraft. Three areas of research come into play to find the proper speed: how fast do existing aircraft sometimes tasked with similar missions fly, how fast do we need to fly to meet the other requirements, and how fast are we capable of flying realistically? Whereas the second two questions are best answered with modeling and simulation, the first one can easily be identified within SWARMing. A short search of one of many sources on the Internet or aircraft data books can be compiled into the chart shown in Figure 24.1.

The plot shows several groups of aircraft. The group of aircraft that are short range and low speed is made up of the older attack aircraft and the F-117. The F-18E and F-22A are higher speed, but not significantly faster. The strategic bombers, including the B-1B, B-2, and B-52, are all much longer range, but only moderate cruise speeds. It seems unlikely that there would be much interest in an aircraft that had a shorter range or lower speed than current state-of-the-art aircraft. This plot, along with some of the other information gathered, was presented to the customer for feedback. For this particular case, the customer indicated an interest in developing a vehicle and required technology for operation primarily in the shaded region of the plot.

From past experience, the speed of an aircraft shapes the aircraft and sizes the engines, but the required range and payload drives the total size of the aircraft. Figure 24.1 also suggests that the difference between a long-range strike aircraft and other aircraft is around 1000 nm. The customer specified a range of 3000 nm in the requirements. Would a radius of 2700 nm still reach the majority of the intended targets or would 3300 nm give significantly greater ability to prosecute targets? Using a great circle mapping tool and selecting an arbitrary base of operation, we obtain a map such as the one shown in Figure 24.2. Operating radii of 1000, 3000, and 6000 nm are shown for comparison.

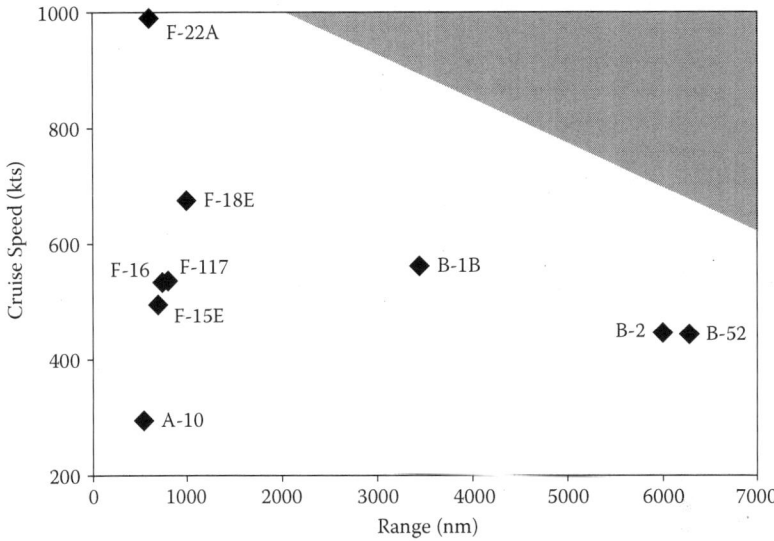

Figure 24.1   Range and speed of existing aircraft.

The 3000-nm operating radius appears to cover large portions of Asia, eastern Africa, and the Middle East. This operating area overlaps with the areas of several other bases as marked on the map. It appears that a slight increase in the range of the LRS would enable the system to cover a greater portion of the Middle East and portions of Asia, but otherwise has little benefit here. Reducing the range would likely result in being able to meet other requirements with greater ease as well as potentially reducing the cost; however, the customer then would require a greater amount of forward basing in order to meet its operating requirements. Such a trade-off is likely very sensitive to political considerations. Identifying and speaking to such considerations may be essential to a successful proposed solution.

The customer's requirements do not specify whether or not the range is refueled or unrefueled. This brings up the question of how much refueling is reasonable for a mission. Continuing with a basing out of Diego Garcia and assuming a strike in the Middle Eastern Theater, we should look into the availability of tankers. Military Web pages identify that Al Udeid Airbase in Qatar is the main U.S. operating base for the 349th Expeditionary Air Refueling Squadron flying KC-135 refueling tankers. Publicly available satellite pictures (such as Google Earth (Google 2008)) of Al Udeid in 2008 show six KC-135 tankers on the parking ramp. The United States Air Force's Fact Sheets report that the KC-135 supports both the probe and the drogue refueling systems used by the U.S. Air Force, Navy, and Marines. Each can hold up to 200,000 pounds of fuel and transfer 6500 pounds of fuel per minute (USAF KC-135 2008). Refueling in mid-air at maximum rate up to 65,000 pounds

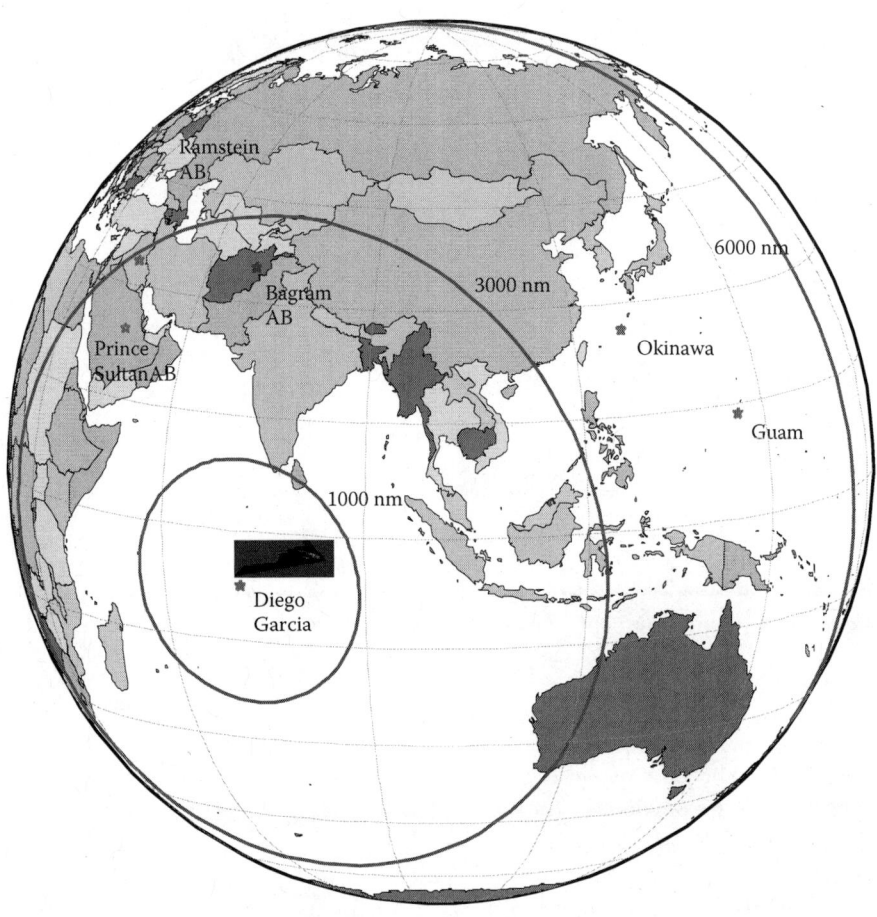

Figure 24.2   Operating radii from Diego Garcia.

would require 10 minutes, not including the time to meet up and connect with the refueling tanker.

Further SWARMing helped identify other implied requirements and effects of the mission requirements. Special facilities might be required for Stealth aircraft. Recent aircraft acquisitions have shown that Stealth is generally desired at the air vehicle level to increase survivability. When systems are more survivable, a lower sortie rate (Figure 24.3) may be required because individual aircraft may be more effective. Unfortunately, the use of Stealth on aircraft can impact the logistics footprint significantly and may reduce the maximum speed of the aircraft. These multidimensional trades will focus future modeling and simulation efforts as well as the trades of interest. SWARMing should continue throughout the system design process to further enhance the understanding of the problem and as needed to fill in information needed for further investigation.

Illustrative Comparison of Weapon-Delivery Potential				
Aircraft	Payload	Daily Sortie Rate	Weapons Delivered per Day	Weapons Delivered in 10 Days
1 × F-15E	3 × GBU-24	1.75	5	53
24 × F-15E	3 × GBU-24	1.75	126	1260
1 × B-2	16 × JDAM[a]	0.33	5	53
16 × B-2	16 × JDAM	0.33	84	840
1 × B-2	16 × JDAM	0.5	8	80
16 × B-2	16 × JDAM	0.5	128	1280

[a]JDAM = Joint Direct Attack Munition.

Figure 24.3 Table of sortie rates.

Management and Planning Tools

The next steps of the system design process are grouped together into a series of management and planning tools. These tools are used to develop, organize, and prioritize customer requirements and engineering characteristics. Each tool is summarized briefly here and discussed in more detail in Chapter 19 ("Systems Design").

Affinity Diagram

Affinity diagram is a process of organized brainstorming. For the customer's requirements (sometimes referred to as "What's") or the engineering characteristics (referred to as the "How's"), a group should take turns identifying what it is that the customer or engineer values in one- or two-word phrases. These phrases can be written on cards and organized into groups based on their relationships or affinities.

Exercise 2

Either on paper or electronically, create an affinity diagram to brainstorm the engineering characteristics for the LRS problem. Then arrange them into logical categories and identify overlaps, intersections, and gaps that exist.

Exercise 2: Instructor's Response

For a system as potentially complicated as the one presented here, the affinity diagram may very likely be quite large. It would not be unreasonable to use an entire wall. However, due to space restrictions, only a subset is shown in Figure 24.4. These have already been organized into one of four groupings: performance, attributes, effectiveness, and economics.

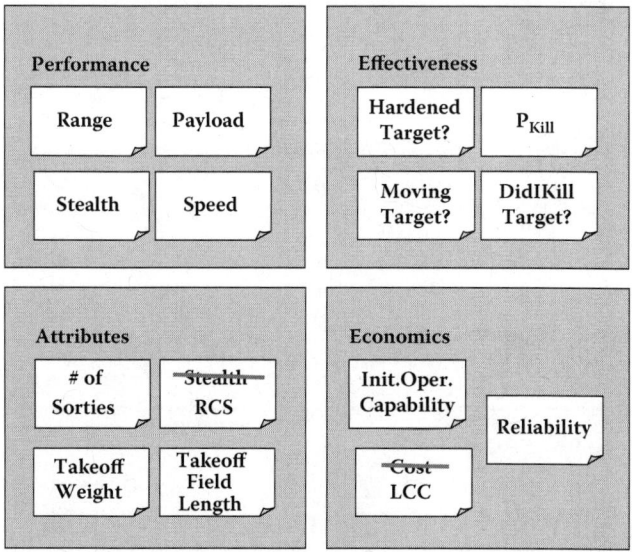

Figure 24.4 Affinity diagram.

Note that "Stealth" appears under both Performance and Attributes. While ideally each item will be under only one heading, it is possible that some items are so related to multiple categories that it is acceptable to list them multiple times. When cross-listing items, be sure that if you update it in one place, that you remember to update it elsewhere as well.

Interrelationship Digraph

Few, if any, of the characteristics identified in the affinity diagram are independent of the others. Just as individual characteristics may fall under multiple categories, there may be complex interrelationships between the characteristics. Mapping the connections helps to ensure that all appropriate trades are considered. This process can quickly become difficult to manage. Traditionally, this might have been done on a bulletin board with string and tacks. More recently, the process is performed within one of the many available graphing software packages.

Exercise 3

Either on paper or electronically, create an interrelationship digraph (Figure 24.5) to determine the interactions between the engineering characteristics that were identified in Exercise 2.

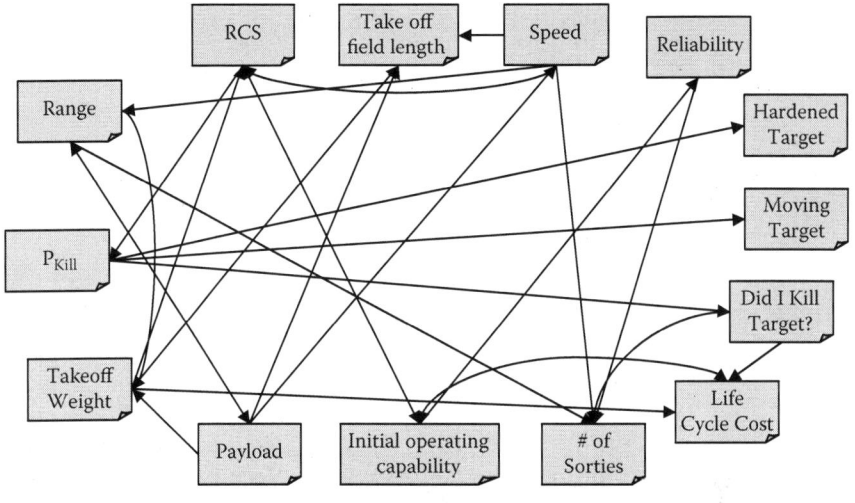

Figure 24.5 Interrelationship digraph.

Exercise 3: Instructor's Response

Considering the small size of the aforementioned affinity diagram, the interrelationship digraph is considerably more complex. Only a few of the requirements have one or two links to other requirements. Takeoff Weight has the most links, at five. Experienced aircraft designers will know that the weight of an aircraft is tightly managed because of the large effect it has on other attributes and also because of the number of design considerations that affect it.

Tree Diagram

The tree diagram breaks down broad goals and requirements into increasing levels of detailed actions. It is ideal for identifying and describing the requirement decomposition process. It can also be used to logically decompose the groups determined by the affinity diagram. Tree diagrams can be created in an outline format or using one of many graphing or mind mapping software packages available.

Exercise 4

Either on paper or electronically, create a tree diagram to identify the root drivers of the customer requirements.

Exercise 4: Instructor's Response

The tree diagram in Figure 24.6 starts by listing at the top level Long Range Strike Capability and decomposing it into all the components needed for that. These include items like Strike Assets, Logistics, and Basing. Strike Assets is further decomposed into the various types of weapons available as well as the various types of strike platforms available. Again, the full tree diagram cannot be shown due to space limitations.

Prioritization Matrix

A prioritization matrix (Figure 24.7) allows for an initial qualitative analysis of the relative importance of different customer requirements or engineering characteristics of the LRS system. The customer requirements developed in the affinity diagram and refined in the interrelationship digraph are listed along the top and left side of a grid. The lower half below and including the diagonal is blanked out and the remaining grids are filled in by relating the importance of the left and top elements. Each value should be read as "the element on the left is X times as important as the element on the top."

Exercise 5

Either on paper or electronically, use a prioritization matrix with 1/9, 1/3, 1, 3, and 9 importance scores to help determine the weighting on the engineering characteristics.

Exercise 5: Instructor's Response

The examples from Exercises 2 and 3 are continued here. Once the matrix has been completed, a quick scan can identify the most or least important elements. An item with large values in its row and small values in its column will be much more important than an item with small values in its row and large values in its column. Figure 24.7 shows that speed is likely the most important aspect of the design while the initial operating capability is least important. Such a prioritization suggests that there were more performance-oriented engineers than contracts people present when the prioritization matrix was completed. As with all of these tasks, having a balanced group of people present when creating these artifacts yields the best results.

Process Decision Program Chart

The Process Decision Program Chart (Figure 24.8; often abbreviated at PDPC) is a method for identifying the risks associated with a mission or project and methods for dealing with those risks. It is split into three columns. The left column lists the

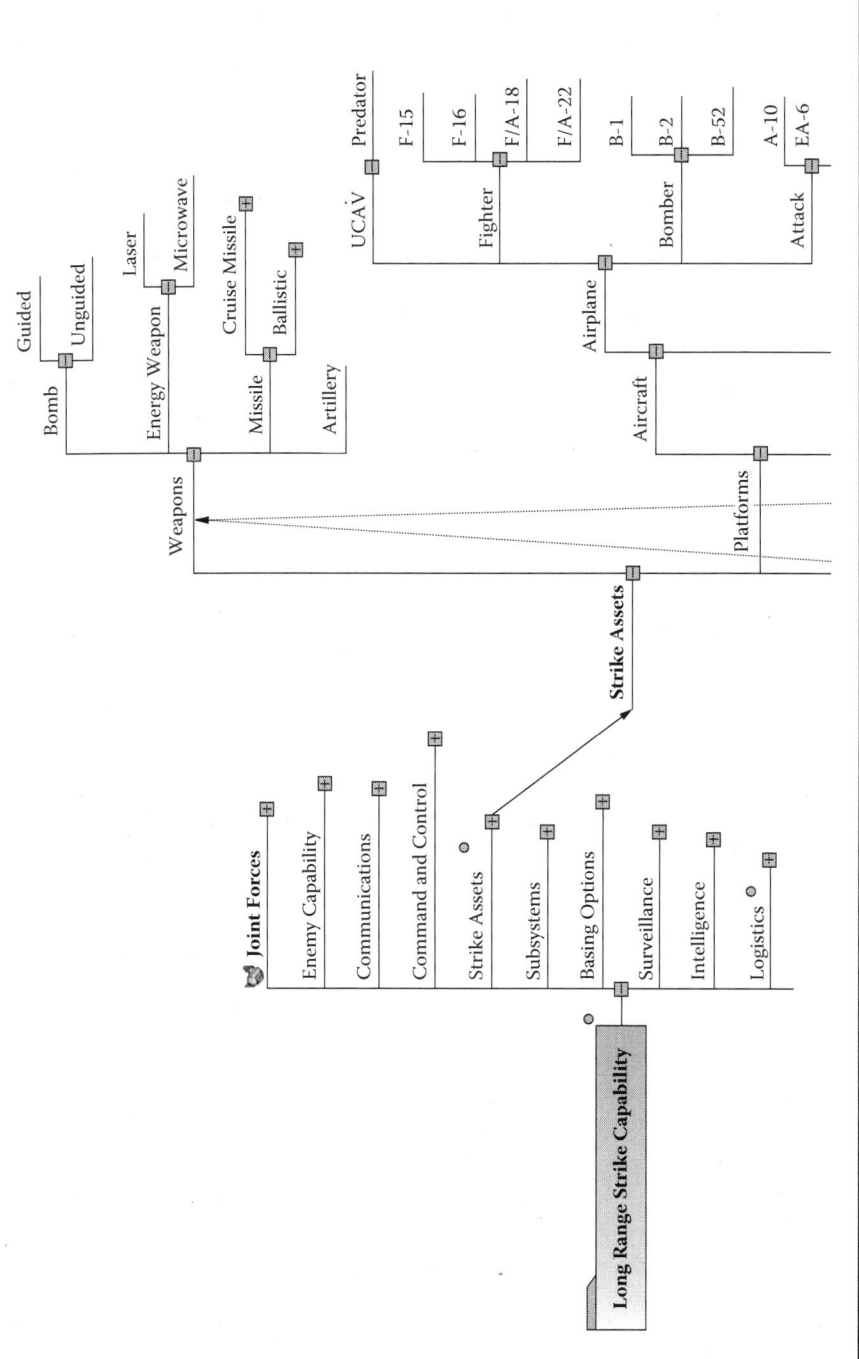

Figure 24.6   Tree diagram.

	Range	Speed	Payload	RCS	TOFL	Takeoff Weight	Sortie Rate	Reliability	IOC	Life Cycle Cost	Prob of Kill	Hardened Target	Moving Target	Target Assessment
Range		1	3	1	1/9	3	1	1	3	1	1/3	3	1	1
Speed			9	3	9	3	3	3	9	3	1	3	3	3
Payload				1/3	3	1	1/3	1/3	3	1/3	1/3	1	1	1
RCS					3	1	1	1	3	1	1	1/3	1	1
TOFL						1/9	1/9	1/9	1	1/3	1/3	1/9	1/9	1/9
Takeoff Weight							1/3	1	3	1	1/3	1/3	1/3	1/3
Sortie Rate								1	9	1	1	1	1	1
Reliability									3	1	1	1	1	1
IOC										1/3	1/3	1/3	1/3	1/3
Life Cycle Cost											1	1	1/3	1/3
Prob of Kill												3	1/3	1/9
Hardened Target													1/3	1/9
Moving Target														1/9
Target Assessment														

Figure 24.7   Prioritization matrix.

plan of events or tasks that must be completed to reach success. The center column lists the risks associated with each of the events in the left column. The right column lists the countermeasures available to minimize these risks. The PDPC can be created easily either digitally or with index cards on a bulletin board.

Exercise 6

Either on paper or electronically, create a Process Decision Program Chart to identify critical risks and countermeasures to completing the notional Long Range Strike mission.

Exercise 6: Instructor's Response

The Plan column in Figure 24.8 essentially follows the mission profile of an aircraft. Targets are found and identified and assigned to platforms by a commander. The platform flies to the target, delivers its weapons, flies back and repairs and re-arms for its next mission. Here, Locate Targets is split up into two subtasks. Each of the tasks is connected to one or more Risks. In a full, exhaustive PDPC, there would likely be hundreds or even thousands of possible risks associated with many tens of Plan tasks. Each of the Risks is then connected to one or more Countermeasures, with the exception of one. This is because there have not been any suggested design changes that can help if there are no targets. Because this identifies a victory condition, it is reasonable that no countermeasures are present.

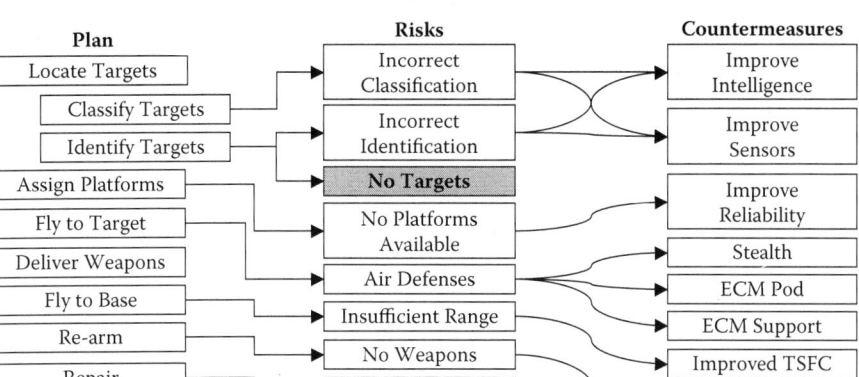

Figure 24.8 Process decision program chart.

Activity Network

An Activity Network bridges from the engineering processes to the program management process. It is very similar to a Gantt chart and can serve the same purpose. For engineering purposes, it is essential for identifying which groups must have work products completed for other groups to use their results. Generally, this would be created within project management software, but it is relatively easy to create using spreadsheet software as well.

Exercise 7

Either on paper or electronically, create an Activity Network (Figure 24.9) to begin laying out a timeframe for the development of the notional LRS system, incorporating the challenges that have been identified in previous examples.

Exercise 7: Instructor's Response

The Activity Network in Figure 24.9 shows the spread of SWARMing, the problem definition, and the analysis of alternatives. This example is relatively simple but captures the essence of an Activity Network.

Quality Function Deployment

The Quality Function Deployment (QFD) matrix, also sometimes known as a House of Quality, combines several of the products created so far with a relationship matrix between the customer requirements and engineering characteristics.

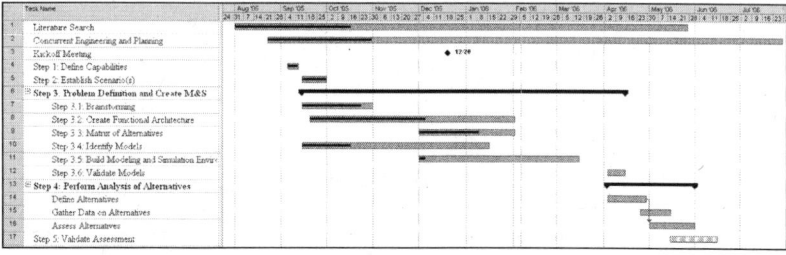

Figure 24.9 Activity network.

It also allows for the easy comparison of alternatives and existing concepts on a qualitative scale as well as prioritizing the engineering characteristics. Because the process of its development is rather involved, it would be helpful to review the appropriate section of Chapter 19 ("Systems Design"). As described there, the QFD itself is a set of qualitative models and model transforms of the requirements and engineering characteristics, and the relationships between them. Much of the QFD can be created in a spreadsheet program, but using specialized QFD software can speed up the process.

Exercise 8

Use the previous exercises to construct the QFD. Use the brainstorming portions to fill out the customer requirements on the left and engineering characteristics on the top. Use the interrelationship digraph to construct the "roof." Brainstorm alternatives to baseline against the customer requirements and engineering characteristics. Complete the center of the QFD using a 0-1-3-9 ranking scheme.

Exercise 8: Instructor's Response

The same engineering characteristics developed in Exercise 2 are listed at the top of the QFD shown in Figure 24.10. The Customer requirements were developed with a similar affinity diagram process. The roof correlates the engineering characteristics with a strong positive, weak positive, weak negative, or strong negative relationship. The tightest relationships are among the Performance and Attributes. The QFD software used here utilizes a triangle to indicate a weak relationship, an open circle to indicate a medium relationship, and a filled circle to indicate a strong relationship.

This QFD also includes measures of technical risk and quantitative goals for some of the engineering characteristics. These features are helpful, but not mandatory to the QFD process. Of particular interest are Weighted and Risk Weighted Importance. These are both shown enlarged in Figure 24.11. At this point, it makes sense to check to see if the weightings on the customer requirements and the center

		Engineering Characteristos (How's)													Competitive Assessment
		Performance			Attributes			Economics			Effectiveness				

Long Range Strike — Engineering Characteristos (How's): Range, Speed, Payload, RCS, Takeoff Field Length, Takeoff Weight, Sortie Rate, Reliability, IOC, Life Cycle Cost, Probability of Kill, Hardered Target Defeat, Moving Target Defeat, Target Assessment.

Customer Importance column and Competitive Assessment legend:
// Tomahawk
◊ F/A-22
⋅⋅ B-2
Ø LRS
scale 0 1 2 3 4 5

Direction of Improvement: ↑ ↑ ↑ ↓ ↓ ↓ ↓ ↓ ↑ ↓ ↑ ↑ ↑ ↑

Customer Requirements (What's)			Cust. Imp.	Range	Speed	Payload	RCS	Takeoff Field Length	Takeoff Weight	Sortie Rate	Reliability	IOC	Life Cycle Cost	Probability of Kill	Hardered Target Defeat	Moving Target Defeat	Target Assessment
Life Cycle	Upgradability	2.0		△	●	○		△	△			●	△	●			
	Agile Acquisition	3.0	△	○	△	○					△	●	○				
	Affordability	4.0	○	●		△		●	△	●	△	●				△
Performance	Survivability	5.0	○	●	△	●		○			●	○		○	△	
	High Success Rate	4.0		●		●	△		○	●		○	●	○	●	●	
	Operate at Night Weather	2.0		○	○	●			●		△	△	●		△		
	Precision	4.0			●							●	●	○			
	Versatile/Threats	5.0	●	●	●	△		△	△		○	△	●	●	△	△	
	Responsiveness	4.0	○	●	○	○	△	○	●	○		○				△	
	Regeneration	3.0		●	△	○			●	●		●				△	
Operations	Interoperability	3.0			●						●	●		△	●	●	
	Phys. Compat.	2.0		△	●		●	●				△		△			
	Maintainability	2.0		○	●				●	●		○					

| How Much | | | 3000 nm | | 20,000lb | | 10,000 ft | | | 0.98 | 2025 | | | | | |

Competitive Assessment:
// Tomahawk 5
◊ F/A-22 4
⋅⋅ B-2 3
Ø LRS 2
1
0

Technical Risk (1–5)	2	4	1	3	1	3	4	3	4	5	4	4	5	3
Weighted Importance	87.	244	179	162	28.	88.	120	150	79.	199	152	96.	137	84.
Relative Importance														
Risk Weighted Importance	174	976	179	486	28.	264	480	450	316	995	608	384	685	252
Relative Risk Weighted Importance														

Figure 24.10 Long range strike QFD.

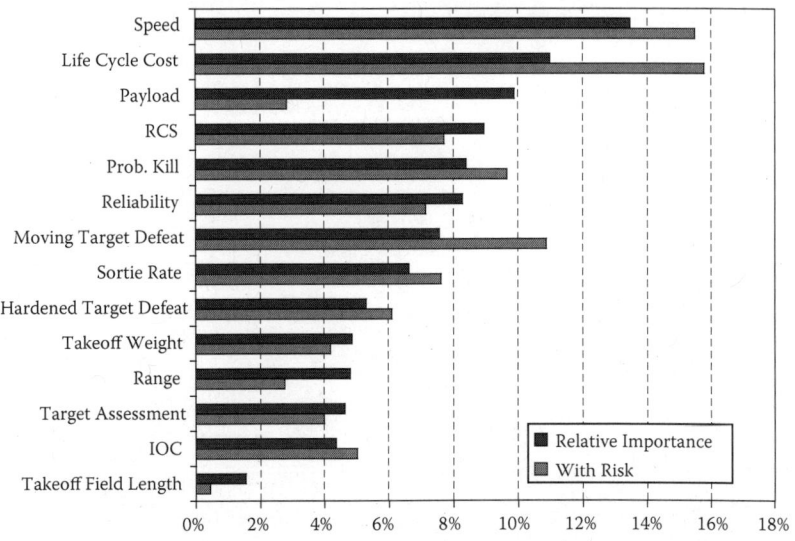

Figure 24.11 LRS engineering characteristics importance ranking.

matrix have resulted in importance rankings that are intuitive. The most important design conditions are the speed and life-cycle cost. These will most likely be the most technically difficult to achieve, and several of the more important customer requirements depend on them. Takeoff Field Length is at the bottom. This is reasonable because most aircraft with engines powerful enough to reach high speeds will also have more than enough thrust to take off in less than two miles of runway. What is surprising is how low IOC is ranked. This suggests that, while meeting the deadline for operations is important, this requirement is not very important in satisfying other customer requirements.

The method shown here is called Quality Function Deployment but the Deployment portion of the method is not obvious. The intention is that the set of Engineering Characteristics at this level is related to even lower engineering characteristics. In this situation, physical characteristics such as Wingspan and Fuel Capacity as well as subsystem characteristics such as Engine Thrust and Advanced Sensors take the role of the engineering characteristics while the previous set now defines the requirements. The second-level deployed QFD is shown in Figure 24.12.

Creation of Alternatives

Once the problem has been better defined and understood, it is necessary to generate a set of alternatives that may each potentially satisfy the customer's requirements. Alternatives should cover as broad a space as possible to help ensure that no options have been overlooked.

Air Vehicle	Customer Importance	Physical Characteristics					Mission Systems				Reliability				Propulsion		
		Wingspan	Aspect Ratio	Internal Weapons	Fuel Capacity	Number of Components	RAM Coatings	Radar Tech Level	Adv. Sensors	Weapon Types	MMH/FH	MTTR	MTBF	Turnaround Time	Engine Diameter	Engine Thrust	Engine TSFC
Performance — Range	88.0		◉	○	◉			△		△	△				○	△	◉
Speed	244	△	△	◉	△		○			△				△	◉	◉	△
Payload	179			◉	△					◉				△			△
Attributes — RCS	162	○	○	◉		△	◉	○	△	◉					◉		
TOFL	28.0	○														◉	
Takeoff Weight	88.0	○	△	△	◉					△					○	◉	○
Sortie Rate	120			△	△		◉			△	◉	◉	○	○			△
Economics — Reliability	150			△		◉	○	○	△	△	◉	△	◉				
IOC	79.0					○	△	○	△						△		
Life-Cycle Cost	199.0					◉	○	◉	△		◉	○	○		△	◉	
Effectiveness — Probability of Kill	152.0		○				◉	◉	△	◉				△			
Hardened Target Defeat	96.0						△	△		◉							
Moving Target Defeat	137						△	◉		◉							
Target Assessment	84.0							◉	◉								
How Much / Organization Difficulty / Weighted Importance / Relative Importance		10 78.0	16 10.0	63 43.0	21 27.0	35 40.0	59 97.0	65 05.0	14 08.0	72 24.0	43 09.0	18 27.0	23 07.0	93 5.0	44 60.0	51 19.0	15 99.0

Figure 24.12   LRS QFD 2.

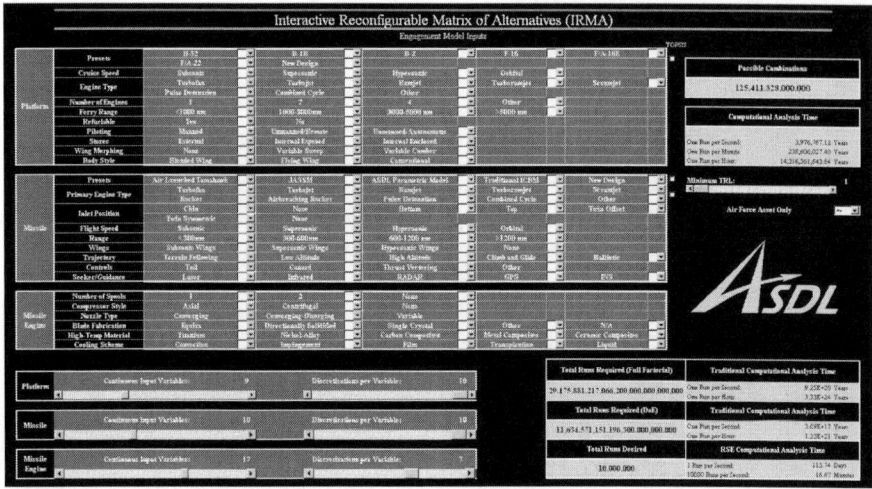

Figure 24.13   Long range strike interactive reconfigurable matrix of alternatives.

Functional and Physical Decomposition

The key to solving many hard engineering problems is to decompose the problem into smaller problems that can be solved. There are two common ways to decompose a problem: physical decomposition or functional decomposition. A physical decomposition splits a system into its physical parts, much like a modular exploded view of a schematic. A functional decomposition instead splits the functions that a system has to perform in order to fulfill its intended purpose. Ideally, a problem should be decomposed both ways, and then the two would be solved together iteratively. For this case study, the systems were decomposed physically into an aircraft, missile, and missile engine. Each of these was then further decomposed into the associated characteristics. For the aircraft, this included physical characteristics such as how it was piloted, the number of engines, and presence of wing morphing. The missile's decomposition included the type of control system used, seeker type, and position of the inlet. The missile engine included the number of spools, nozzle type, and the cooling scheme. The full physical decomposition is shown as part of Figure 24.13.

Matrix of Alternatives

In many instances, several alternatives are proposed for comparison. Unfortunately, these alternatives are rarely created by any structured method, but rather are based on either what a company has already developed or on educated guesses at what potential concepts may be. The functional or physical decomposition is an excellent resource to use when generating concepts more rigorously, using Fritz Zwicky's

Morphological Analysis Method (Zwicky 1969). The individual tasks of the functional decomposition are listed in a vertical column next to an empty matrix, split into broad categories for organization. Next, alternative subsystems, approaches, or solutions to each of the tasks are listed out horizontally next to it, filling the matrix. This creates what is known as a Matrix of Alternatives or Morphological Matrix. Like any other brainstorming process, it is best to include all options and then eliminate those that are not feasible later. To generate a single concept, an individual simply selects one item from each of the rows such that each function has a solution that satisfies it. The total number of combinations possible in a Matrix of Alternatives is the product of the number of options available in each row. The process is identical if using the physical decomposition.

Exercise 9

Either on paper or electronically, create a Matrix of Alternatives based on a functional or physical decomposition of the notional LRS system. Calculate the total number of concepts possible. Note that the response to this exercise is in the next section.

IRMA

A variation on the Matrix of Alternatives known as the Interactive Reconfigurable Matrix of Alternatives (IRMA) is used to improve the usability and incorporate some multiattribute decision making tools. An IRMA focuses on allowing a group to visually make traceable decisions together for each of the rows in order to achieve a better set of concepts. The layout is similar to the Matrix of Alternatives, but also includes a simple calculation to display the number of total possible combinations remaining as decisions are made. As decisions are made within each row, a compatibility matrix in the background eliminates options through the IRMA that require negated options or are incompatible with selected options. Additionally, each row can be equipped with a simple multiattribute decision making tool to bring qualitative or quantitative information to the decision-making process. Finally, the use of flat filters across the entire IRMA can eliminate those options which are considered too risky from a technological or programmatic standpoint (Engler et al. 2007).

Exercise 9: Instructor's Response

The Matrix of Alternatives used for the Platform-Missile-Missile Engine breakdown is shown in Figure 24.13. The second column from the left shows the physical decomposition created earlier. Not all rows have the same number of possible alternatives. Some, such as the Missile's Primary Engine Type, have as many as ten options while others, such as the Platform's Refuelable quality have just two. The total number of possible combinations for this IRMA is on the order of 125 trillion. A number this size casts doubt on how just three or four concepts may have been generated in the past. After a single family of concepts is identified by selecting one item in each row (a thread through the matrix), it is important to remember that within that family are an infinite number of actual designs based on the varying dimensions associated with physical design. At the base of the IRMA is an additional set of slide bars to help remind the designer of the size of the design space, assuming that a simple design space exploration is performed.

Quantitative Modeling

There are many ways to explore the design space associated with each of the families of concepts selected. The simplest approach is to select a point based on previous experience, existing baselines, or a competitor's product and test to see if this design meets your requirements. If it does, then no further design is required. If not, some modification is performed and the design is tested again. The question arises regarding how one tests a design. For some systems, where prototypes are relatively cheap and easy to produce, a physical system may be created for each case and tested. Such approaches are generally specific to a field and outside the scope of this text. For more complex or expensive systems, a virtual prototype is created within a modeling and simulation (M&S) environment. Due to the time required to develop and run any modeling and simulation environment, no exercises are presented here. Instead, a discussion of what methods were used for the LRS problem may provide insight into one of the many ways to down-select to a single solution.

For the LRS system, several levels of modeling are required. Because the system has already been broken up into a platform, missile, and missile engine, it is likely that a separate model will be required for each of these. Each model was likely originally developed by a separate group as a stand-alone tool, and they may all be of varying fidelity. A relatively simple, commercially available engine cycle program known as GasTurb (Kurzke 2008) is used to calculate performance of several engine types including turbojet, turbofans, and ramjets. A missile sizing model is built using the information and trends from Gene Fleeman's book on tactical missile design (Fleeman 2006). NASA Langley's Flight Optimization System (FLOPS) aircraft sizing and synthesis code is used to calculate baseline LRS performance (McCullers 1995). These three codes make it easy to understand the performance of each of the components in the larger LRS system, but they do not capture the interactions between them or their combined effectiveness.

The customer stated more specifically that there was an interest in performing all trades at the level of overall system effectiveness. In order to capture the effectiveness of a system, the aircraft-missile-engine combination must be analyzed within an appropriate scenario. This drives the need for a higher-level engagement modeling environment. A simple "Scud-hunt"-type mission was developed inside an agent-based simulation under the guidance of the customer and industry partners. In this, long-range strike aircraft armed with a powered weapon were sent against an opposing force equipped with surface-to-air missiles and time-critical tactical ballistic missiles. Within this environment, the missile engine drove the fuel consumption and available thrust of the missile. The missile's range and speed were combined with the range and speed of the aircraft to allow the aircraft to intelligently decide when to release its weapon. The behavior and number of aircraft could be varied by a series of parametric inputs to control the commander's tactics. Similarly, the opponent's capabilities and behavior were variable, to test which cases were easier or harder for the LRS solution. Within this setup, all four models could be run for a single design and specific scenario, yielding results describing the number of targets eliminated and number of LRS aircraft lost.

Surrogate Modeling

The four-program modeling environment allowed an engineer to test a single design condition. However, due to the long runtime and necessity to manually transfer results from one model to another, it was not well suited to true design space exploration on the scale desired. What was needed were much faster models that could be easily linked together. The concept of surrogate models, as discussed in Chapter 20 ("Advanced Design Methods"), is that a high-fidelity numerical regression is performed to accurately capture the trends of a more detailed model within a set of ranges. These would allow a single case to be run fully integrated in a small fraction of a second instead of the nearly 30 minutes required to run each model manually and combine them. In addition to increasing the speed at which individual designs can be tested, the continuous nature of the surrogate models enables a wide variety of options for graphical display and evaluation. There are two approaches to surrogate modeling of multiple individual models. The first is to use a set of inputs at the base level and run all of the models to generate final results for regression. This would create a single monolithic surrogate model. In past experience, this is more difficult to integrate and is only helpful when there is a very large set of intermediary variables between the models. The other option is to create surrogate models for each individual code. This allows greater reusability of the surrogate models as well as giving some visibility to the intermediate values for later error checking. The latter method was selected for the LRS problem.

For many problems, response surface equations (RSEs) based on multivariate second-order Taylor-series polynomial expansions are more than sufficient to capture the physics of the problem. This approach was used for both the aircraft and missile models with accuracies of more than 95 percent. The polynomial-type surrogates are not able to capture plateau-like behavior or irregular and shifting peaks that are often prevalent in agent-based simulation. Because the engine modeling included three different engine architectures, the same behaviors were present in the transitions from one engine type to another. For these, single-layer perceptron neural network surrogate models were used. Neural networks require significantly more data and time to regress, but have the benefit of being able to theoretically approximate any function with enough data. When applied to the engagement and engine models, these surrogates were also greater than 95 percent accurate to the original models.

Selection of Alternatives

With an extremely fast modeling and simulation environment available in the form of surrogate models, it was possible to test many more than the handful of points that may have been traditionally looked into. This opens up several new methods for design space exploration. The simplest leap would be to wrap a simple optimization tool around these equations and let it identify the best possible design through massive iteration. If the space is multimodal, it may be possible to end up with a suboptimal design. Another alternative is to use Monte Carlo simulation to generate a large number of design cases and scenarios (on the order of tens of thousands to millions) and use a series of constraints on each of the objectives and requirements to narrow the space incrementally until only a small handful of the best cases and points remain. These cases should then be run through the full modeling and simulation environment to ensure that their surrogate values are accurate and for further investigation. At this point it is relatively simple to select a data point with the customer present and still have the full analysis available at an instant.

Summary

A notional LRS system consisting of an aircraft carrying a powered missile has been investigated here. An intensive literature search, known as SWARMing, was used to understand and define the customer's explicit and implicit requirements. Through a series of systems engineering tools, the requirements were distilled into specific phrases and prioritized. These were translated into engineering characteristics through the Quality Function Deployment process. Using a Matrix of Alternatives, several families of concepts were defined for further exploration. A modeling and simulation environment consisting of an aircraft sizing model, a missile sizing model, an engine performance model, and a simple agent-based engagement model was used to create surrogate models. Exercising these surrogate models within an optimizer or Monte Carlo filtering allowed the selection of a small set of concepts to be presented to the customer.

References

Engler, W.O., Biltgen, P.T., and Mavris, D.N. January 2007. Concept Selection Using an Interactive Reconfigurable Matrix of Alternatives (IRMA). *45th AIAA Aerospace Sciences Meeting and Exhibit*, Reno, NV.

Fleeman, E. 2006. *Tactical Missile Design*, 2nd ed. Reston: American Institute of Aeronautics and Astronautics.

Google. 2008. Google Earth. http://earth.google.com (accessed 26 June 2008).

Kurzke, J. May 14, 2008. The GasTurb Program. http://www.gasturb.de (accessed June 26, 2008).

Mavris, D.N. Architecture-Based Modeling and Simulation in Support of the Accelerated Capabilities Development and Delivery (ACD&D) Program. Final Report, December 31, 2004.

McCullers, L.A. 1995. *Flight Optimization System,* User's Guide Version 5.7. Langley: NASA Langley Research Center.

USAF (United States Air Force) Link. April 2008. KC-135 STRATOTANKER. http://www.af.mil/factsheets/factsheet.asp?id=110 (accessed June 26, 2008).

Zwicky, F. 1969. *Discovery, Invention, Research—Through the Morphological Approach.* Toronto: The Macmillan Company.

Chapter 25

Remote Monitoring

This case study will challenge the class to draw from what they have learned in all four of the previous parts of the book. The case study considers the development of a system architecture in response to a set of technology-specific requirements. Architecture-based assessment informs the suitability of the architecture against the requirements to suggest improvements to the architecture itself.

Key Concepts

MDA™ architecture development
   Computation-independent model
   Platform-independent model
   Platform-specific model
Domain models
Architecture-based assessment

Background

This chapter presents a case study in which a system architecture is developed that can be used for the technical assessment of system alternatives. At the end of the case study, the architecture will be used to make a defendable judgment of the feasibility of each system alternative. The aim of this assessment is to identify anything about the alternative and its role in the architecture that will prevent the system from doing what it was intended to do. The relationship between architecture-based assessment, and system, and architecture, which have been previously defined in this book, is defined as follows:

> A *system* is a combination of interacting elements integrated to realize properties, behaviors, and capabilities that achieve one or more stated purposes. The *architecture* of a system is a specification of the parts and connectors of the system and the rules for the interactions of the parts using the connectors. *Architecture-based assessment* is the

concordant assessment of the capabilities and interoperability of a family of systems.

This case study will consider the suitability of the developed architecture against the system requirements through the use of MDA™. The context chosen for this case study is a remote monitoring system for a healthcare provider.

Class Exercise

The following exercise and instructor response can be used as reference material for an instructor to create similar exercises that are of interest to a class that is using this book for academic instruction or professional training. The exercise in this case study was performed by a graduate-level systems architecture class of 11 students at Loughborough University in the spring semester 2008. Of the four case studies in this textbook (which the same class was assessed by), this was the most challenging because of the breadth of concepts that had been covered by the end of the course and the many interpretations that the problem statement could be given. However, class members actually found this the simplest exercise because, by the end of the course, they had learned to think in new ways about architecture and systems engineering. This was the last of the in-class assessments. In what follows are the exact instructions, materials, and instructor's response to the exercise. The instructions include the responsibility of the class to organize itself to work in small groups. The practice of architecture and systems engineering is as much about personal and professional relationships as is it about technical details.

Instructions to Class

- The purpose of this exercise is to assess the ability of the class to practice creating a system architecture that can be used for the technical assessment of system alternatives.
- The class will then use the architecture to make a defendable judgment of the feasibility each system alternative, that is:
  - Can you identify anything about the alternative and its role in the architecture that will prevent the system from doing what it was intended to do?

The class will use the following procedure to respond to the exercise:

- Organize yourselves.
  - Designate one member of the class as the chief architect.
- Use the requirements narrative for the system concept to
  - Create a systems architecture to be used for assessments
  - Understand alternative solutions from a systems viewpoint
  - Discover problems with the alternatives

■ The chief architect will lead the class in a 15-minute presentation to the instructor on the architecture and the assessment of alternatives. The presentation will be at the start of the course summary.

Class Assessment

The class will be assessed as follows:

■ The class will be assessed as a whole based on
  – The utility of your systems architecture. That is, how useful was it in supporting your assessment?
  – The correctness of your assessment. That is, did you identify a real issue for the system alternative?
  – How defendable is your assessment of the alternative?
■ The assessment will be provided to the class at the end of the class presentation.
  – For each student, the mark for this assessment will be substituted for any one previous mark that is lower.

Remote Monitoring System Concept

After hospital treatment, some medical patients need extended periods of monitoring with equipment such as ECG cardiac behavior and EEG for brainwaves. The typical period of monitoring is between 24 hours and 1 week. Requiring the patient to remain in the hospital for extended monitoring is expensive due to room costs. Frequently, the monitoring is for the purpose of diagnosis and assessment rather than any life-threatening reason. In these cases, it has been suggested that there should be a way to monitor patients from their home. The patient would need the freedom to move about in his/her home without interfering with the monitoring process. And at times, patients may also need to leave their homes.

The remote monitoring system (RMS) would use the same type of attachable electrode sensors as used in the hospital but would add a technology for remote operation that would link into the patient's home computer or cell phone for Internet access.

The healthcare provider requires one of three specified wireless technologies to implement the RMS and enable it to link into the Internet. These wireless technologies are the Bluetooth 1.2 and 2.0 technologies and another type of cellular technology based on GPRS. The RMS implementation must also account for the Internet service provider used in the patient's home.

Technical Details

Further details of the system requirements, including acronym explanation and relevant technical standards, are outlined below:

■ ECG: the electrocardiograph is used to measure cardiac behavior. Electrode sensors are placed on the surface of the body. For simplicity, assume the nominal rate of sampled data to be 1 KHz.
■ EEG: the electroencephalograph is used to measure the electrical activity of the brain. Electrode sensors are placed on the scalp. Assume the nominal rate of sampled data to be 100 kHz.
■ Bluetooth: an industrial specification and technology for wireless personal area networks.
  – Bluetooth 1.2 is backward compatible with 1.1 and supports transmission speeds of up to 700 kbps.
  – Bluetooth 2.0 supports transmission rates of up to 3.0 Mbps.
■ GPRS: the General Service Radio Packet is a packet-oriented Mobile Data Service that provides data rates from 56 to 114 kbps.
■ Both technologies are implemented using microchip transceivers with embedded microprocessors.

Architecture Development

The RMS requirements have been stated as a narrative with reference to certain standards to enable the proposed system to operate with technology already deployed in the organization. Note that while the standards are explicit requirements, the overall concept of the RMS implies a number of implicit requirements. Through the use of logical modeling, as outlined in this book, and other requirements discovery techniques, these requirements can be captured through the conversion of the requirements prose into logical models that are congruent and have conceptual integrity.

The Developed Architecture

Utilizing an MDA approach, the first activity for developing the architecture is the creation of a domain model through the identification of domains and their dependencies. This was achieved through brainstorming the domains (not functionality!) that the system needed to meet the requirements. These domains were derived directly and through interpretation of the requirements. The domains and their dependencies are shown in Figure 25.1.

This initial domain model can be further refined through an iterative process of evaluation and updating. A key part of this refinement was the separation of

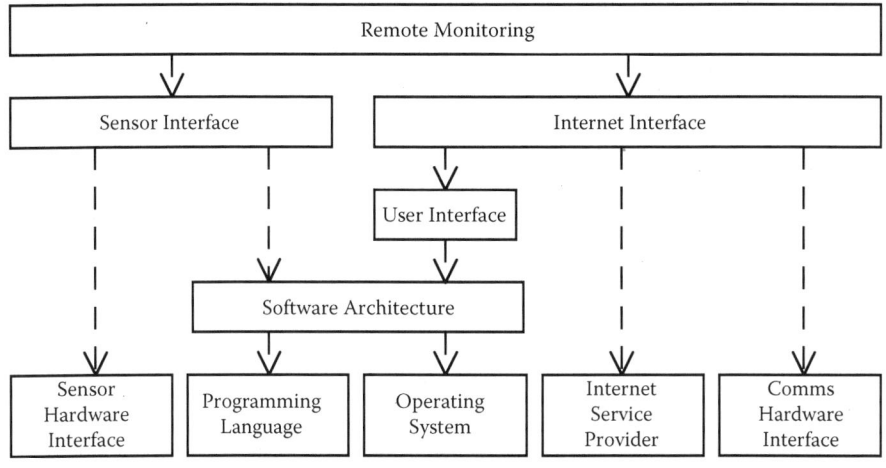

Figure 25.1   Dependencies in the UML package diagram.

domains into the four domain types specified by Kennedy Carter and representing these as layers in the domain model. The four domain types are:

- Application—purpose of the system from the user's point of view.
- Service—basic services, independent of application (e.g., interfaces, logging, alarms, etc.)
- Software—provides the execution environment for all the application and service domains.
- Implementation—represents preexisting technologies and components.

The domain model and the four Kennedy Carter domain layers are shown in Figure 25.2.

These four types relate to the three specified MDA model types: the Computationally Independent Model, or CIM; the Platform Independent Model, or PIM; and the Platform Specific Model, or PSM. These viewpoints are analogous to the domain-type layers, each moving from a level of abstraction down to an implementation level. Figure 25.3 helps illustrate the relationship between the three viewpoints of MDA and the four Kennedy Carter domain types.

This relationship helps reinforce the concept of moving down the domain model from a level of abstraction to an implementation level. Note that no solution has been specified in this model; it simply captures the domains in the overall architecture of the system. The more traditional systems engineering cycle could now be employed to further define and specify these domains to provide the actual implementation of the system. However, the requirements narrative for the RMS specifies a number of alternate technologies that we can now consider to understand the

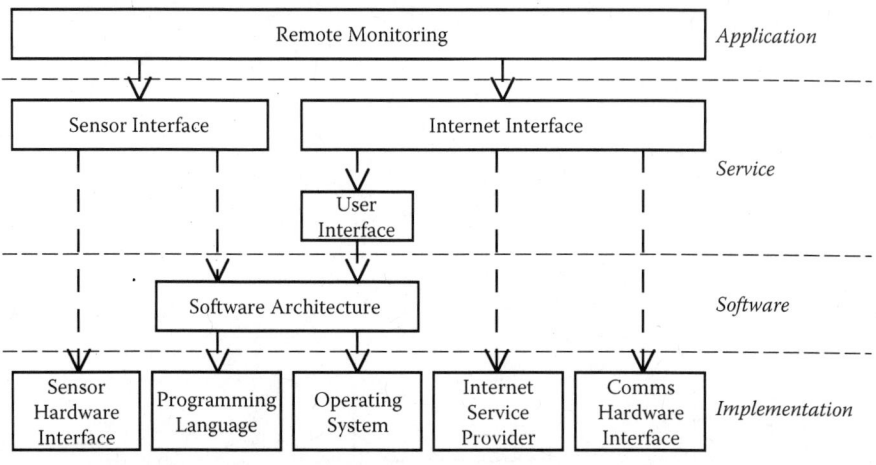

Figure 25.2   MDA domain model.

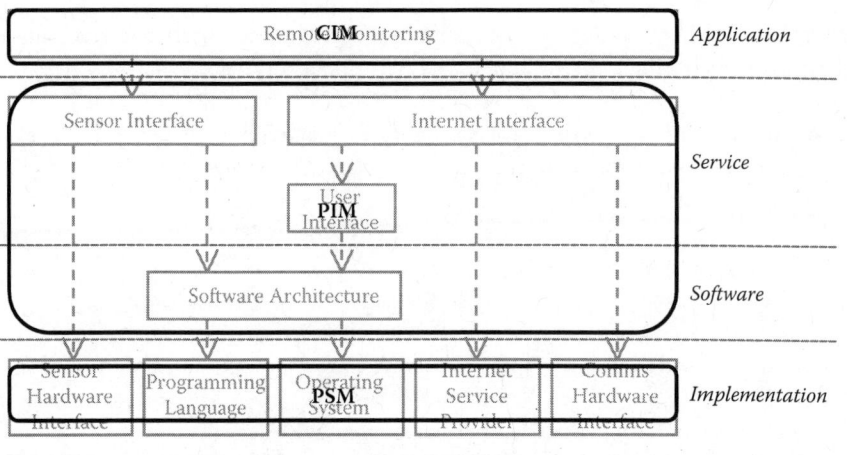

Figure 25.3   Relation to MDA-specified model types.

implications of these choices on the overall system through interpretation within the architecture.

Architecture-Based Assessment

With a domain model developed that could, were the domains specified appropriately, fulfill the requirements of the RMS, we can now perform an assessment of the architecture. The requirements for the RMS specify that one of three wireless technologies must be used to implement the RMS and enable it to link into the internet. This requirement impacts the "Comms Hardware Interface" domain, which is where the wireless technology would reside. The requirements specify that this wireless link will link the RMS to either the patient's cell phone or home computer. There are three variables of interest: the range of the wireless technology (to link the RMS to the internet interface), the bandwidth of the wireless technology, and the bandwidth of the device providing a connection to the Internet (the cell phone or the home computer). To bound the problem, we consider the bandwidth issue first. The specified wireless technologies have a bandwidth of between 56 and 3000 kbps. The ECG and EEG sensors have a nominal rate of sampled data between 1 and 100 kHz. Assuming 8-bit encoding, this results in an output of between 8 and 800 kbps. The final consideration in terms of bandwidth limitations is the Internet connection speed itself, which varies depending on the type of connection, the physical distance from the telephone exchange, etc. These bandwidths are illustrated on the domain model as shown in Figure 25.4.

The illustration of the architecture with the bandwidths shown highlights the uncertainty present in the system—namely, the bandwidth output of the Sensor Interface (which depends on the bit encoding level used; the example shown is for 8-bit encoding) and the bandwidth of the Internet service provider. Assuming the bandwidth of the Internet service provider is equal to or greater than the chosen wireless technology, and taking the proposed 8-bit encoding, a solution matrix can be generated, which is shown in Figure 25.5.

The matrix shown in Figure 25.5 suggests that only the 3000-kbps Bluetooth 2.0 technology is suitable, having the required bandwidth for both the minimum and maximum data bandwidths of the EEG and ECG sensors. This still assumes, of course, that the Internet service provider can match this bandwidth. Consulting the domain

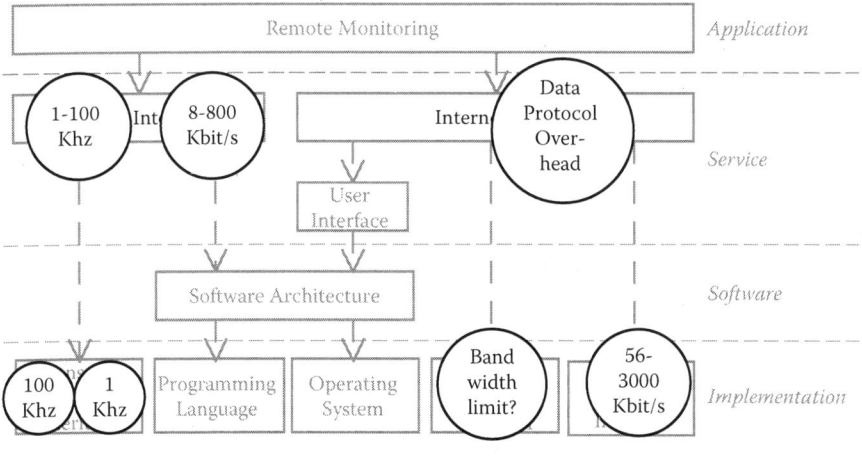

Figure 25.4 Domain model throughput assessment.

	GPRS	Bluetooth 1.2	Bluetooth 2.0
	56-114 kbps	700 kbps	3000 kbps
Min rate 8 kbps	✓	✓	✓
Max rate 800 kbps	✗	✗	✓

Figure 25.5 Wireless technology solution space.

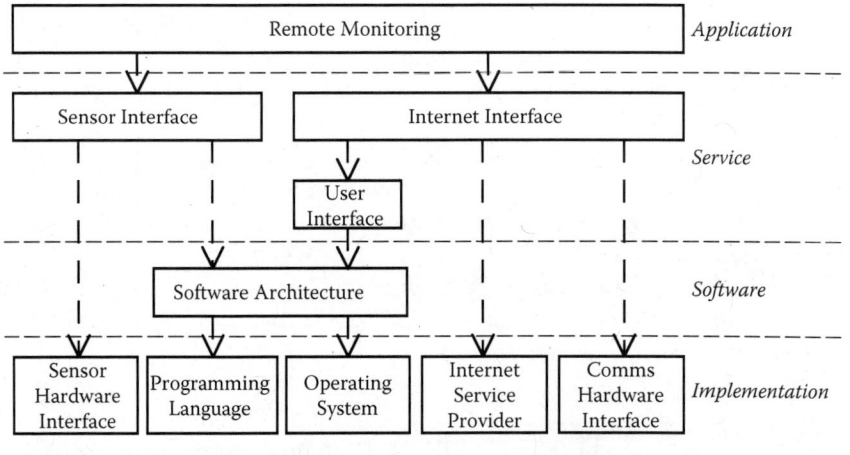

Figure 25.6 Domain model: software architecture highlighted.

model again, however, highlights a key domain that interfaces between the EEG/ECG sensors (in the sensor hardware interface domain) and the Internet service provider domain. That domain is the Software Architecture domain, as shown in Figure 25.6.

The Software Architecture domain is the key to understanding how the RMS architecture operates and how the architecture can cope with the apparent bandwidth limitations. Recall that the issue is the balancing of the bandwidth requirements of the sensors to the available bandwidth of the wireless technology specified by the healthcare provider. In Figure 25.6 it can be seen that the Software Architecture domain is the key link between these two bandwidths. Utilizing a domain-specific technology such as signal processing or data compression, the actual data bandwidth presented to the Comms Hardware Interface domain (the wireless link) can be significantly reduced. With a 10:1 compression ratio, the solution space is shifted as shown in Figure 25.7.

It is unlikely that the healthcare provider will want a continuous data feed of the EEG and ECG signals, but rather will only want to know of anomalies and

	GPRS 56-114 kbps	Bluetooth 1.2 700 kbps	Bluetooth 2.0 3000 kbps
Min rate 0.8 kbps	✓	✓	✓
Max rate 80 kbps	?	✓	✓

Figure 25.7   Solution space with data compression.

have the option of viewing a live data stream intermittently. This raises other considerations, such as data logging and how the system handles requests for information from the healthcare provider. At the higher architectural level where we can consider the interface between the RMS and the healthcare providers system with which it will interact, the question of bandwidth becomes much more significant; it is quite obvious that with multiple RMS deployed, maintaining enough dedicated bandwidth to cope with continuous data streams will be prohibitive. Conceptually, the variable data rate achievable through the Software Architecture domain provides a much better system; it can be imagined how this would enable the RMS to respond to fluctuating bandwidth availability, not just at the wireless link but between it and the healthcare providers recipient system across the Internet.

The conclusion from all of this is that any of the wireless technologies can be used to implement the system, provided the software architecture can sufficiently reduce the data throughput. An optimal solution would focus on a software architecture capable of providing a data stream that responds to available system bandwidth negating the need to pick a "winner" from the multiple specified wireless technologies.

This still leaves the issue of range, between the RMS and the Internet interface, unanswered. Consider the MDA domain model again shown in Figure 25.2. Note that the Comms Hardware Interface and Internet Service Provider domains together link the rest of the RMS to the Internet. Any wireless device is in danger of being moved out of range and hence unable to communicate; and for the RMS, if the Comms Hardware Interface domain is not connected; then the RMS cannot communicate back to the healthcare provider over the Internet. For an RMS, this could be a critical issue, as the system will need to be able to make a connection should an anomaly be detected by the EEG/ECG sensors. Additionally, it is very possible that the patient may not have an Internet connection available in his or her home for the RMS to connect to. Of the three wireless technologies specified by the healthcare provider, only GPRS has the possibility of being "always connected," as it links into the mobile phone network and does not depend on a stationary Internet access point in the patient's home. However, GPRS has the lowest bandwidth and is also the most expensive network (as the mobile phone carrier charges per megabyte of data carried as opposed to telephone Internet providers, which charge a flat rate for a very high bandwidth limit). An optimal system solution may be an RMS that can connect to the patient's home Internet connection (if available) to utilize its higher bandwidth, but which also has built-in GPRS allowing an "always connected" capability (e.g., due to moving out of range of the home Internet connection, if there is one, or because of power cuts, etc.). This level of redundancy makes sense from an architectural perspective as the Internet connection is in the system's environment and hence outside of the RMS's control. This suggests utilization of both GPRS and Bluetooth wireless technologies to provide the suggested redundancy.

Case Study Summary

Recall the relationship between architecture based assessment and system and architecture, repeated below:

> A *system* is a combination of interacting elements integrated to realize properties, behaviors, and capabilities that achieve one or more stated purposes. The *architecture* of a system is a specification of the parts and connectors of the system and the rules for the interactions of the parts using the connectors. *Architecture-based assessment* is the concordant assessment of the capabilities and interoperability of a family of systems.

In summary, the architecture developed, which was shown in MDA using the Kennedy Carter domain types, presents a clear, simple picture of the domains and their interactions. For the assessment of the architecture, this simple picture allows the capabilities to be tagged onto the domains, as shown with the bandwidths, to allow an assessment of the capabilities of the system with the technology limitations prescribed by the stakeholders. The domains explicitly partition systems, which in turn allows an understanding of the interoperability of the domains and the impact of the capabilities of the domains, as was shown with the Software Architecture domain's impact on the bandwidth issue. Throughout this process the clear picture of the architecture provides a communication medium between the development team and helps ensure a common understanding of the architecture, not only within the team, but also with stakeholders. This entire assessment was conducted only considering the prescribed technologies. Hence, the resulting architecture is mostly implementation free and later it can be developed as a PSM with the appropriate technologies as chosen by the domain specialists.

Index